MACHINE DESIGN

Irving J. Levinson, P.E.

Provost, Highland Lakes Campus
Oakland Community College
Bloomfield, Michigan

RESTON PUBLISHING COMPANY, INC.
Reston, Virginia
A PRENTICE-HALL COMPANY

Library of Congress Cataloging in Publication Data

Levinson, Irving J
 Machine design.

 Includes index.
 1. Machinery—Design. I. Title.
TJ230.L559   621.8'15   77-4421
ISBN 0-87909-461-3

© 1978 by
Reston Publishing Company, Inc.
A Prentice-Hall Company
Reston, Virginia   22090

10 9 8 7 6 5 4 3 2 1

Printed in the United States of America.

IN MEMORIAM

Irving J. Levinson devoted many long hours of the last four years of his life toward the completion of his text on *Machine Design*. Uppermost in his thoughts during those four years was the desire that the final manuscript be of the highest quality. The completed text justifies his desires. It is hoped that this text will serve to quench a thirst for knowledge of those interested in the subject matter. An indebtedness is acknowledged to the author not only for this book, but also for the many other works he has produced.

Matthew I. Fox
PRESIDENT, RESTON PUBLISHING CO., INC.

CONTENTS

PREFACE

This is a book that is concerned with the fundamentals of mechanical machine design—any single one of the chapters could be enlarged to fill voluminous volumes—each too heavy to carry—that would be out of date on the eve of their publication. Knowledge is erupting, understanding is broadening, and more thinkers are thinking simply because there are more thinkers. It is virtually impossible for any single individual to have total awareness—to know and to understand the entire gamut of all mechanical design criteria. If, however, through some unexplained miracle, such an individual were to exist at this very moment, he would vanish tomorrow.

So—this is a book that is concerned with the fundamentals of mechanical machine design; this concern does not preclude or exclude, as the discussion progresses from topic to topic, an attempt at a realistic and contemporary approach to theory, problems, and projects. No one will be asked, for instance, to conjure up a Babylonian granite chipper sharpener, a Viking-style retractable oarlock, or a machine to manufacture long narrow neckties.

It is assumed that the reader has a basic—or at least a reviewable—knowledge of drafting, statics, dynamics, and strength of materials. While these subjects are re-examined in some detail, little time can be afforded to fundamental language translations and to long extended derivations. Terms like stress, strain, force, moment, tolerance, and fatigue should not sound foreign to the reader.

Both problems and suggested design projects appear at the end of each chapter—the problems usually have definite answers and these answers are given as an aid to understanding; hints to the solutions of these problems are also supplied. Similarly, the design projects also have answers; these answers, however, will vary from engineer to engineer depending upon his assumptions and his imagination. The *best solution*, if such an entity exists, to any design project, is left to the creativity of the doer—and rightly so.

Irving J. Levinson

ACKNOWLEDGMENT

Acknowledgment is given to William J. Patton with the deepest gratitude for his generosity and efforts in the completion of the author's manuscript. Mr. Patton's thoughtfulness, integrity, and knowledge have served as a tribute to the late Irving J. Levinson.

Marilyn T. Levinson

CHAPTER 1

STRESS AND STRAIN: A REVIEW OF FUNDAMENTAL CONCEPTS

ma-chine (mə shēn') *n.* An apparatus consisting of interrelated parts with separate functions, used in the performance of some kind of work ... a device which transmits and modifies force or motion ... also called a *simple machine* ; any of six or more elementary mechanisms, as the lever, wheel and axle, pulley, screw, wedge, and inclined plane ... also called a *complex machine* ; a combination of simple machines.

1-1 INTRODUCTION

In the beginning God created ... man; and man devised, contrived, invented, developed, constructed, and mass-produced machines— machines to cultivate his fields, grind his grain, fight his battles, transport him from place to place on the face of the earth, and allow him to probe the universe—machines that bridge the gap from simple to complex, from safety pins to spacecraft. However ingenious and intricate these machines were, are, or will be, their mechanical segments are always reducible to no more than six simple mechanisms: levers, wheels and axles, pulleys, screws, wedges, and inclined planes.

Since it is not essential that the digestive process be understood to enjoy food, it is similarly not essential that the physical basis of things be understood to design workable machines. After all, archaeological digs and the carbon-dating process have placed an age of over 20,000 years on some rather crude but functional man-formed tools, an age of 12,000 years on implements to till the soil, an age of over 5000 years on the use of gears and the alloying of metals. Thus possessed with an inquisitive, inquiring, and inventive mind, man during his own development surely questioned "why" things worked, "why" he could not lift himself by pulling on his own boot straps. Truths were slow in coming—if man has existed on earth in his present form for 25,000 years, he had to wait 24,600 years (just yesterday) for the mathematical truths of mechanics—the analyses of the lever, wheel and axle, pulley, screw, wedge, and inclined plane—the explanation of the interrelationships between: force and motion, work and energy, impulse and momentum, stress and strain. Man had to wait for the development of his own genius in the forms of da Vinci, Galileo, Descartes, Newton, Euler, Coulomb, Hooke, Young, Cauchy, Navier, and Poisson —all our contemporaries relative to the archaeological scale of time.

Today's machine designer is, in truth, a machine redesigner. He acknowledges and appreciates that the arts and sciences of the past, present, and future are entwined so as to be inseparable. He knows that the pollution-free internal combustion engine of tomorrow will have evolved from design changes of today's engine. Why the emphases on redesign? For one or more of the following possible reasons: a more economical use of existing raw materials, the improvement and development of energy sources, the incorporation of new materials of construction, boundaries that are imposed which involve unusual temperatures and pressures, new surrounding atmospheres (or lack of atmospheres), operational dangers and failures in what has been designed, and general aesthetics—next year's model must look different.

The machine design process, as it exists today—and as it will probably exist tomorrow—is an admixture of both science and art; fortunately as time goes by, the "science' will overshadow the "art." The development of acceptable mathematical theories, when coupled with the ever-increasing precision in the act of measurement, contributes to the expanding coffers of knowledge. Research papers, engineering journals, and manufacturers' literature are the sources that the practicing engineer invariably draws upon. Thus, the experts surround us—scientists, engineers, inventors, and manufacturers each contributing a dot or two to the never-to-be-finished portrait of understanding, started centuries ago and painted in the fashion of pointillism—we are advancing from the unknown.

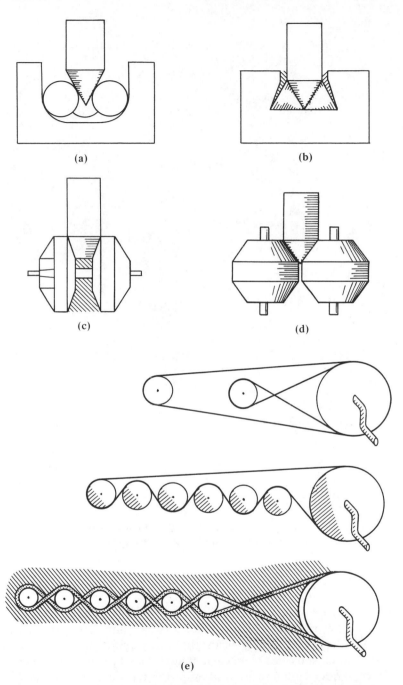

(a)

(b)

(c)

(d)

(e)

Figure 1-1 / Mechanisms of all sorts and sizes were devised by Leonardo da Vinci: typical are the ball and roller bearings which did not come into use until the 1900's—some 400 years after their inception. Civilization had to discover the need and the materials. (Sketched after drawings by Leonardo da Vinci)

Figure 1-2   /   A replica of the original sewing machine patented by I. M. Singer, circa 1851. (Courtesy of the Singer Sewing Machine Co.)

1-2   SIMPLE STRESS

Stress, a vector quantity which is dimensionally similar to the term *pressure*, is the magnitude of force that acts on a unit area; *uniform stress*, an idealized quantity, simply means that every unit of area at a particular cross section is subjected to an equal magnitude of force. If the magnitude of force varies (a more realistic situation) at a given cross-sectional area, the resulting stress is defined simply as *nonuniform stress*.

   Fig. 1-3(a) shows a notched prismatic bar of cross-sectional area A acted upon by an axial tensile force F; assume the dimensions of the

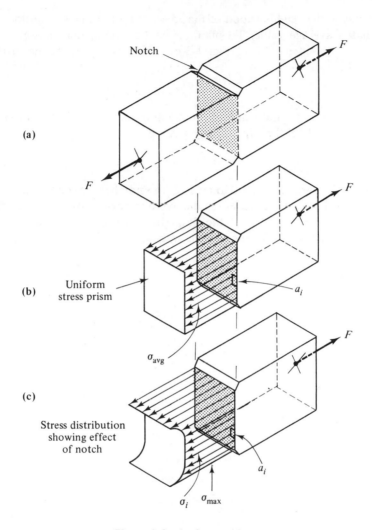

Figure 1-3 / Stress prisms.

notch to be negligible as compared to the cross-sectional area of the bar. By dividing the force F by the area A perpendicular to the axis of the force, the average or theoretical uniform stress σ_{avg} is obtained.

$$\sigma_{avg} = \frac{F}{A} \qquad (1\text{-}1)$$

When plotted, as in Fig. 1-3(b), a rectangular stress prism is obtained. Experience based upon precise measurements tells us that a notch—or, for that matter, any internal or external geometric discontinuity—causes the stress in the neighborhood of that discontinuity to be greater than the calculated average value. To maintain static equilibrium, however, the volume of the stress prism, Fig. 1-3(c), must exactly equal the product $\sigma_{avg} A$, thus

$$F = \sum f_i = \sum \sigma_i a_i = \sigma_{avg} A \qquad (1\text{-}2)$$

where f_i is an elemental value of force acting on a unit of area a_i, and σ_i is the magnitude of the stress at any given point.

EXAMPLE 1-1

An axial force $F = 1000$ lb acts on the notched prismatic bar shown in Fig. 1-4(a). Determine: (a) the average normal stress acting on the notched cross section of the bar; (b) and (c), the maximum value of the axial stress for the two theoretical stress distributions shown in

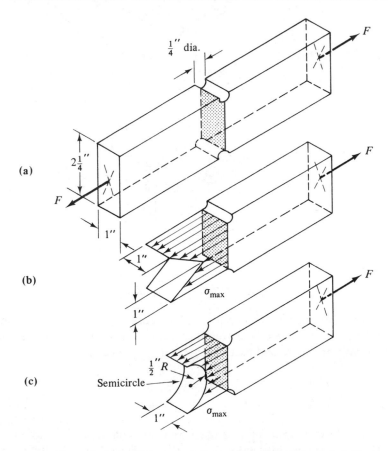

Figure 1-4 / Stress prisms for a notched bar (Example 1-1).

Figs. 1-4(b) and (c) where the minimum stress in each instance is assumed to be 200 psi.

solution

(a) The average stress across a normal section is given by Eq. 1-1:

$$\sigma_{avg} = \frac{F}{A} = \frac{1000}{1 \times 2} \frac{\text{lb}}{\text{in}^2} = 500 \text{ psi}$$

(b) $F = 1000$ lb must equal the volume of the stress prism (Eq. 1-2); this stress prism consists of rectangular volume less a triangular volume.

$$F = 1000 \text{ lb} = \text{(volume of rectangular prism)}$$
$$- \text{(volume of triangular prism)}$$
$$= \sigma_{max} \times 2 \text{ in.} \times 1 \text{ in.} - [\tfrac{1}{2} \times 1 \text{ in.} \times 2 \text{ in.} \times (\sigma_{max} - 200)]$$
$$2\sigma_{max} - \sigma_{max} - 200 = 1000 \text{ psi}$$
$$\sigma_{max} = 1200 \text{ psi}$$

(c) $F = 1000$ lb again must equal the volume of the stress prism.

$$F = 1000 \text{ lb} = \text{(volume of rectangular prism)}$$
$$- \text{(volume of semicircular prism)}$$
$$= \sigma_{max} \times 2 \text{ in.} \times 1 \text{ in.}$$
$$- [\tfrac{1}{2} \times \pi \times \tfrac{1}{2} \text{ in.} \times (\sigma_{max} - 200) \times 1 \text{ in.}]$$
$$2\sigma_{max} - \frac{\pi}{4} \sigma_{max} + 50\pi = 1000$$
$$\sigma_{max} = 694 \text{ psi}$$

1-3 AXIAL, SHEAR, AND BEARING STRESSES: DIRECT STRESSES

Direct stresses are, by the generally accepted definition, those stresses that can be determined by simply dividing *force* by *area*. These direct stresses fall into three fundamental categories—those stresses caused by forces that: axially push or pull, exert a tearing action, or tend to crush or create a contact pressure. These three direct stresses are respectively termed *axial*, *shear*, and *bearing stresses*.

Figure 1-5(a), a pinned clevis attached to a bolted pipe clamp, illustrates how the axial pulling force F can exert, at various locations within the assembly, the three forms of direct stress. The shaded cross

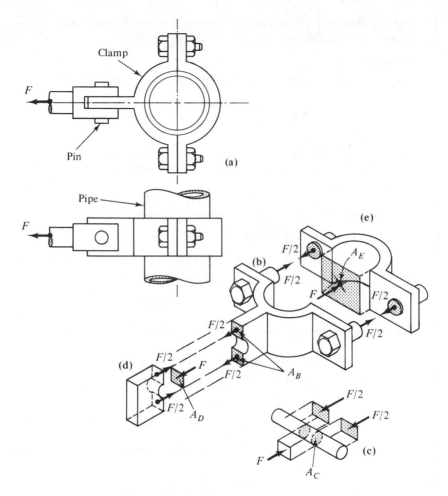

Figure 1-5 / Direct stresses in a clevis.

section shown in Fig. 1-5(b) must support the axial tensile load $(F/2 + F/2) = F$, which has been transmitted by the hinge pin; the axial stress, considered to be uniform, is found by dividing the magnitude of the force F by the quantity of area A_B that is normal (perpendicular) to the axis of the force.

The clevis hinge-pin shown in the free-body of Fig. 1-5(c) is subjected to forces that tend to shear or tear the center section of the pin from its ends. Assuming that the two circular shaded areas A_C within the pin equally support the load F, the shear stress τ would be equal to the magnitude of F divided by the quantity of area $2A_C$. The hinge-pin, supported by the shank of the clamp, bears on the drilled hole—the pin creates a pressure force and tends to elongate the hole, as in Fig. 1-5(d). The *bearing stress* σ_D is found by simple arithmetic to equal the magnitude of the force F divided by the *projected area* of the hole, which is, in this instance, the product of the diameter of the pin and the width of the shank.

A free-body diagram, Fig. 1-5(e), shows the aft section of the clamp to bear against the pipe; again, the bearing stress σ_E is equal to the magnitude of the force divided by the projected area A_E.

The bolts are in tension, and their respective axial stress would be computed by dividing the magnitude of force that each must support by the least area in the bolt cross section. This computation is made assuming that prior to loading, the bolts are just "snugged" tight. Pretightening would, of course, result in an initial tensile stress or *prestress* in the bolts. If the clamp is assumed to be fairly rigid (nonyielding), it can be assumed that the stress in each bolt would be that caused by the pretightening and not affected by the load F, unless, of course, the load F exceeds the sum of the pretightening forces in the bolt, in which case the clamp halves would separate. Greater detail is given to this topic in Chapter 5; however, a brief explanation of bolt forces seems to be worthy of a brief discussion at this point.

Consider a situation where an elastic bolt tightened to an initial tension F_1 joins two rigid (nonyielding) members A and B, as in Fig. 1-6(a). In Fig. 1-6(b), F_1 is the initial load in the bolt caused by pre-tightening, and P is the pressure force between the plates. Now an external force F_2 is applied, as in Fig. 1-6(c); the pressure force between the plates diminishes to P_i while the bolt force remains F_1 since its elastic deformation (stretch) has not changed. Separation between the plates A and B will occur when the pressure force P_i approaches and finally equals zero— only then will the bolt force increase, because the bolt undergoes further

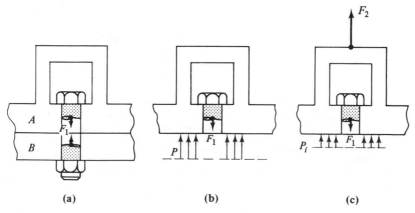

(a)   (b)   (c)

Figure 1-6 / Bolt stresses.

elongation. The graph of Fig. 1-7 gives values of acceptable safe loads for American Standard bolts of various sizes. The design of connections and joints is discussed in greater detail in Chapter 5. It is worthy to note, however, that bolts are easily and inexpensively replaced—good design dictates that they should be the "weak link" in a combination of machine components.

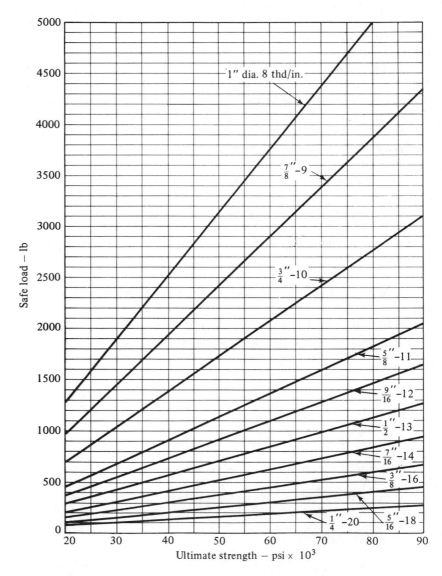

Figure 1-7 / Safe loads of American Standard bolts as a function of ultimate strength.

EXAMPLE 1-2

The six bolts are to be the "weakest link" in the assembly shown in Fig. 1-8. Assume the permissible tensile stress σ_T in the leg of the Tee to be 15,000 psi, the permissible bearing stress σ in the slip-pin to be 44,000 psi, and the allowable shear stress in the slip-pin to be 33,000 psi. The load F is symmetrically applied; each end of the pin is subjected to a force of $F/2$. Find: (a) the load F; (b) the width w; and (c) the appropriate bolt size and stress-grade. To provide for the "weak link" in the assembly, design the bolts to each support $F/6$ and the Tee and pin to withstand at least a 10% overload.

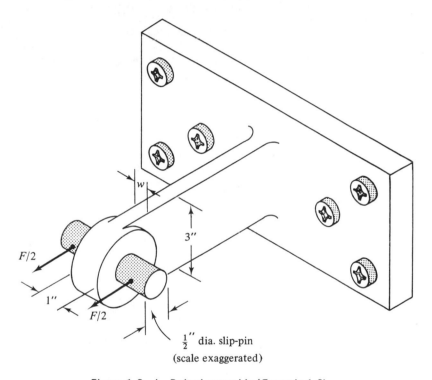

$\frac{1}{2}''$ dia. slip-pin
(scale exaggerated)

Figure 1-8 / Bolted assembly (Example 1-2).

solution

(a) Either the bearing stress or the shear stress in the slip-pin will govern the design, hence:

Bearing:

$$F + 0.1F = \text{(bearing stress)} \times \text{(projected bearing area)}$$
$$1.1F = 44{,}000 \text{ lb/in.}^2 \times 1 \text{ in.} \times 0.5 \text{ in.}$$
$$F = 20{,}000 \text{ lb}$$

Shear:

$$F + 0.1F = (\text{shear stress}) \times (\text{shear area})$$
$$1.1F = 33{,}000 \text{ lb/in.}^2 \times 2 \times \pi \times \overline{0.25}^2 \text{ in.}^2$$
$$F = 11{,}800 \text{ lb}$$

Shear governs!

(b)   Next, use Eq. 1-1 to determine the width w of the Tee:

$$F + 0.1F = 1.1 \times 11{,}800 = \sigma A = 15{,}000 \text{ lb/in.}^2 \times 3 \text{ in.} \times w$$
$$w = 1.1 \times 11{,}800/15{,}000 \times 3 = 0.288 \text{ in.}$$

Use $\frac{5}{16}$-in. (0.3125 in.) stock.

(c)   The graph of Fig. 1-7 is used to find the most appropriate bolt size and stress-grade for the assembly, where

$$F/6 = 11{,}800/6 = 1970 \text{ lb}$$

A $\frac{3}{4}$-in.–10 thd stress-grade = 60,000 psi bolt will safely support 2060 lb; since this fail load is greater than the required 1970 lb, the shear strength of the pin and the tensile strength of the Tee must be rechecked to insure that the pin and Tee will not fail, leaving six perfectly good bolts holding nothing.

Tee:

$$\sigma = \text{bolt force/area} = 6 \times 2060/3 \times 0.3125 = 13{,}500 \text{ psi}$$
$$\text{Satisfactory(!) since } 13{,}500 < 15{,}000 \text{ psi.}$$

Pin:

$$\tau = \text{bolt force/area} = 6 \times 2060/2 \times \pi \times \overline{0.25}^2 = 31{,}500 \text{ psi}$$
$$\text{Satisfactory (!) since } 31{,}500 \text{ psi} < 33{,}000 \text{ psi.}$$

Answer:

Use six $\frac{3}{4}$-in.–10, 60,000 psi stress-grade bolts.

1-4   COMPONENTS OF STRESS

Materials are frequently encountered that are strong in tension or compression and weak in shear—the converse may also occur. A material's principal virtue may be its ability to withstand shearing forces. To visualize

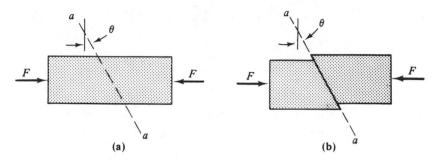

Figure 1-9 / Angular plane of stress.

the phenomenon, consider the prismatic bar in Fig. 1-9(a) that is acted upon by the compressive force F. It is conceivable that shear or slippage may occur across an arbitrary plane a–a inclined at an angle θ as shown in Fig. 1-9(b). Internal shear forces F_s and normal forces F_n that act on an inclined plane within a member can be found by considering the free-body diagram of Fig. 1-10(b), where the components F_n and F_s must vectorially add and equate to F by the principles of statics. Then

$$F_n = F \cos \theta \qquad \text{and} \qquad F_s = F \sin \theta \qquad (1\text{-}3)$$

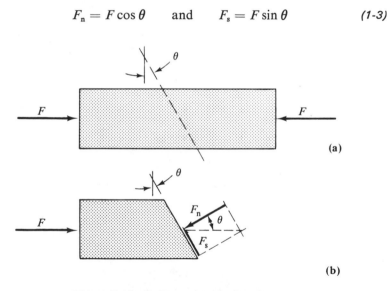

Figure 1-10 / Forces on angular planes.

The respective normal and shear stress σ_θ and τ_θ are found by dividing the force vectors F_n and F_s by the oblique area $A/\cos \theta$ where A is the transverse area, the area normal to the axis of the force.

$$\sigma_\theta = \frac{F_n}{A/\cos \theta} = \frac{F \cos \theta}{A/\cos \theta} = \sigma \cos^2 \theta \qquad (1\text{-}4)$$

and

$$\tau_\theta = \frac{F_s}{A/\cos \theta} = \frac{F \sin \theta}{A/\cos \theta} = \frac{\sigma}{2} \sin 2\theta \qquad (1\text{-}5)$$

The normal stress is a maximum at $\theta = 0°$ and equal to σ or simply F/A. The shear stress has a maximum value of $\sigma/2$ at $2\theta = \pm 90°$ or at $\theta = \pm 45°$; the shear stress is zero on planes that are normal to the axis of the force.

EXAMPLE 1-3

Two mild steel bars having rectangular cross sections are silver-soldered along their inclined planes as shown in Fig. 1-11. (a) Determine the safe tensile load F if the permissible working stresses at the joint in tension and shear are 900 psi and 2000 psi, respectively. (b) Find the efficiency of the joint if the stress in the steel is not to exceed 18,000 psi.

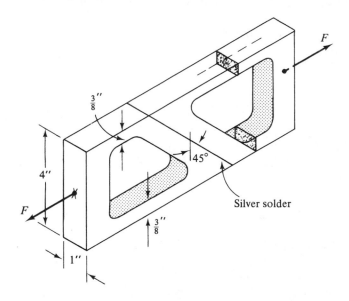

Figure 1-11 / Silver-soldered joint (Example 1-3).

solution

(a) In this situation, tension governs since $\sigma_{\theta \, max} < \tau_{\theta \, max}$

$$900 < (2000/2) \text{ psi}$$

Therefore,

$$F = \tau_\theta A_\theta = 2000 \times \frac{4 \times 1}{\cos 45°} = \frac{8000}{0.707} = 11,300 \text{ lb}$$

The steel is capable of sustaining a *direct* tensile load F' equal to the applied force divided by the magnitude of the shaded transverse area shown:

$$F' = 18,000 \times 2 \times 1 \times \tfrac{3}{8} = 13,500 \text{ lb}$$

(b)   The efficiency of the joint is

$$\text{Eff} = \frac{F}{F'} \times 100 = \frac{11,300}{13,500} \times 100 = 83.7\%$$

1-5   TWO DIMENSIONAL ORTHOGONAL FORCE SYSTEMS: GENERALIZED PLANE STRESS

The effects on a body by a system of two-dimensional forces—forces in one plane that act at right angles to one another—can be investigated by considering the state of equilibrium existing on the object shown in Fig. 1-12(a). The *state of stress* at a minute point P within the body and the forces that act at this point as a function of stress are shown in Fig. 1-12(b). A state of internal equilibrium must exist, that is, the vector polygon of forces must close, as in Fig. 1-12(d). Since all forces are a function of the oblique area A, Fig. 1-12(c), it can be shown that

$$\sigma_\theta = \frac{\sigma_x + \sigma_y}{2} + \frac{\sigma_x - \sigma_y}{2} \cos 2\theta \qquad (1\text{-}6)$$

and

$$\tau_\theta = \frac{\sigma_x - \sigma_y}{2} \sin 2\theta \qquad (1\text{-}7)$$

The maximum and minimum values of σ and τ derived from these equations are called *principal* stresses σ_{p_1} and σ_{p_2}. These principal stresses occur at

$$\cos 2\theta = \pm 1$$
$$2\theta = 0°, 180°$$
$$\theta = 0°, 90°$$

thus

$$\sigma_{p_1} = \sigma_x \text{ at } = 0°$$

and

$$\sigma_{p_2} = \sigma_y \text{ at } = 90°$$

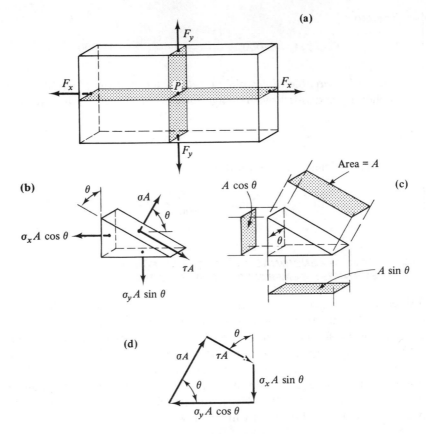

Figure 1-12 / Biaxial stresses.

Whether σ_{p_1} and σ_{p_2} are maximums or minimums depends, of course, on both the magnitudes and directions of σ_x and σ_y. No planes, however, within the body will have normal stresses greater than σ_x or σ_y.

The shear stress τ is a maximum at $\sin 2\theta = \pm 1$; hence

$$\sin 2\theta = \pm 1$$
$$2\theta = 90°, 270°$$
$$\theta = 45°, 135°$$

Thus, in simple biaxial stress, the maximum or minimum shear stress occurs on planes inclined at 45° with the principal normal planes—the orthogonal x- and y-axes.

$$\tau_{\max} = \frac{\sigma_x - \sigma_y}{2} \quad \text{at } \theta = 45$$

and
$$\tau_{min} = -\frac{\sigma_x - \sigma_y}{2} \quad \text{at } \theta = 135$$

It is interesting to observe that when $\sigma_x = \sigma_y$, all planes within the body are free of shearing stress.

Equations (1-6) and (1-7), fortunately, have a very simple graphical interpretation which is attributed to Otto Mohr who developed the concept early in the 19th century. A set of coordinate axes are drawn as illustrated in Fig. 1-13(a). These horizontal and vertical axes represent normal and shearing stresses, respectively, and must not be confused with an x- and y-coordinate system. While the sign convention is arbitrary, once an assumption regarding plus and minus signs is made, the convention must be adhered to throughout the analyses. It is arbitrarily assumed, therefore, that the normal stresses, tension and compression, will be positive and negative, respectively. A shear stress that produces a clockwise moment about a point on the element will be considered positive; conversely, a counterclockwise moment will be produced by a negative shear stress. The following steps, then, will progressively create Mohr's circle:

1 / Select an appropriate scale for the set of coordinate axes [Fig. 1-13(a)], and locate the normal stresses [Fig. 1-13(b)] σ_x and σ_y at points A and B on the coordinate axes [Fig. 1-13(c)]. If $\sigma_y > \sigma_x$, then σ_y would be furthest to the right. Either or both of the normal stresses could be negative, denoting one or both to be compressive.

2 / Locate point 0, the midpoint between σ_x and σ_y, as shown in Fig. 1-13(c). With 0 as a center, draw a circle that intersects the normal axis at σ_x and σ_y; the radius of this circle is equal to $(\sigma_x - \sigma_y)/2$.

3 / The intersection of any diameter a–a, as in Fig. 1-13(d), with the circumference of the circle will give values of σ_θ and τ_θ for any oblique plane inclined at an angle θ as defined in Fig. 1-13(b). Note that θ is the angle between a normal to the oblique plane and the x-axis.

4 / In the element, σ_x and σ_y are 90° to one another—in the circle, σ_x and σ_y are 180° to one another; all angles measured in the circle are double those measured on the element. Angular directions, however, are similar; clockwise rotation on the element means clockwise on the circle, and counterclockwise rotation on the element means counterclockwise on the circle.

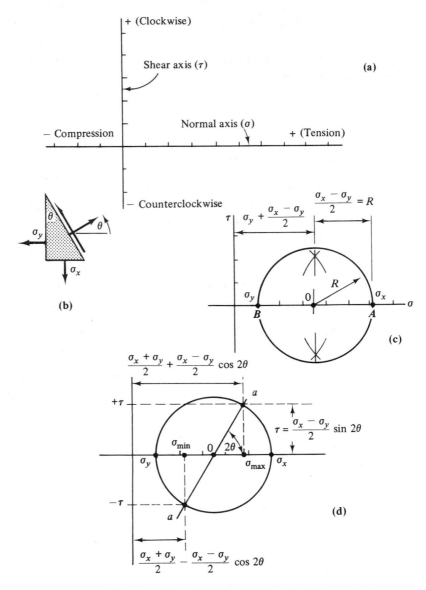

Figure 1-13 / *Mohr's circle for stresses.*

5 / Values of stress, σ and τ, assume the values defined in Eqs. (1-6) and (1-7), as shown by the dimensioning of Mohr's circle. Coordinates of all points on the circle will give values of stress on every oblique plane within the element.

EXAMPLE 1-4

A plate of uniform thickness is subjected to the forces shown in Fig. 1-14(a). Use Mohr's circle to determine the normal and shearing stresses that act on the shaded plane inclined as shown.

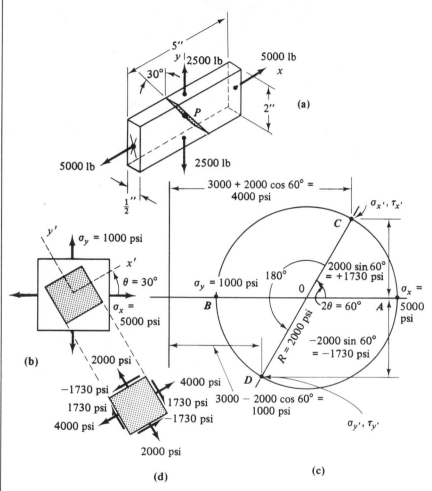

Figure 1-14 / Stresses on a plane (Example 1-4).

solution

The stresses σ_x and σ_y are first computed:

$$\sigma_x = \frac{F_x}{A_x} = \frac{5000}{2 \times 0.5} = 5000 \text{ psi}$$

$$\sigma_y = \frac{F_y}{A_y} = \frac{2500}{5 \times 0.5} = 1000 \text{ psi}$$

An element with its associated state of stress [Fig. 1-14(b)] is isolated, and Mohr's circle is constructed as illustrated in Fig. 1-14(c); the radius of Mohr's circle is seen to be 2000 psi. A second element, the shaded cube of Fig. 1-14(d), is redrawn turned counterclockwise 30°—the element's coordinate axes will be arbitrarily called its x'- y'-axes. The intersection of radius OC with the circle at C defines the state of stress on the parallel faces of the element that are normal to the x'-axis:

$$\sigma_{x'} = 3000 + 2000 \cos 60° = +4000 \text{ psi}$$

$$\tau_{x'} = 2000 \sin 60° = +1730 \text{ psi}$$

The y'-axis is $(90 + 30)°$ counterclockwise from the x-axis. In Mohr's circle the state of stress on faces normal to the y'-axis is at $(180 + 60)°$ counterclockwise from the radius OA on the circle and is defined by point D.

$$\sigma_{y'} = 3000 - 2000 \cos 60° = +2000 \text{ psi}$$

$$\tau_{y'} = -2000 \sin 60° = -1730 \text{ psi}$$

Note: The sign convention previously defined was adhered to, and the element, Fig. 1-14(d), is in static equilibrium.

EXAMPLE 1-5

To aid in the understanding of the significance of signs, positive and negative, as they relate to the direction of stress, the previous example is repeated with F_x acting in compression, as shown in Fig. 1-15(a). The general steps to the solution will remain the same—only the arithmetic will change.

solution
Proceed as in Example 1-4; follow the analyses through the successive steps in the figure.

$$\sigma_x = \frac{F_x}{A_x} = \frac{-5000}{2 \times 0.5} = -5000 \text{ psi}$$

$$\sigma_y = \frac{F_y}{A_y} = \frac{2500}{5 \times 0.5} = 1000 \text{ psi}$$

The coordinates of point C, which now are down and to the left, are:

$$\sigma_{x'} = -2000 - 3000 \cos 60° = -3500 \text{ psi (compression)}$$

$$\tau_{x'} = -3000 \sin 60° = -2600 \text{ psi (counterclockwise)}$$

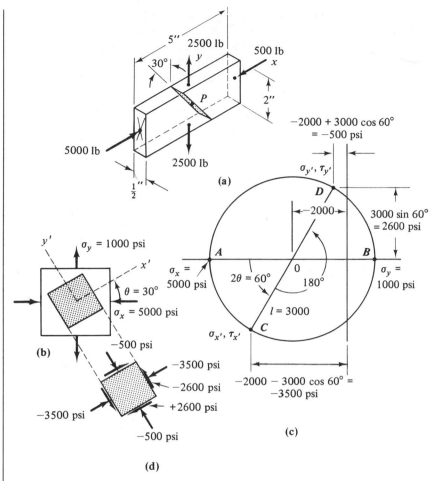

Figure 1-15 / Example 1-5.

And similarly, the coordinates of point D are

$$\sigma_{y'} = -2000 + 3000 \cos 60° = -500 \text{ psi (compression)}$$
$$\tau_{y'} = +3000 \sin 60° = 2600 \text{ psi (clockwise)}$$

The state of stress is shown in Fig. 1-15(d).

The ease of the geometrical approach as a method of determining the stress at a point can be fully appreciated when the method is applied to *generalized plane stress*—the case of biaxial and shearing stresses acting simultaneously.

To illustrate the construction of Mohr's circle for generalized plane stress, the "solution" to Example 1-4 will be considered as "the problem" with the object being to find the principal normal stresses that act. An additional question will also be asked: What are the magnitudes and directions of the principal (maximum) shearing stresses?

The stress at a point in Fig. 1-16(a) consists of tensile stresses on faces normal to the x- and y-axes and shearing stresses which are posi-

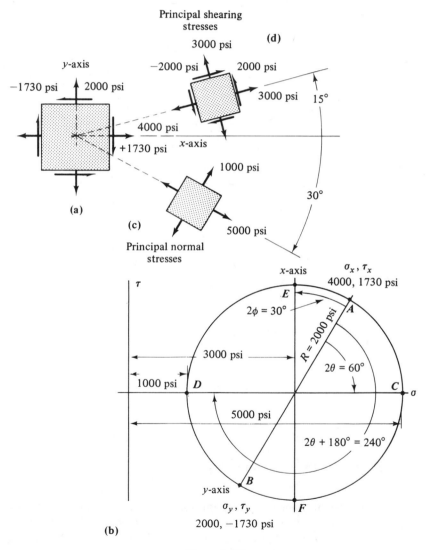

Figure 1-16

tive on planes normal to the x-axis and negative on planes normal to the y-axis. To construct Mohr's circle, locate the coordinates (σ_x, τ_x) and (σ_y, τ_y)—numerically these would be (4000, 1730 psi) and (2000, −1730 psi), respectively, at A and B, as shown in Fig. 1-16(b). Next, the diameter AB is drawn, and the circle is constructed. Numerical values are readily apparent; $\overline{OC} = 1000$ psi, $\overline{AC} = 1730$ psi, and

$$\overline{OA} = \overline{OB} = \sqrt{\overline{OC^2} + \overline{CA^2}} = \sqrt{1000^2 + 1730^2} = 2000 \text{ psi}$$

The principal normal stresses, their maximum and minimum values, are represented by points D and E, thus

At D: $\sigma_{x'} = 3000 + \overline{OA} = 3000 + 2000 = 5000$ psi

At E: $\sigma_{y'} = 3000 - \overline{OB} = 3000 - 2000 = 1000$ psi

To arrive at point C which represents $\sigma_{x'}$, the radius line $\overline{OA}$ of Mohr's circle must be rotated through an angle 2θ clockwise; similarly $\overline{OA}$ must be rotated through a clockwise angle of $(180 + 2\theta)°$ to arrive at point D.

$$2\theta = \arctan \frac{1730}{1000} = 60°$$

Since rotation in the circle is twice that in the element, the state of principal stress is drawn as indicated in Fig. 1-16(c); the x'-axis is 30° clockwise, and the y'-axis is $(180 + 2\theta)/2 = 120°$ clockwise, both measured angularly from the original x-axis. Compare the orientation with that described in Example 1-4.

Points E and F define the principal shear stresses; note that the maximum and minimums are equal in magnitude but opposite in sign. Note also that the normal stresses $\sigma_{x''}$ and $\sigma_{y''}$ associated with the principal shear stress are equal in both magnitude and direction. The element and its rotation are illustrated in Fig. 1-16(d), where $\phi = 15°$.

Symbolically, the equation that governs the principal values of plane stress is

$$\sigma_{\substack{p_{max} \\ min}} = \frac{\sigma_x + \sigma_y}{2} \pm \sqrt{\left(\frac{\sigma_x - \sigma_y}{2}\right)^2 + (\tau_{xy})^2} \qquad (1\text{-}8)$$

on planes inclined at

$$\theta_{p_1} = \frac{1}{2} \arctan \left(\frac{2\tau_{xy}}{\sigma_x - \sigma_y}\right) \qquad (1\text{-}9)$$

and $$\theta_{p_2} = \theta_{p_1} + 90° \qquad (1\text{-}10)$$

The principal values of the shearing stress are given by

$$\tau_{\substack{max \\ min}} = \pm\sqrt{\left(\frac{\sigma_x - \sigma_y}{2}\right)^2 + (\tau_{xy})^2} \qquad (1\text{-}11)$$

on planes inclined at

$$2\theta_{s_1} = 90° - 2\theta_{p_1} \qquad (1\text{-}12)$$

and

$$\theta_{s_2} = \theta_{s_1} + 90° \qquad (1\text{-}13)$$

In the preceding equations, τ_{xy} is defined as the shear stress acting on planes that are normal to the x-axis; τ_{yx}, which is always equal to $-\tau_{xy}$ because moments must balance, is the shear stress acting on planes that are normal to the y-axis. The graphical representation of these stresses is depicted in the sketches shown in Fig. 1-17.

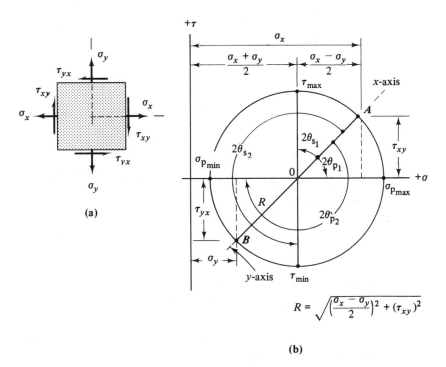

(a)

(b)

$$R = \sqrt{\left(\frac{\sigma_x - \sigma_y}{2}\right)^2 + (\tau_{xy})^2}$$

Figure 1-17 / Principal stresses.

EXAMPLE 1-6

An element within a body is subjected to the state of stress shown in Fig. 1-18(a). Draw Mohr's circle for the state of stress, and find: (a) the magnitudes of the principal normal stresses and the respective

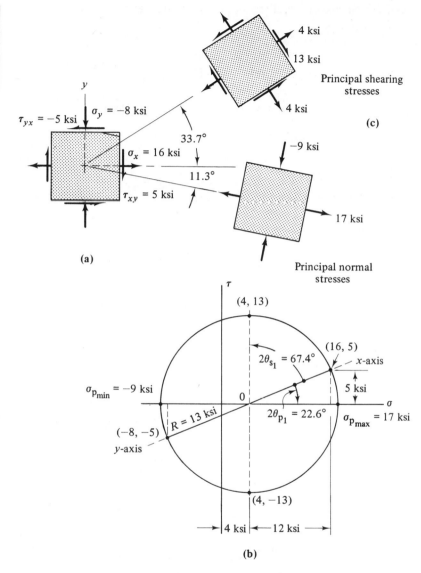

Figure 1-18 / Combined shear and direct stresses (Example 1-6).

orientation of the element; and (b) the principal shearing stresses and their associated orientation.

solution

(a) The radius of Mohr's circle is

$$R = \sqrt{12^2 + 5^2} = 13 \text{ ksi}$$

The principal normal stresses as shown in the circle are

$$\sigma_{p\text{max}} = 4 + 13 = +17 \text{ ksi} \quad \text{(tension!)}$$

at $\qquad \theta_{p_1} = \dfrac{1}{2} \text{ arc tan} \dfrac{5}{12} = 11.3°$ (clockwise from the positive x-axis)

and $\qquad \sigma_{p\text{min}} = 4 - 13 = -9 \text{ ksi} \quad \text{(compression!)}$

at $\qquad \theta_{p_2} = 90° + 11.3° = 101.3°$ (clockwise from the positive x-axis)

The orientation of the element is shown in Fig. 1-18(c).
(b)   The principal shearing stresses, as given in Mohr's circle, are

$$\tau_{\substack{\text{max} \\ \text{min}}} = \pm 13 \text{ ksi}$$

at $\qquad \theta_{s_1} = 45° - \theta_{p_1} = 33.7°$ (counterclockwise from the positive x-axis)

and $\qquad \theta_{s_2} = \theta_{s_1} + 90° = 123.7°$ (counterclockwise from the positive x-axis)

The normal stresses on the planes of principal shear are seen to be $+7$ ksi at both τ_{max} and τ_{min}.

alternate solution
Equations (1-8) through (1-13) afford a direct mathematical solution; unfortunately, comprehension of the physical nature of the solution is lost in what cannot be seen.

$$\sigma_{\substack{p\text{max} \\ \text{min}}} = \frac{\sigma_x - \sigma_y}{2} \pm \sqrt{\left(\frac{\sigma_x - \sigma_y}{2}\right)^2 + (\tau_{xy})^2}$$

$$= \frac{16 - 8}{2} \pm \sqrt{\left(\frac{16 + 8}{2}\right)^2 + (5)^2} = 4 \pm 13 = 17, -9 \text{ ksi}$$

$$\theta_{p_1} = \frac{1}{2} \text{ arc tan} \frac{2\tau_{xy}}{\sigma_x - \sigma_y} = \frac{1}{2} \text{ arc tan} \frac{2 \times 5}{24} = 11.3°$$

$$\theta_{p_2} = 90° + 11.3° = 101.3°$$

$$\tau_{\substack{\text{max} \\ \text{min}}} = \pm \sqrt{\left(\frac{\sigma_x - \sigma_y}{2}\right)^2 + (\tau_{xy})^2}$$

$$= \pm 13 \text{ ksi}$$

$$\theta_{s_1} = 45° - \theta_{p_1} = 45° - 11.3° = 33.7°$$
$$\theta_{s_2} = \theta_{s_1} + 90° = 33.7° + 90° = 123.7°$$

The answers are the same; the visualization and interpretation are much more difficult.

1-6 TORSIONAL STRESS AND STRAIN

Reaction must follow action—it is a pattern of life as well as a law of physics. Thus, stress must coexist with strain, and strain must accompany stress. When an object is pulled or pushed by a force or forces, its dimensions must change; when torques or couples act on a member, the member twists. In circular sections, the relationships between the torque T, shear stress τ, and angle of twist θ are easily derived; the algebraic relationships for maximum values are:

$$\tau_{\max} = \frac{TD}{2J} \qquad (1\text{-}14)$$

and

$$\theta = \frac{TL}{GJ} \qquad (1\text{-}15)$$

where G is the shear modulus of elasticity, and J is the polar moment of inertia. An assumption, proven valid and required in the development of the equations, states that *circular sections within a body will remain circular during elastic deformation*, while the planes along diameters such as *abcd* in Fig. 1-19(b) will uniformly and proportionally warp as shown. A microscopic view of point P, in Fig. 1-19(c), which lies on the surface of a thin wafer, illustrates the angular deformation associated with the twisting. The shear stresses τ_{xy} and τ_{yx} alone act at P. Principal stresses that are hidden within the element are exposed through the use of Mohr's circle, as in Figs. 1-19(d) and 1-19(e). Since the radius of Mohr's circle has a magnitude of τ_{xy}, it follows that, for pure shear

$$|\tau_{xy}| = |\tau_{yx}| = |\sigma_{P_{\max}}| = |\sigma_{P_{\min}}| \qquad (1\text{-}16)$$

An elastic constant K, called the *spring rate*, is defined as the ratio of the applied torque T to the angle of twist θ. The units of K are usually expressed in in.-lb per radian. Table 1-1 gives values of shear stress, angular strain, and spring rate for several common torsional sections. Numerical substitutions have been made for the applicable polar moments of inertia. Unfortunately, values of stress and strain for noncircular sections are not easily derived—these equations are obtained through a combination of advance theory of elasticity and experimentation.

The "spring-rate concept" applied to torsion bar analysis affords an interesting yet simple approach to problem solving. Consider the stepped shaft shown in Fig. 1-20(a). The torque T, which is transmitted

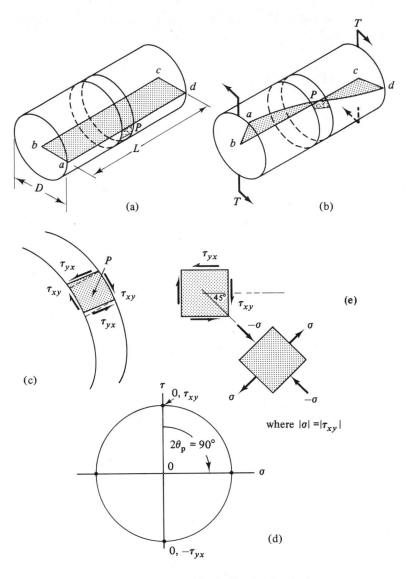

Figure 1-19 / Torsional stress.

equally through both segments, will cause a total angular strain of

$$\theta = \theta_A + \theta_B$$

Since $\theta_A = T/K_A$ and since $\theta_B = T/K_B$, the equivalent spring rate K_e for the

TABLE 1-1

TORSIONAL STRESS, ANGLE OF TWIST, AND TORSIONAL SPRING CONSTANTS

cross section	shear stress, τ psi τ = psi, T = torque in in.-lb	angle of twist, θ rad G = shear modulus, psi L = length, in.	spring constant k, in.-lb/rad
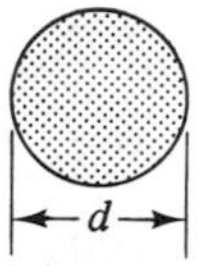	$\dfrac{16}{\pi}\dfrac{T}{d^3}$	$\dfrac{32}{\pi}\dfrac{1}{d^4}\dfrac{TL}{G}$	$\dfrac{\pi}{32}d^4\dfrac{G}{L}$
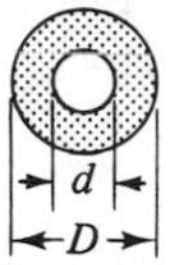	$\dfrac{16}{\pi}\dfrac{DT}{(D^4-d^4)}$	$\dfrac{32}{\pi}\dfrac{1}{(D^4-d^4)}\dfrac{TL}{G}$	$\dfrac{\pi}{32}(D^4-d^4)\dfrac{G}{L}$
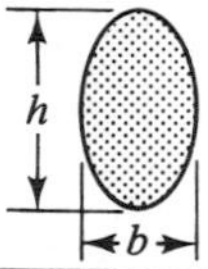	$\dfrac{16}{\pi}\dfrac{T}{b^2h}$ $(h>b)$	$\dfrac{16}{\pi}\dfrac{(b^2+h^2)}{b^3h^3}\dfrac{TL}{G}$	$\dfrac{\pi}{16}\dfrac{b^3h^3}{(b^2+h^2)}\dfrac{G}{L}$
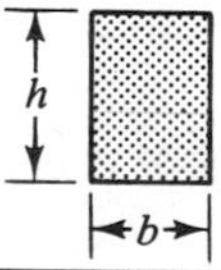	$4.5\dfrac{T}{b^2h}$	$3.5\dfrac{(b^2+h^2)}{b^3h^3}\dfrac{TL}{G}$ for $h=2b$	$0.29\dfrac{b^3h^3}{(b^2+h^2)}\dfrac{G}{L}$
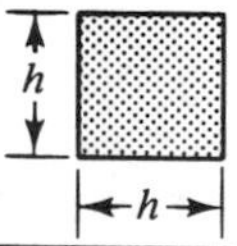	$4.5\dfrac{T}{h^3}$	$7.2\dfrac{1}{h^4}\dfrac{TL}{G}$	$0.141h^4\dfrac{G}{L}$
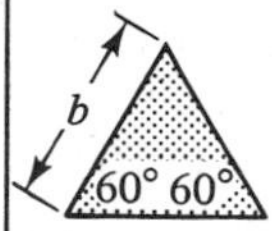	$20\dfrac{T}{b^3}$	$46.2\dfrac{1}{b^4}\dfrac{TL}{G}$	$0.02b^4\dfrac{G}{L}$
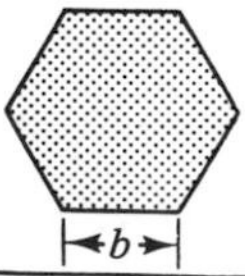	$1.1\dfrac{T}{b^3}$	$0.97\dfrac{1}{b^4}\dfrac{TL}{G}$	$1.03b^4\dfrac{G}{L}$

system would be

$$K_e = \frac{T}{\theta} = \frac{T}{\theta_A + \theta_B} = \frac{T}{\dfrac{T}{K_A}+\dfrac{T}{K_B}} = \frac{1}{\dfrac{1}{K_A}+\dfrac{1}{K_B}} \qquad (1\text{-}17)$$

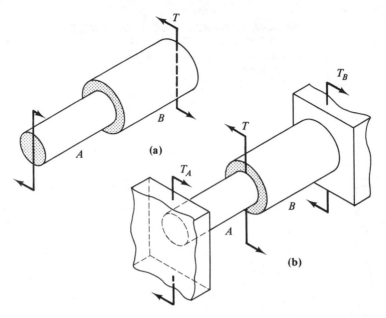

Figure 1-20 / Stepped shaft in torsion.

A shaft of n segments in *series* would have an equivalent spring rate of

$$K_e = \cfrac{1}{\cfrac{1}{K_1} + \cfrac{1}{K_2} + \cdots + \cfrac{1}{K_n}}$$

Note: By definition, a series system is one in which all segments sustain the same torque T.

Consider, next, the same two shafts now held between left and right rigid supports, Fig. 1-20(b). The torque T, applied at the junction of A and B is distributed (the more rigid the shaft, the greater share of the load it will carry) between segments A and B. Both shafts in this *parallel* system twist equally, thus

$$T = T_A + T_B \qquad \text{and} \qquad \theta_A = \theta_B$$

The equivalent spring rate would be

$$K_e = \frac{T}{\theta} = \frac{T_A + T_B}{\theta} = \frac{K_A\theta + K_B\theta}{\theta} = K_A + K_B \qquad (1\text{-}18)$$

EXAMPLE 1-7

A combination of connected solid steel ($G = 12 \times 10^6$ psi) torsion bars are acted on by a torque T, as shown in Fig. 1-21. The bars are held between left and right rigid supports. Determine: (a) the equivalent spring rate; (b) the angle of twist in degrees of each segment if $T = 5000$ in.-lb; and (c) the maximum stress in each of the three segments.

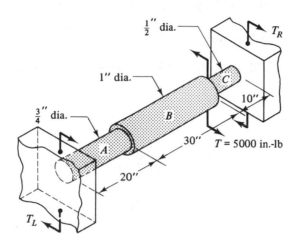

Figure 1-21 / Example 1-7.

solution

(a) For this particular torque location, segments A and B are in series, and their combination is in parallel with segment C.

$$K_e = \cfrac{1}{\cfrac{1}{K_A} + \cfrac{1}{K_B}} + K_C = \frac{K_A \times K_B}{K_A + K_B} + K_C$$

where

$$K_A = \frac{G}{L} \frac{\pi D^4}{32} = \frac{12 \times 10^6 \times \pi \times \overline{0.75}^4}{20 \times 32}$$

$$= 18.6 \times 10^3 \text{ in.-lb/rad}$$

$$K_B = \frac{12 \times 10^6 \times \pi \times 1^4}{30 \times 32} = 39.3 \times 10^3 \text{ in.-lb/rad}$$

$$K_C = \frac{12 \times 10^6 \times \pi \times \overline{0.5}^4}{10 \times 32} = 7.36 \times 10^3 \text{ in.-lb/rad}$$

Hence
$$K_e = \left[\frac{18.6 \times 39.3}{18.6 + 39.3} + 7.36\right] \times 10^3$$

$$= 20.0 \times 10^3 \text{ in.-lb/rad}$$

(b) The angle of twist θ_C at the junction of segments B and C, the point of application of the torque, is

$$\theta_C = \frac{T}{K_e} = \frac{5000}{20 \times 10^3} = 0.25 \text{ rad} \times \frac{360°}{2\pi \text{ rad}} = 14.3°$$

The torque T_R is next computed.

$$T_R = K_C\,\theta_C = 7.36 \times 10^3 \times 0.25 = 1840 \text{ in.-lb}$$

Since moments must balance,

$$T_L = T - T_R = 5000 - 1840 = 3160 \text{ in.-lb}$$

Thus
$$\theta_A = \frac{T_L}{K_A} = \frac{3160}{18.6 \times 10^3} = 0.17 \text{ rad or } 9.74°$$

and
$$\theta_B = \frac{T_L}{K_B} = \frac{3160}{39.3 \times 10^3} = 0.08 \text{ rad or } 4.58°$$

Check $\theta_A + \theta_B \overset{?}{=} \theta_C; \quad 0.17 + 0.08 = 0.25; \quad$ yes!

(c) The stress in each segment is next computed:

For A: $\tau_{max} = \dfrac{16T_A}{\pi D^3} = \dfrac{16 \times 3160}{\pi \times 0.75^3} = 38,300$ psi

For B: $\tau_{max} = \dfrac{16 \times 3160}{\pi \times 1^3} = 16,100$ psi

For C: $\tau_{max} = \dfrac{16 \times 1840}{\pi \times 0.5^3} = 75,000$ psi

1-7   AXIAL STRESS AND STRAIN

As with torsion, axial stress and strain are physical actions and reactions; when a force acts on the prismatic body (Fig. 1-22) the body's dimensions —length, width, and thickness—change. The strain ϵ, a measure of *unit*

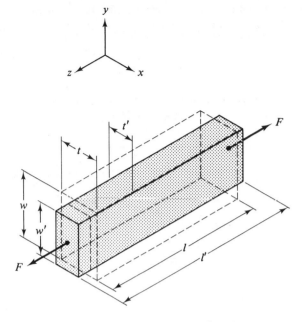

Figure 1-22 / Elements of strain.

change in length Δ/L along each of the coordinate axes, is defined as

$$\epsilon_x = (l' - l)/l$$
$$\epsilon_y = (w' - w)/w \qquad (1\text{-}19)$$
$$\epsilon_z = (t' - t)/t$$

Strain, as dimensional analysis indicates, is a number void of units; a strain of 0.01 can equally be interpreted to mean a change of 0.01 inches per inch or 0.01 feet per foot.

Early in the nineteenth century, the French mathematician Poisson demonstrated that in elastic materials, the ratio of lateral strain to axial strain (that strain along the axis of the force) is a constant μ. For the member shown in Fig. 1-22, Poisson's ratio would be defined as

$$\mu = \frac{\epsilon_y}{\epsilon_x} = \frac{\epsilon_z}{\epsilon_x} \qquad (1\text{-}20)$$

a numerical value which could be positive or negative. In most metals, Poisson's ratio numerically falls between the limits of $\frac{1}{4}$ and $\frac{1}{3}$.

A second important elastic constant that is a function of the physical properties of the given material is its *modulus of elasticity E*, which is defined as the ratio of axial stress to the accompanying axial strain.

$$E = \frac{\text{stress}}{\text{strain}} = \frac{F/A}{\Delta/L} = \frac{FL}{\Delta A} \tag{1-21}$$

Experimentally determined stress–strain diagrams, similar to that shown in Fig. 1-23, are used to find the modulus E. Precise measurements of

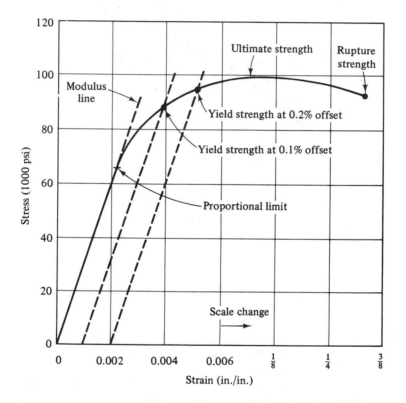

Figure 1-23 / Stress–strain diagram.

strain are made as a function of an increasing measurable load on a standardized "tensile bar." Stress is computed based on the original cross-sectional area, and the typical graph, Fig. 1-23, is plotted. Unfortunately, materials do not behave as the physicist would pray they should, so, often, the modulus of elasticity is a fictitious quantity. Most design limits stay well below the point where stress and strain cease to be proportional, simply because most materials are subjected to frequent load reversals—from tension to compression to tension to compression and so on. By not allowing the permissible stress to exceed the *proportional limit*, it is assumed that the material will elastically return to its original length when the load is removed—the material will recover. In static loading, much usable strength exists within the material beyond its elastic limit, and acceptable strengths are based on a certain amount of knowledgeable permanent deformation. These acceptable strengths are represented by the intersection of the stress-strain diagram with lines drawn parallel to the modulus line as shown—these permissible strengths are referred to as "yield strength at 0.1% offset, or at 0.2% offset."

Mohr's circle for the state of stress at point P is illustrated in Fig. 1-24; as can be seen, a material may be subjected to an axial load yet fail in shear. Thus, total design analysis requires total component analysis.

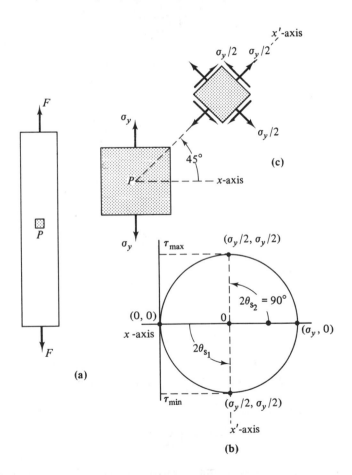

Figure 1-24

The concept of spring rate affords, as was demonstrated in the discussion of torsional stress and strain, a comfortable approach to the analysis of axial loading and in particular to statically indeterminate situations. From Eq. (1-22), the axial spring rate is defined as

$$K = \frac{F}{\Delta} = \frac{EA}{L} \qquad (1\text{-}22)$$

where E, A, and L are, respectively, the modulus of elasticity, the cross-sectional area normal to the axis of the force, and the length.

In a *series system*, Fig. 1-25(a), the force is transmitted equally through each member; the equivalent spring rate is, therefore,

$$K_e = \cfrac{1}{\cfrac{1}{K_A} + \cfrac{1}{K_B}}$$

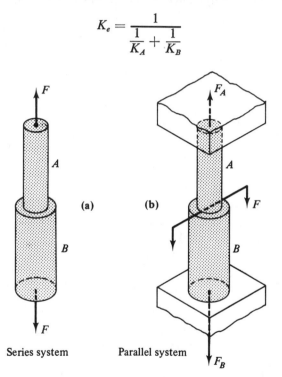

(a)   (b)

Series system   Parallel system

Figure 1-25 / Series and parallel systems.

and for n segments in series

$$K_e = 1/\sum \frac{1}{K_n}$$

In a *parallel system*, Fig. 1-25(b), the deflection is apportioned between segments A and B; the spring rate is

$$K_e = K_A + K_B$$

and for n segments in parallel

$$K_e = \sum K_n \qquad (1\text{-}23)$$

EXAMPLE 1-8

The system shown in Fig. 1-26 consists of four solid cylindrical steel members ($E = 30 \times 10^6$ psi, and $\mu = 0.30$) stacked as illustrated; assume that S_1 and S_2 are nonyielding supports. Find: (a) the total deformation Δ of the system; (b) the stress in each member; and (c) the diametrical expansion of each member.

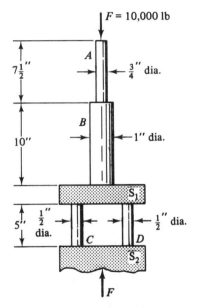

Figure 1-26 | A stacked system of members (Example 1-8).

solution

(a) The equivalent spring-rate concept (this is one of several approaches that could be employed) will be used. Members A and B are in series with the parallel combination of C and D; hence

$$K_e = \cfrac{1}{\cfrac{1}{K_A} + \cfrac{1}{K_B} + \cfrac{1}{K_C + K_D}}$$

where $K_A = \dfrac{EA}{L} = \dfrac{30 \times 10^6 \times \pi \times \overline{0.75}^2}{4 \times 7.5} = 1.77 \times 10^6 \ \text{lb/in.}$

$K_B = \dfrac{30 \times 10^6 \times \pi \times 1^2}{4 \times 10} = 2.36 \times 10^6 \ \text{lb/in.}$

$K_C = K_D = \dfrac{30 \times 10^6 \times \pi \times \overline{0.5}^2}{4 \times 5} = 1.18 \times 10^6 \ \text{lb/in.}$

Therefore, $K_e = \dfrac{10^6}{\dfrac{1}{1.77} + \dfrac{1}{2.36} + \dfrac{1}{2 \times 1.18}} = 0.708 \times 10^6 \ \text{lb/in.}$

The deflection at the point of application of the load F is

$$\Delta = \frac{F}{K_e} = \frac{10,000}{0.708 \times 10^6} = 0.0141 \ \text{in.}$$

(b)   The direct compressive stresses in each member are next found:

$$\sigma_A = \frac{F}{A} = \frac{10,000}{\pi \times \overline{0.75}^2/4} = 22,600 \ \text{psi}$$

$$\sigma_B = \frac{10,000}{\pi \times 1^2/4} = 12,700 \ \text{psi}$$

$$\sigma_C = \sigma_D = \frac{F/2}{A} = \frac{5000}{\pi \times \overline{0.5}^2/4} = 25,500 \ \text{psi}$$

(c)   The axial strain in each segment is found, and lastly, the diametrical expansions.

$$\epsilon_A = \frac{\sigma_A}{E} = \frac{22,600}{30 \times 10^6} = 7.53 \times 10^{-4} \ \text{in./in.}$$

$$\epsilon_B = \frac{12,700}{30 \times 10^6} = 4.23 \times 10^{-4} \ \text{in./in.}$$

$$\epsilon_C = \epsilon_D = \frac{25,500}{30 \times 10^6} = 8.50 \times 10^{-4} \ \text{in./in.}$$

By definition, the symbolic diametrical strain is

$$\epsilon_{\text{dia}} = \mu \epsilon_{\text{axial}}$$

where the numerical substitutions give

$$\Delta_A = \epsilon_{\text{dia}} \times \text{dia} = 0.3 \times 7.53 \times 10^{-4} \times 0.75$$
$$= 1.69 \times 10^{-4} \text{ in.}$$
$$\Delta_B = 0.3 \times 4.23 \times 10^{-4} \times 1 = 1.27 \times 10^{-4} \text{ in.}$$
$$\Delta_C = \Delta_D = 0.3 \times 8.50 \times 10^{-4} \times 0.5 = 1.28 \times 10^{-4} \text{ in.}$$

1-8   THERMAL STRESS AND STRAIN

Nature has provided the designer with a most troublesome problem—when most materials are heated or cooled, their dimensions change—they expand or contract. To add to the dilemma, not only are the various elements that comprise a machine or structure subjected to various different temperatures, the materials themselves expand or contract at different rates depending upon their composition. While the effects of natural expansion and contraction cannot be totally controlled, they can be minimized by allowing the various components to expand more or less freely without geometric interference. Bridge designers have, for years, used "floating construction." One of the ends of the expanse is pin-connected to its supports, while the other end floats on sets of rollers. The expansion–contraction problem perplexed pendulum clock-makers for centuries—a slight change in temperature caused a change in the length of the pendulum, which in turn changed its period of oscillation. Novel compensating devices were invented, and then one day, a nickel alloy, Invar, was discovered whose thermal dimensional change was negligible. By that time, the electric clock was in vogue.

Fortunately, for the sake of computation, if none other, the expansion rate as a function of temperature is linear in the usable (elastic) region of most materials. A constant α, the coefficient of thermal expansion, is defined as the ratio of strain ϵ to temperature change.

$$\alpha = \frac{\epsilon}{\Delta T} = \frac{\delta}{L \times \Delta T} \qquad (1\text{-}24)$$

Solving Eq. (1-24) for deformation δ gives

$$\delta = \alpha L \times \Delta T \qquad (1\text{-}25)$$

In American engineering practice, the customary units of α are inches per inch per degree Fahrenheit. Steel, for example, has a coefficient of thermal expansion of 6.5×10^{-6} in./in./°F, which means that all dimensions—length, width, thickness, and diameter—will change by 0.0000065 in./in. for every degree of temperature change. While this is a small number, it cannot be ignored: a mile of steel railroad track that undergoes a temperature change of 100°F will change its length by

$$\delta = \alpha L \times \Delta T = 6.5 \times 10^{-6} \times 5280 \times 12 \times 100 = 41.2 \text{ in.}$$

If totally constrained, this expansion or contraction would constitute a stress of 19,500 psi in the rail, since

$$\delta = \frac{FL}{EA} = \frac{\sigma L}{E} = \alpha L \, \Delta T$$

$$\sigma = \alpha \times \Delta T \times E = 6.5 \times 10^{-6} \times 100 \times 30 \times 10^6 = 19,500 \text{ psi}$$

Superposition, a powerful mathematical concept that will find frequent application throughout this book, is an approach that ably lends itself to problems involving thermal stress and strain. Superposition simply means "to mathematically superimpose—to solve separately and combine results." For example, a beam will sag or deflect under a given load. If the load is doubled, the deflection will double (the solutions were added), superimposed upon one another.

EXAMPLE 1-9

An aluminum rod is fastened to a nonyielding support at A as shown in Fig. 1-27. The free end can move 0.02 in. before touching a similar support at B. Determine: (a) the temperature increase to cause the free end to contact B; and (b) the stress in the rod if its temperature increases 100°F.

solution

The following data are obtained from the Appendix:

$$E = 10 \times 10^6 \text{ psi}$$

$$\alpha = 13.1 \times 10^{-6} \text{ in./in./°F}$$

(a)   The change in temperature ΔT as a function of thermal deformation δ_T, α, and L is

$$\Delta T = \frac{\delta_T}{\alpha L} = \frac{0.02}{13.1 \times 10^{-6} \times 20} = 76.3°F$$

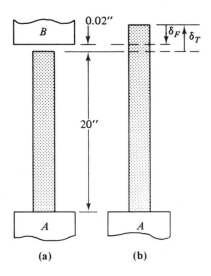

Figure 1-27 / Thermal strain (Example 1-9).

(b) Two situations will be superimposed; first the bar will be allowed to freely expand δ_T, Fig. 1-27(b); the support at B is imagined to be nonexistent. Then the support will be shoved in place; to forcibly accomplish this, the rod will deform δ_F. Geometry will give the relationship between δ_T and δ_F as shown in the figure.

$$\delta_T = \delta_F + 0.02$$

$$\alpha L\, \Delta T = \frac{FL}{EA} + 0.02 = \frac{\sigma L}{E} + 0.02$$

Substitution of data gives

$$13.1 \times 10^{-6} \times 20 \times 100 = \frac{20 \times \sigma}{10 \times 10^6} + 0.02$$

and
$$\sigma = \frac{(0.026 - 0.02) \times 10 \times 10^6}{20} = 3000 \text{ psi}$$

EXAMPLE 1-10

A steel bar and an aluminum bar, each secured to a rigid support, are fastened at their free ends by a 1-in.-diameter pin, as shown in Fig. 1-28. Determine the shearing stress on the pin if the temperature drops 50°F. The bars are initially free of stress.

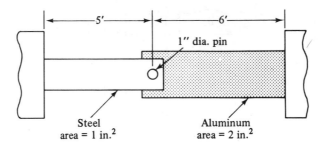

Figure 1-28 / Shear stress in a pin due to thermal strain (Example 1-10).

solution

With one constraint removed, the right-hand wall for instance, the joined and stress-free members would contract $\delta_{T_s} + \delta_{T_{al}}$. To pull the members back to the wall would mean to forcibly stretch them $\delta_{F_s} + \delta_{F_{al}}$; hence

$$\delta_{F_s} + \delta_{F_{al}} = \delta_{T_s} + \delta_{T_{al}}$$

$$\left(\frac{FL}{EA}\right)_s + \left(\frac{FL}{EA}\right)_{al} = (\alpha L \, \Delta T)_s + \alpha L \, \Delta T_{al}$$

Substitution of data (see Appendix) and factoring gives

$$\frac{12F}{10^6}\left[\frac{5}{30 \times 1} + \frac{6}{10 \times 2}\right] = 50 \times 12 \times 10^{-6}[(6.5 \times 5) + (13.1 \times 6)]$$

$$F = 11,900 \text{ lb}$$

Finally, the shear stress in the pin is

$$\tau = \frac{F}{A} = \frac{11,900}{\pi \times \overline{0.5^2}} = 15,200 \text{ psi}$$

While thermal stresses cannot be totally avoided, the designer can exercise some judicious control over them. He can avoid, wherever possible, rigid constraints, and he can allow materials to "float." He can also build in intentional flexibility—the bellows and the expansion loops are examples of this intentional floppiness. He can wisely use gaskets and flexible seals to afford a control over pressure leakages and the loosening of bolts. And he can consider the importance of insulation to retard heat flow, thereby lessening temperature differentials that occur between "hot" and "cold" components.

1-9   NORMAL AND SHEAR STRESSES IN SIMPLE BEAMS

Simple beams are defined as members, initially straight and of uniform cross section supported only at the ends of the beam, that are acted upon by forces normal to the axis of the member. A simple beam is also assumed to have a symmetrical cross section with reference to the axis of the force or forces. Two basic types of stress act within the beam. One stress is caused by internal bending moments M that simultaneously cause some fibers to elongate and others to contract. The second type of stress is caused by transverse shearing forces V which tend to cause slippage between axial planes. The shaded portion of the beam shown in Fig. 1-29(a) and magnified in Fig. 1-29(b) illustrates how a fiber situated at a distance y from the neutral plane (a plane within the beam whose length is unchanged) elongates as a function of the curvature $\Delta\theta$ under the action of the moment. The maximum elongation occurs at a distance c_t (with the subscript denoting that the action is associated with a tensile stress). Simultaneously, the maximum contraction (compression) occurs at a distance c_c from the neutral plane. It can be shown that the neutral plane (or *neutral axis* as it is often called) lies at the centroid of the cross section —an important point to remember.

Since it is assumed that plane a–a simply rotates about O to a'–a', it follows that the stress distribution, Fig. 1-29(c), varies proportionally with distance from the neutral plane. The value of the normal stress as a function of y is

$$\sigma = \frac{My}{I} \tag{1-26}$$

Where I is the moment of inertia of the cross-sectional area with reference to the axis of the neutral plane and M, the bending moment at that particular point in the beam, the maximum values of tensile and compressive stresses are, respectively

$$\sigma_t = \frac{Mc_t}{I} \tag{1-27}$$

and

$$\sigma_c = \frac{Mc_c}{I} \tag{1-28}$$

The volumes of the tensile and compressive stress prisms, Fig. 1-29(c), are dimensionally equal to forces F. These forces equate and, to maintain static equilibrium, form a couple that balances the moment M.

Equations (1-27) and (1-28) are often written in terms of *section*

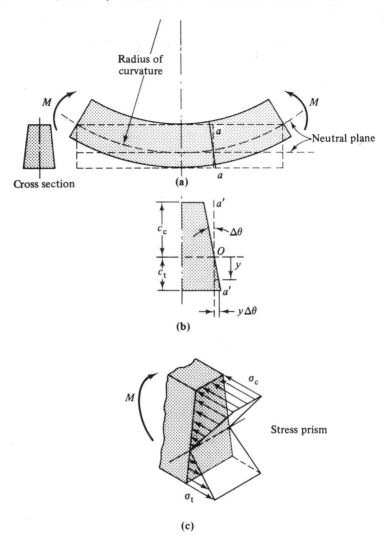

Figure 1-29 / Flexural stresses.

moduli Z_t and Z_c

$$\sigma_t = \frac{M}{Z_t} \tag{1-29}$$

and

$$\sigma_c = \frac{M}{Z_c} \tag{1-30}$$

where Z is the ratio I/c; comparative magnitudes of section moduli provide a rapid method of making a relative check of section strength. Table 1-2 lists some properties of frequently encountered beam sections; reference

TABLE 1-2

MOMENT OF INERTIA AND SECTION
MODULUS OF BEAM SECTIONS

	cross section	moment of inertia I_x	section modulus Z_x
Square	$x -$ b	$\dfrac{b^4}{12}$	$\dfrac{b^3}{6}$
Rectangle	$x -$ h b	$\dfrac{bh^3}{12}$	$\dfrac{bh^2}{6}$
Hollow rectangle	$x-$ h_1 b_1 h b	$\dfrac{bh^3 - b_1 h_1^3}{12}$	$\dfrac{bh^3 - b_1 h_1^3}{6h}$
Equal rectangles	$x - - - -$ $h_1 - - h$ b	$\dfrac{b(h^3 - h_1^3)}{12}$	$\dfrac{b(h^3 - h_1^3)}{6h}$
Circle	$x - - - - - -$ d	$\dfrac{\pi d^4}{64}$	$\dfrac{\pi d^3}{32}$
Hollow circle	$x - - - -$ d_1 d	$\dfrac{\pi(d^4 - d_1^4)}{64}$	$\dfrac{\pi(d^4 - d_1^4)}{32d}$

can be made to engineering handbooks for a complete listing of properties of areas.

As mentioned, a second type of stress, that caused by tendency toward *horizontal shear*, occurs in beam loading. This shearing tendency can be described by imagining the beam to consist of a number of thin layers piled one on top of the other as shown in Fig. 1-30(a). A load acting on this beam will cause the beam to sag, Fig. 1-30(b), with the layers slipping as shown. In a solid beam, Fig. 1-30(c), this shearing tendency still exists. By examining a magnified view of a segment P of the beam, Fig. 1-30(d), it can be seen that the shear force is caused by the difference in normal forces F_1 and F_2. When developed, the equation governing the

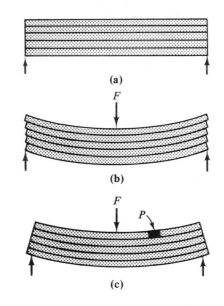

(a)

(b)

(c)

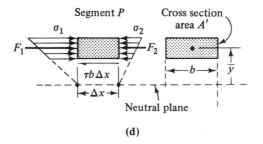

(d)

Figure 1-30 / Shear due to bending.

shear stress τ at a point within the beam is

$$\tau = \frac{V}{Ib} A'\bar{y} \qquad (1\text{-}31)$$

where $V =$ Magnitude of the transverse shear force at the point within
the beam
$I =$ Area moment of inertia about the neutral plane
$b =$ Width of the material at the shear plane under investigation
$A'\bar{y} =$ First moment of the cross-sectional area lying above the
plane at which the shear stress is to be computed

In Eq. (1-31), V and I are constants for any given section; the shear stress is, therefore, a maximum when the quantity $A'\bar{y}/b$ is a maximum. In a rectangular beam, Fig. 1-31(a), the shear stress is a maximum

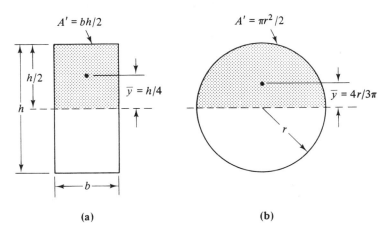

(a)   (b)

Figure 1-31 | *Maximum shear stress occurs at the neutral plane.*

at the neutral plane and is equal to

$$\tau_{\max} = \frac{V}{Ib} A'\bar{y} = \frac{V}{b^2 h^3/12} \times \frac{bh}{2} \times \frac{h}{4} = \frac{3V}{2A} \qquad (1\text{-}32)$$

where A is the area of the entire cross section. In a circular section, Fig. 1-31(b), the maximum shear stress is again at the neutral plane and is equal to

$$\tau_{\max} = \frac{4V}{3A} \qquad (1\text{-}33)$$

EXAMPLE 1-11

A box beam supports a distributed and a concentrated load as shown in Fig. 1-32(a). Find: (a) the maximum tensile and compressive stresses acting in the beam; (b) the maximum shearing stress within the beam; (c) the state of stress at point P which lies at $x = 4$ ft and 2 in. down from the top; and (d) the principal normal and shearing stresses.

solution

(a) and (b)   Since the beam is symmetrically loaded, the reactions R_1 and R_2 share the load equally.

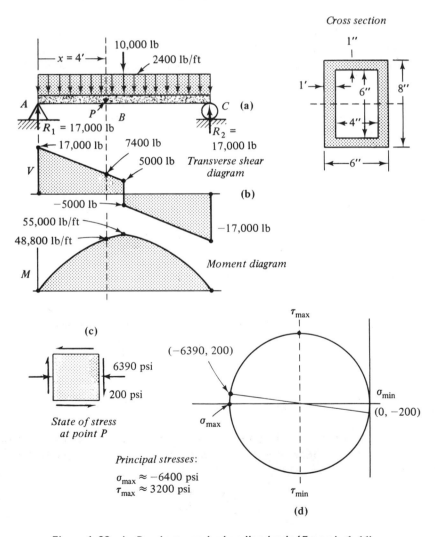

Figure 1-32 / Box beam under bending loads (Example 1-11).

$$R_1 = R_2 = \frac{(2400 \times 10) + 10,000}{2} = 17,000 \text{ lb}$$

The *transverse shear* and *moment* diagrams are next plotted as shown in Fig. 1-32(b). Since *the area of the shear diagram between any two points on the beam is equal to the difference in moment between these two points*, it follows that

$$M_B = \frac{17,000 + 5000}{2} \times 5 = 55,000 \text{ lb-ft}$$

Next, the moment of inertia with reference to the neutral plane is determined.

$$I_{NA} = \frac{1}{12} bh^3 - \frac{1}{12} b'h'^3 = \frac{1}{12}[(6 \times 8^3) - (4 \times 6^3)] = 184 \text{ in.}^4$$

The first moment of the beam's cross-sectional area lying above the neutral plane is next computed.

$$A'\bar{y} = A_1\bar{y}_1 - A_2\bar{y}_2 = (4 \times 6 \times 2) - (4 \times 3 \times 1.5) = 30 \text{ in.}^3$$

Finally, the normal and shearing stresses are found. Since the neutral plane is an axis of symmetry, the maximum tensile and maximum compressive stresses are equal.

$$\sigma = \frac{M_c}{I} = 55,000 \text{ lb-ft} \times 12 \text{ in.} \times 4 \text{ in.} \times \frac{1}{184 \text{ in.}^4}$$

$$= 14,300 \text{ psi} \quad \text{at } B$$

and $\quad \tau = \frac{V}{Ib} A'\bar{y} = \frac{17,000 \text{ lb}}{184 \text{ in.}^4 \times 2\text{in.}} \times 30 \text{ in.}^3$

$$= 1390 \text{ psi} \quad \text{at } A \text{ and } C$$

(c) and (d)   The moment M and transverse shear V at P are, respectively:

$$V_{x=4} = 17,000 - (2400 \times 4) = 7400 \text{ lb}$$
$$M_{x=4} = (17,000 \times 4) - (2400 \times 4 \times 2) = 48,800 \text{ lb-ft}$$

The magnitudes of the normal and shearing stresses are, therefore,

$$|\sigma| = \frac{My}{I} = \frac{48,800 \times 12 \times 2}{184} = 6390 \text{ psi}$$

$$|\tau| = \frac{V}{Ib} A'\bar{y} = \frac{7400}{184 \times 2} \times [(6 \times 2 \times 1) - (4 \times 1 \times 0.5)] = 200 \text{ psi}$$

The state of stress at P is illustrated in Fig. 1-32(c); Mohr's circle is then drawn, Fig. 1-32(d); and the principal stresses as defined in the circle are found either analytically or graphically.

1-10 STRESSES IN CURVED BEAMS SUBJECTED TO PURE MOMENTS

Because of the variation in arc length, stresses in curved beams are not proportional to distance from the neutral plane as they are in straight beams. A typical comparative stress distribution on the shaded portion of a composite straight and curved beam is shown in Fig. 1-33. The essential

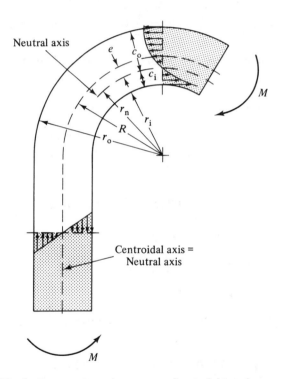

Figure 1-33 / Stress pattern in a composite straight and curved beam.

difference in the curved beam is the shift toward the center of curvature of the neutral plane. This shift e is a function of radius of curvature R and the geometry of the cross-sectional area.

It is not always readily apparent whether the maximum stress will occur at the inside or outside fibers if the beam section lacks symmetry. In a rectangle, circle, or equal-flanged beam, however, the maximum stress will always occur at the inside fiber. Equations, which can be derived, that give values of stress σ_i and σ_o at the inside and outside fibers are

$$\sigma_i = \frac{Mc_i}{Aer_i} \qquad (1\text{-}34)$$

and

$$\sigma_o = \frac{Mc_o}{Aer_o}$$

where A is the cross-sectional area of the beam. The remaining symbols are defined in Fig. 1-33. Table 1-3 gives values of r_n, e, and R for five common beam sections that are likely to be encountered in machine frames, clamps, and hooks.

EXAMPLE 1-12

The member, Fig. 1-34(a), is subjected to a pure moment as shown. Determine the maximum tension, compression, and shear stress at planes *abcd* and *efgh* if: (a) $R = 3$ in.; and (b) $R = 21$ in.

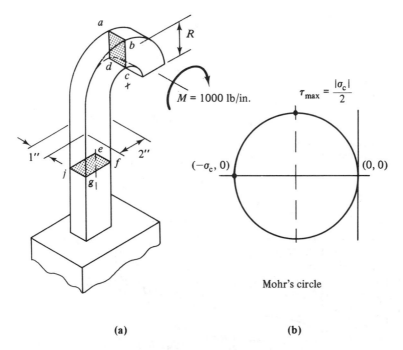

(a)   (b)

Figure 1-34 | Curved beam subject to a bending moment (Example 1-12).

TABLE 1-3

VALUES OF THE CONSTANTS r_n, e, AND R FOR CURVED BEAMS

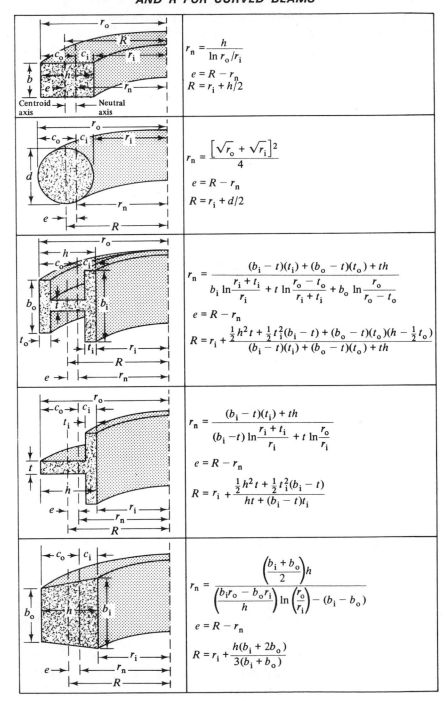

$$r_n = \frac{h}{\ln r_o / r_i}$$

$$e = R - r_n$$

$$R = r_i + h/2$$

$$r_n = \frac{\left[\sqrt{r_o} + \sqrt{r_i}\right]^2}{4}$$

$$e = R - r_n$$

$$R = r_i + d/2$$

$$r_n = \frac{(b_i - t)(t_i) + (b_o - t)(t_o) + th}{b_i \ln \dfrac{r_i + t_i}{r_i} + t \ln \dfrac{r_o - t_o}{r_i + t_i} + b_o \ln \dfrac{r_o}{r_o - t_o}}$$

$$e = R - r_n$$

$$R = r_i + \frac{\frac{1}{2}h^2 t + \frac{1}{2}t_i^2(b_i - t) + (b_o - t)(t_o)(h - \frac{1}{2}t_o)}{(b_i - t)(t_i) + (b_o - t)(t_o) + th}$$

$$r_n = \frac{(b_i - t)(t_i) + th}{(b_i - t)\ln \dfrac{r_i + t_i}{r_i} + t \ln \dfrac{r_o}{r_i}}$$

$$e = R - r_n$$

$$R = r_i + \frac{\frac{1}{2}h^2 t + \frac{1}{2}t_i^2(b_i - t)}{ht + (b_i - t)t_i}$$

$$r_n = \frac{\left(\dfrac{b_i + b_o}{2}\right)h}{\left(\dfrac{b_i r_o - b_o r_i}{h}\right)\ln\left(\dfrac{r_o}{r_i}\right) - (b_i - b_o)}$$

$$e = R - r_n$$

$$R = r_i + \frac{h(b_i + 2b_o)}{3(b_i + b_o)}$$

solution

(a) The constants r_n, e, h_i, and h_o are first computed for the curved section:

$$r_n = \frac{h}{\ln r_o/r_i} = \frac{2}{\ln 4/2} = \frac{2}{0.69315} = 2.8854 \text{ in.}$$

$$e = R - r_n = 3 - 2.8854 = 0.1146 \text{ in.}$$

$$c_i = r_n - r_i = 2.8854 - 2 = 0.8854 \text{ in.}$$

$$c_o = r_o - r_n = 4 - 2.8854 = 1.1146 \text{ in.}$$

The stress at the inner fiber is

$$\sigma_i = \frac{Mc_i}{Aer_i} = \frac{1000 \times 0.8854}{1 \times 2 \times 0.1146 \times 2} = 1930 \text{ psi}$$

The stress at the outer fiber is

$$\sigma_o = \frac{Mc_o}{Aer_o} = \frac{1000 \times 1.1146}{1 \times 2 \times 0.1146 \times 4} = 1216 \text{ psi}$$

From Mohr's circle, Fig. 1-34(b), the maximum shear stress is equal to

$$\tau_{max} = \frac{\sigma_{max}}{2} = \frac{1930}{2} = 965 \text{ psi}$$

In the straight section of the beam, tensile and compressive stresses are equal.

$$\sigma_t = \sigma_c = \left| \frac{Mc}{I} \right| = \frac{1000 \times 1}{\frac{1}{12} \times 1 \times 2^3} = 1500 \text{ psi}$$

It is interesting to note that even for the small radius of curvature, $R = 3$ in., the percent difference in maximum stresses—curved beams versus straight beams—is not a large number.

$$\% \text{ difference} = \frac{1930 - 1500}{1930} \times 100 = 22\%$$

(b) It would be expected that for a large radius of curvature, σ_o and σ_i would more nearly equal the stress in the straight portion of the member. Proceeding as before:

$$r_n = \frac{h}{\ln r_o/r_i} = \frac{2}{\ln r_o/r_i} = \frac{2}{\ln 22/20} = \frac{2}{0.09531} = 20.9842 \text{ in.}$$

$$e = R - r_n = 21 - 20.9842 = 0.01584 \text{ in.}$$

$$c_i = r_n - r_i = 20.9842 - 20 = 0.9842 \text{ in.}$$

$$c_o = r_o - r_n = 22 - 20.9842 = 1.0158 \text{ in.}$$

$$\sigma_i = \frac{Mc_i}{Aer_i} = \frac{1000 \times 0.9842}{1 \times 2 \times 0.01584 \times 20} = 1550 \text{ psi}$$

$$\sigma_o = \frac{Mc_o}{Aer_o} = \frac{1000 \times 1.0158}{1 \times 2 \times 0.01584 \times 22} = 1460 \text{ psi}$$

Thus, $\sigma_o \approx \sigma_i \approx 1500$ psi. If the member was considered straight for the large radius, $R = 21$ in., the error in the answer would be negligible.

A second approach to the curved beam problem, which is perhaps not as accurate but is just as adequate, is through the use of graphs of the type shown in Fig. 1-35. Eight typical sections are considered with a concentration factor K given as a function of the ratio R/c. The quantity c is the distance from the centroidal axis of the section to the inside edge (for a nonsymmetrical section, c would have to be determined first). The graphs do point out two important aspects of the problem of stress in curved beams: first, cross-section geometry is not nearly as important as one would imagine, especially at the outer fibers; second, for large radii of curvature, the beam may be considered to be straight.

1-11 COMBINED LOADING

Simplicity of the action of a single load usually does not reign in the usual machine member. Figure 1-36 illustrates how a single force acting on a member can produce direct, torsional, and bending loads at a particular section. Individually, the stresses are easily calculable by the formulas previously described:

$$\text{Direct stress:} \quad \sigma = \frac{P}{A}$$

$$\text{Torsional stress:} \quad \tau = \frac{TD}{2J} \quad \text{(for circular sections)}$$

$$\text{Bending stress:} \quad \sigma = \frac{Mc}{I}$$

$$\text{Beam shear stress:} \quad \tau = \frac{V}{Ib} A'\bar{y}$$

The combined stress is, of course, a function of the addition of the singular stresses—axial stresses caused by various loads may be added by super-

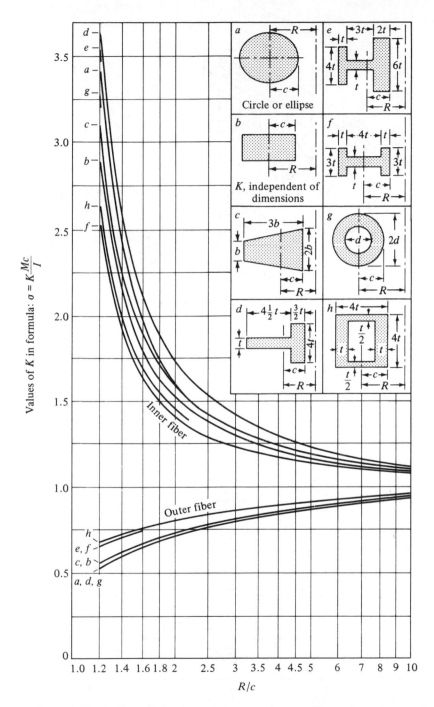

Figure 1-35 / Factor K for determining stress in curved flexural members.

55

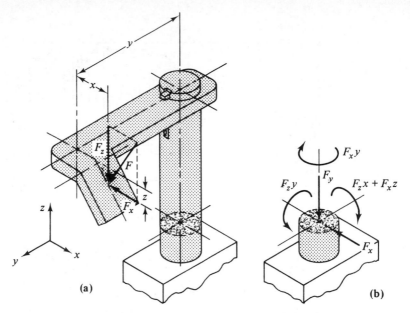

(a)

(b)

Figure 1-36 | Stresses due to eccentric loading.

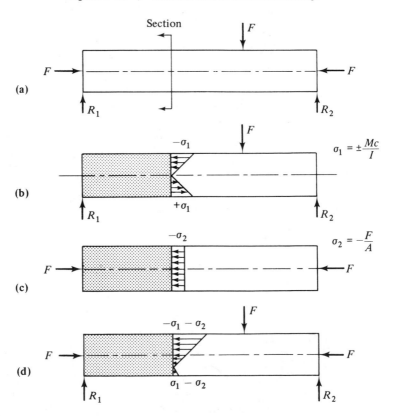

Figure 1-37 | Superposition of stresses.

imposing one upon the other. In a similar manner, shear stress can be added directly to shear stresses; Mohr's circle is employed to find the result of the combination of axial and shear loads.

As an example of direct superposition addition of stress, consider the beam shown in Fig. 1-37(a). The tensile and compressive stresses caused by bending, Fig. 1-37(b), and the stress caused by the compressive load, Fig. 1-37(c), are superimposed to give the final stress distribution shown in Fig. 1-37(d). The maximum axial stress at a particular point is found by

$$\sigma = -\frac{P}{A} \pm \frac{Mc}{I} \qquad (1\text{-}35)$$

Bending in two orthogonal directions is still another example of the superposition of stress. The progressive steps to the final analysis of the stress distribution within the beam shown in Fig. 1-38 are illustrated in the sequence of sketches. A line of zero stress, the neutral axis, is always apparent in this type of loading. Three examples follow that illustrate an appropriate approach to the analysis.

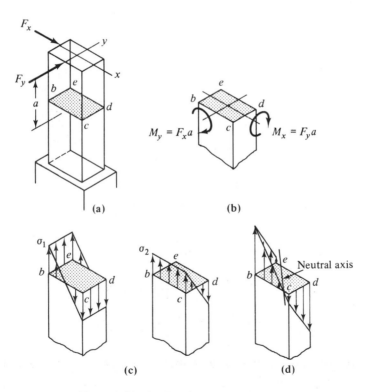

Figure 1-38 / Bending in two directions.

EXAMPLE 1-13

Determine the magnitude and direction of the stresses that act at each of the four corners of the plane *abcd* in Fig. 1-39(a).

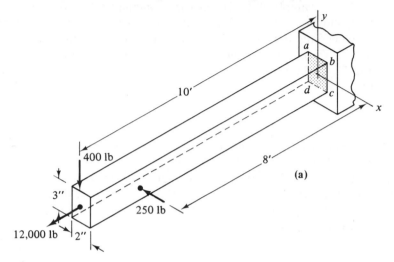

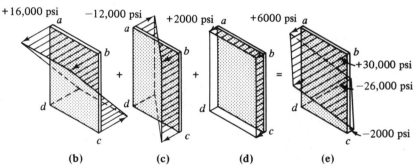

Figure 1-39 / A complex loading (Example 1-13).

solution

The section moduli are first computed.

$$Z_x = \frac{bh^2}{6} = \frac{2 \times 3^2}{6} = 3 \text{ in.}^3$$

$$Z_y = \frac{b^2h}{6} = \frac{2^2 \times 3}{6} = 2 \text{ in.}^3$$

Next, the bending stresses, first about the *x*-axis and then about the *y*-axis, are computed.

$$\sigma_{x\text{-axis}} = \pm \frac{M}{Z_x} = \frac{400 \times 10 \times 12}{3} = \pm 16{,}000 \text{ psi}$$

$$\sigma_{y\text{-axis}} = \pm \frac{M}{Z_y} = \frac{250 \times 8 \times 12}{2} = \pm 12{,}000 \text{ psi}$$

Finally, the direct tensile stress is found.

$$\sigma_{\text{direct}} = +\frac{P}{A} = \frac{12{,}000}{2 \times 3} = 2000 \text{ psi}$$

Thus
$$\sigma_a = 16{,}000 - 12{,}000 + 2000 = 6000 \text{ psi}$$
$$\sigma_b = 16{,}000 + 12{,}000 + 2000 = 30{,}000 \text{ psi}$$
$$\sigma_c = -16{,}000 + 12{,}000 + 2000 = -2000 \text{ psi}$$
$$\sigma_d = -16{,}000 - 12{,}000 + 2000 = -26{,}000 \text{ psi}$$

The composite stress distribution is shown in Fig. 1-39(e).

EXAMPLE 1-14

Two loads act on the 2-in.-diameter shaft as shown in Fig. 1-40(a). Find: (a) the maximum tensile stress in the shaft; and (b) the maximum shear stress caused by bending.

solution

(a) The end reactions are found in the usual manner, and the free body diagram is drawn as shown in Fig. 1-40(b). Two-dimensional shear and moment diagrams are then drawn, Fig. 1-40(c) and (d). Forces and moments are directed magnitudes, and they can be added vectorially; the maximum moment occurs at B and the maximum transverse shear between points A and B.

$$V_{\max} = \sqrt{200^2 + 600^2} = 632 \text{ lb}$$

$$M_B = \sqrt{600^2 + 1800^2} = 1900 \text{ lb-ft}$$

$$M_C = \sqrt{900^2 + 900^2} = 1270 \text{ lb-ft}$$

At B: $\sigma_{\max} = \dfrac{M}{Z} = \dfrac{M}{\pi d^3/32} = \dfrac{32 \times 1900 \times 12}{\pi \times 2^3} = 29{,}000 \text{ psi}$

(b) Between A and B:

$$\tau_{\max} = \frac{V}{Ib} A'\bar{y} = \frac{4}{3}\frac{V}{A} = \frac{4}{3} \times 632 \times \frac{4}{(2)^2} = 268 \text{ psi}$$

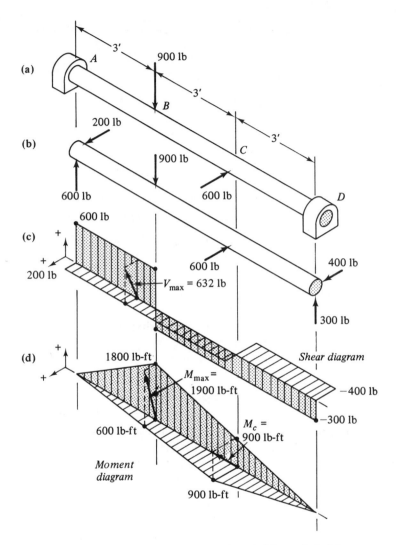

Figure 1-40 / Shaft under biaxial loads (Example 1-14).

EXAMPLE 1-15

The offset cylindrical bar supports a load of 200 lb as shown in Fig. 1-41. Determine the permissible eccentricity y if the maximum permissible tensile stress is not to exceed 10,000 psi.

solution

Two tensile stresses, one caused by bending and the other by the direct load, act within the member at point P. This point is selected as

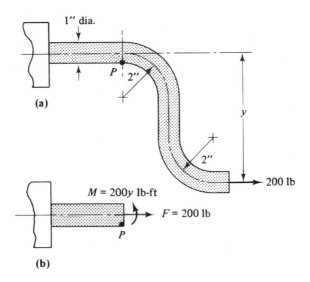

Figure 1-41 / Offset cylindrical bar under load (Example 1-15).

being the most critical, since it is the inside fiber of the curved beam and is at a section where the moment is a maximum.

For the curved beam:

$$r_n = \frac{(\sqrt{r_o} + \sqrt{r_i})^2}{4} = \frac{(\sqrt{2.5} + \sqrt{1.5})^2}{4} = 1.9684 \text{ in.}$$

$$e = R - r_n = 2 - 1.9684 = 0.03159 \text{ in.}$$

$$c_i = r_n - r_i = 1.9684 - 1.5 = 0.4684 \text{ in.}$$

$$A = \frac{\pi d^2}{4} = \frac{\pi \times 1^2}{4} = 0.785 \text{ in.}^2$$

Next, the two tensile stresses are superimposed by direct addition.

$$\sigma = \frac{Mc_i}{Aer_i} + \frac{F}{A}$$

$$10{,}000 = \frac{200 \times y \times 0.4684}{0.785 \times 0.03159 \times 1.5} + \frac{200}{0.785}$$

$$10{,}000 = 2520y + 255$$

$$y = 3.87 \text{ in.}$$

Bending stresses and direct axial stresses are induced when loads are applied *eccentrically*, as illustrated in Fig. 1-42(a). The equivalent loading, Fig. 1-42(b), consists of the bending moments M_x and M_y and the

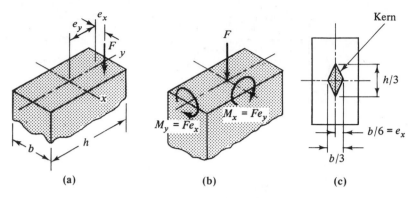

Figure 1-42 / Eccentric loads.

direct load of F. The stress at any corner would be:

$$\sigma = \pm\frac{F}{A} \pm \frac{M_x}{Z_x} \pm \frac{M_y}{Z_y} \qquad (1\text{-}36)$$

Equation (1-36) is an important design relationship to consider when using materials of construction that are weak in tension. The permissible amount of eccentricity for an allowable tensile stress of *zero* can be found by equating the tensile bending stresses to the direct stress.

$$\frac{F}{A} = \frac{M_x}{Z_x} + \frac{M_y}{Z_y}$$

$$\frac{F}{A} = \frac{Fe_y}{bh^2/6} + \frac{Fe_x}{b^2h/6} = \frac{F}{A}\left(\frac{6e_y}{h} + \frac{6e_x}{b}\right)$$

$$1 = \frac{6e_y}{h} + \frac{6e_x}{b}$$

By setting first $e_x = 0$ and then $e_y = 0$, the permitted separate axial eccentricities for the rectangular area are found:

$$e_y = h/6 \qquad \text{and} \qquad e_x = b/6$$

Thus, a load acting anywhere within the shaded area (called the *kern*), Fig. 1-42(c), will not produce a tensile stress in the member. Table 1-4 gives dimensions of the kerns of several frequently encountered sections.

Axial stresses and shear stresses cannot be added directly; they are of different species, like apples and oranges—their sum is fruit. Mohr's circle is the necessary instrument for finding principal stresses as the example that follows illustrates.

TABLE 1-4

KERNS OF COMMON SECTIONS

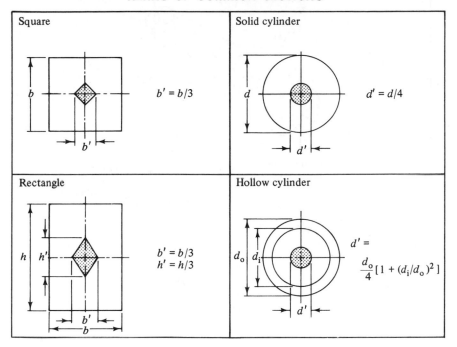

Square	Solid cylinder
$b' = b/3$	$d' = d/4$
Rectangle	Hollow cylinder
$b' = b/3$ $h' = h/3$	$d' = \dfrac{d_o}{4}[1 + (d_i/d_o)^2]$

EXAMPLE 1-16

A ship's propeller mounted on a 3-in.-diameter solid steel shaft, Fig. 1-43(a), produces the torque and pull shown. Find the principal stresses that are developed in the shaft.

solution

The tensile and shearing stresses are computed; Mohr's circle is drawn and the principal stresses determined.

$$\tau = \frac{TD}{2J} = \frac{30{,}000 \times 3}{2 \times \pi \times 3^4/32} = 5660 \text{ psi}$$

$$\sigma = \frac{F}{A} = \frac{20{,}000}{\pi \times 3^2/4} = 2830 \text{ psi}$$

From Mohr's circle

$$R = \sqrt{1415^2 + 5660^2} = 5840 \text{ psi}$$

Hence $\sigma_{max} = 1415 + 5840 = 7255$ psi (tension)

$\sigma_{min} = 1415 - 5840 = -4425$ psi (compression)

$\tau_{\substack{max \\ min}} = 5840$ psi

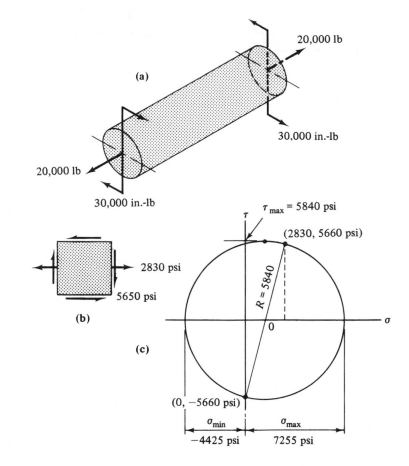

Figure 1-43 / Principal stresses in a propeller shaft (Example 1-16).

1-12 STRESS CONCENTRATION

Any change in section geometry on a structural member, abrupt or gentle, like a hole, dimensional change, or even surface roughness, will cause the stress in the vicinity of the discontinuity to increase beyond the average value. The maximum stress in each instance occurs at the discontinuity. Failure, particularly under cyclic loading, begins at those points of high localized stress. Figure 1-44 illustrates the typical stress pattern for a plate with a hole that nature provides in the material and man attempts to justify. Techniques of experimental stress analysis—strain gage and photoelastic model analyses—coupled with advanced theory of elasticity have

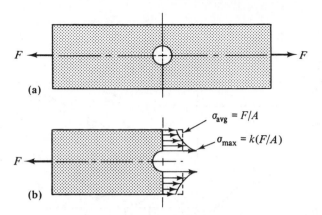

Figure 1-44 / Stress concentration due to a hole.

produced curves[1] that can predict with fair accuracy a stress concentration factor k in terms of the critical dimensions of the member. The actual maximum stress in terms of k for axial, bending, and torsional loads would be, respectively,

$$\sigma = k\frac{F}{A}, \qquad \sigma = k\frac{Mc}{I}, \qquad \text{and} \qquad \tau = k\frac{MD^*}{2J} \qquad (1\text{-}37)$$

*For circular members only.

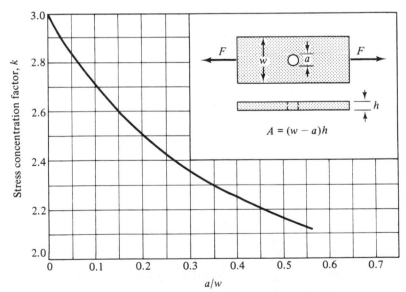

Figure 1-45 / Stress concentration factor k for the plate with hole in tension.

[1]R. E. Peterson, *Stress Concentration Design Factors* (New York, N.Y.: John Wiley & Sons, Inc., 1953).

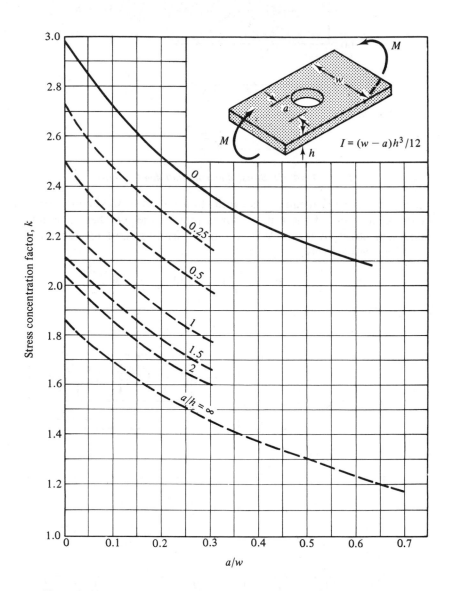

Figure 1-46   /   Stress concentration factor k for plate with hole in bending.

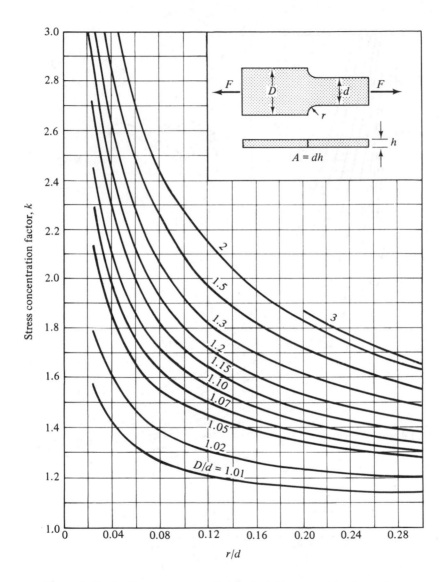

Figure 1-47 / Stress concentration factor k for stepped bar in tension.

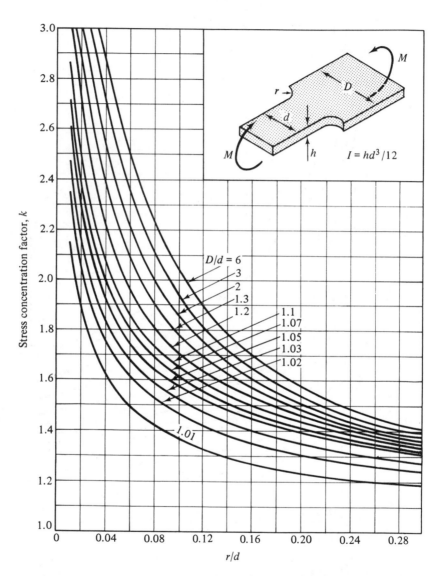

Figure 1-48 / Stress concentration factor k for stepped bar in bending.

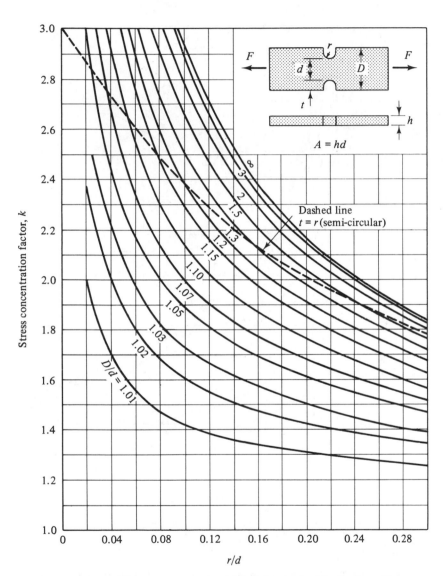

Figure 1-49 / Stress concentration factor k for notched flat bar in tension.

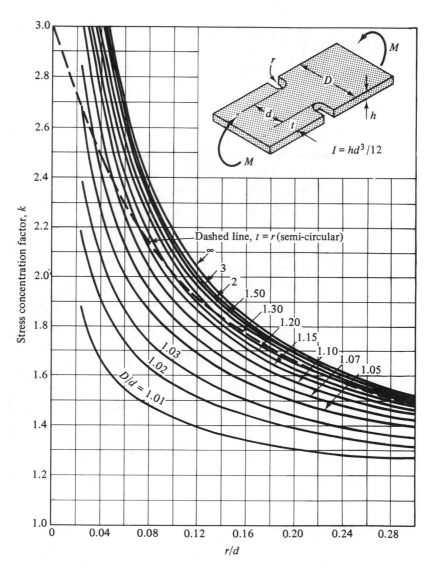

Figure 1-50 / Stress concentration factor k for notched flat bar in bending.

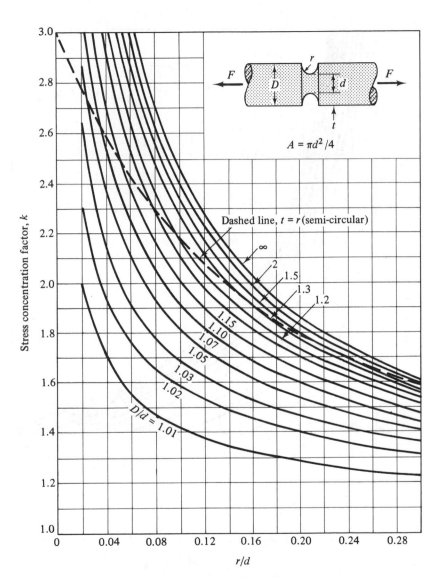

Figure 1-51 / Stress concentration factor k for grooved shaft in tension.

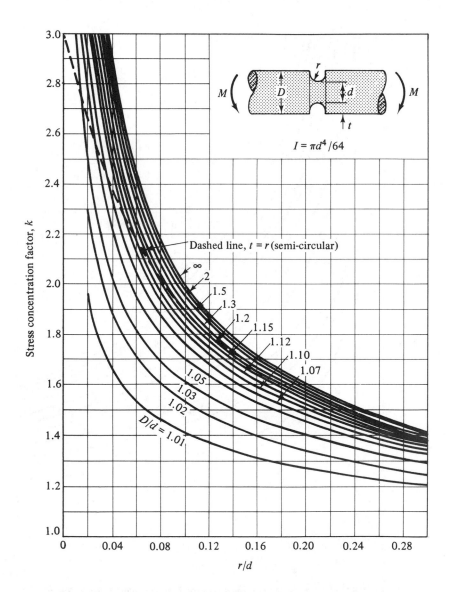

Figure 1-52   /   Stress concentration factor k for grooved shaft in bending.

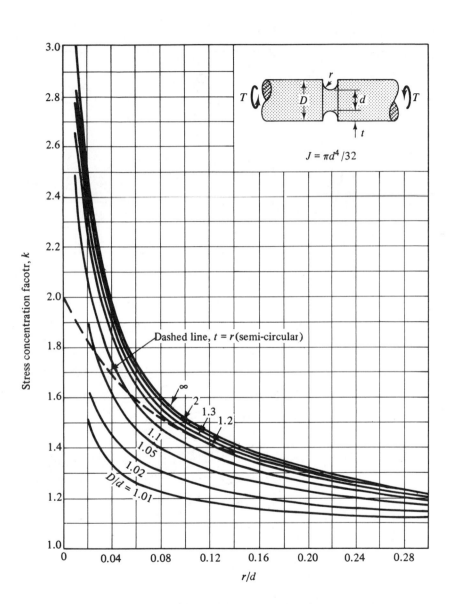

Figure 1-53 / Stress concentration factor k for grooved shaft in torsion.

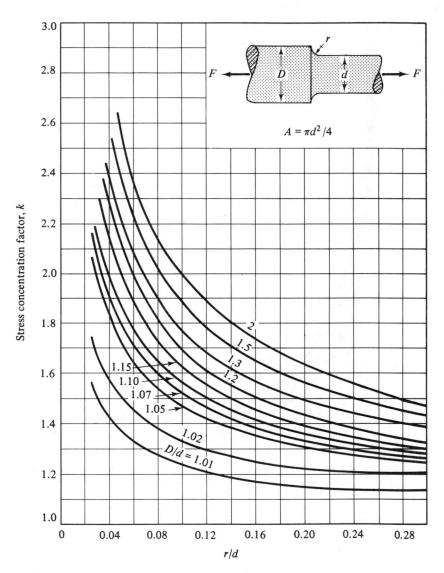

Figure 1-54 / Stress concentration factor k for stepped shaft in tension.

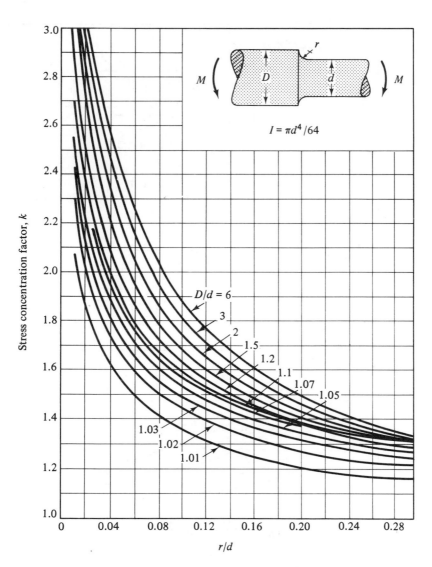

Figure 1-55 / Stress concentration factor k for stepped shaft in bending.

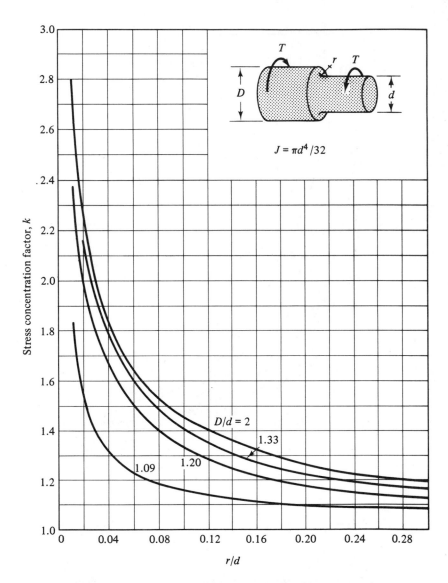

Figure 1-56 / Stress concentration factor k for stepped shaft in torsion.

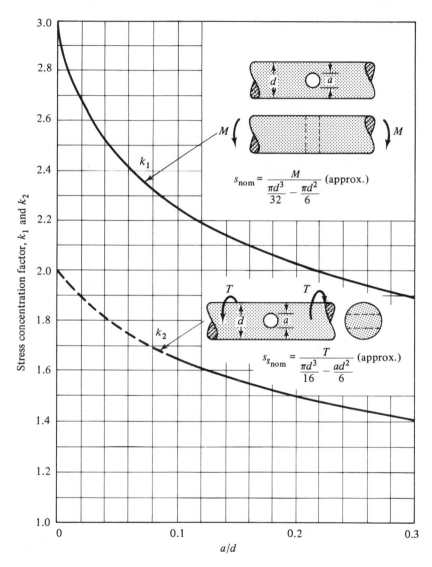

Figure 1-57 / *Stress concentration factors k_1 and k_2 for shaft with transverse hole in bending and torsion.*

EXAMPLE 1-17

The grooved circular shaft shown in Fig. 1-58 is subjected to both a bending moment and an axial force. Determine the maximum tensile stress developed at the weakest section.

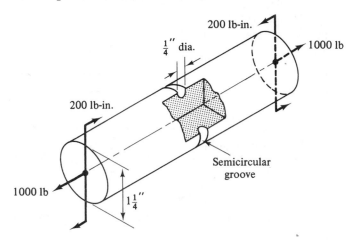

Figure 1-58 / Grooved shaft under load (Example 1-17).

solution

For the section: $r/d = 0.125/1 = 0.125$, and $D/d = 1.25/1 = 1.25$. From the appropriate curves, $k_{direct} \, \sigma = 2.1$, and $k_{bending} \, \sigma = 1.7$; hence

$$\sigma_{direct} = 2.1\frac{F}{A} = 2.1 \times \frac{1000}{\pi 1^2/4} = 2.1 \times 1270 = 2680 \text{ psi}$$

$$\sigma_{bending} = 1.8\frac{M}{Z} = 1.8 \times \frac{200}{\pi 1^3/32} = 1.8 \times 2040 = 3670 \text{ psi}$$

$$\sigma_{max} = 2680 + 3670 = 6350 \text{ psi}$$

Strain is also affected by an abrupt change in section geometry, and while the added stretch or twist is usually considered negligible in most design problems, the increased flexibility does play an important role in the study of torsional vibrations. A sudden change in shaft diameter results in an increased flexibility and is referred to as the *junction effect*.[2] A portion of the larger shaft adjacent to the step is inactive, Fig. 1-59; the magnitude of the junction effect depends upon the ratio of diameters of the sections and the size of the fillet radius. The computations involve adding an increment ΔL to the length of the smaller shaft and subtracting the same

[2]R. Bruce Hopkins, "Angular Deflection of Stepped Shafts," *Machine Design*, July 5, 1962.

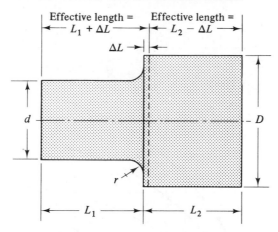

Figure 1-59 / Junction effect.

increment of length from the larger shaft. Curves[3] based on experimentation for two transitions are given in Figs. 1-60 and 1-61.

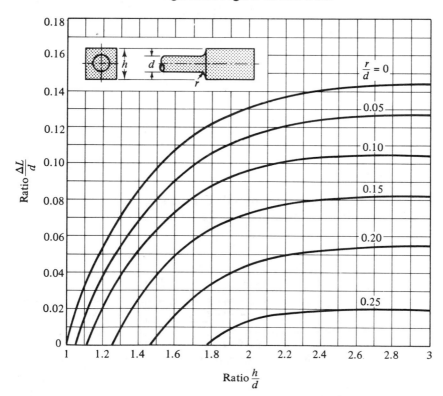

Figure 1-60 / Junction effect for shafts with square and circular cross sections.

---

[3]E. J. Nestorides, *A Handbook on Torsional Vibrations* (London, England: Cambridge University Press, 1900).

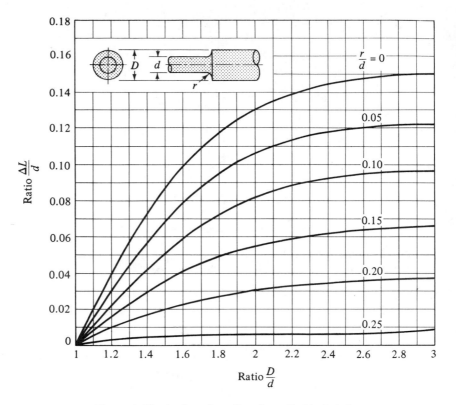

Figure 1-61   /   Junction effect for cylindrical shafts.

EXAMPLE 1-18

Determine: (a) the true equivalent spring constant for the system shown in Fig. 1-62; and (b) the percentage of error if the *junction*

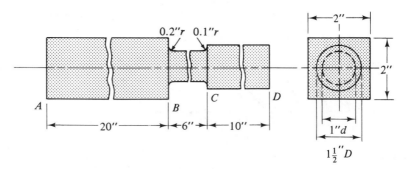

Figure 1-62   /   Stepped shaft (Example 1-18).

effect is ignored. The material is stretch and has a shear modulus of elasticity of 12×10^6 psi.

solution

(a) For the junction at B:

$$h/d = 2/1 = 2 \quad \text{and} \quad r/d = 0.2/1 = 0.2$$

From the graph in Fig. 1-60:

$$\Delta L/d = 0.043 \quad \text{and} \quad \Delta L = 0.043 \text{ in.}$$

For the junction at C:

$$D/d = 1.5/1 = 1.5 \quad \text{and} \quad r/d = 0.1/1 = 0.1$$

From the graph in Fig. 1-61:

$$\Delta L/d = 0.05 \quad \text{and} \quad \Delta L = 0.05 \text{ in.}$$

Therefore, the effective lengths are:

$$L_{AB} = 20 - 0.043 = 19.957 \text{ in.}$$

$$L_{BC} = 6 + 0.043 + 0.05 = 6.093 \text{ in.}$$

$$L_{CD} = 10 - 0.05 = 9.95 \text{ in.}$$

The spring constants are next computed.

$$K_{AB} = 0.141h^4\frac{G}{L} = \frac{0.141 \times 2^4 \times 12 \times 10^6}{19.957} = 1.357 \times 10^6 \text{ in.-lb/rad}$$

$$K_{BC} = \frac{\pi}{32}d^4\frac{G}{L} = \frac{\pi \times 1^4 \times 12 \times 10^6}{32 \times 6.093} = 0.193 \times 10^6 \text{ in.-lb/rad}$$

$$K_{CD} = \frac{\pi}{32}d^4\frac{G}{L} = \frac{\pi \times 1.5^4 \times 12 \times 10^6}{32 \times 9.95} = 0.599 \times 10^6 \text{ in.-lb/rad}$$

The corrected equivalent spring constant for the shaft is

$$K_e = \frac{1}{\dfrac{1}{K_{AB}} + \dfrac{1}{K_{BC}} + \dfrac{1}{K_{CD}}} = \frac{10^6}{\dfrac{1}{1.357} + \dfrac{1}{0.193} + \dfrac{1}{0.599}}$$

$$= 0.132 \times 10^6 \text{ in.-lb/rad}$$

(b)   If the junction effect is neglected, the individual spring constants would be

$$K'_{AB} = 1.357 \times 10^6 \times \frac{20}{19.957} = 1.360 \times 10^6 \text{ in.-lb/rad}$$

$$K'_{BC} = 0.193 \times 10^6 \times \frac{6}{6.093} = 0.190 \times 10^6 \text{ in.-lb/rad}$$

$$K'_{CD} = 0.599 \times 10^6 \times \frac{10}{9.95} = 0.602 \times 10^6 \text{ in.-lb/rad}$$

$$K'_e = \frac{10^6}{\dfrac{1}{1.360} + \dfrac{1}{0.190} + \dfrac{1}{0.602}} = 0.131 \times 10^6 \text{ in.-lb/rad}$$

$$\% \text{ error} = \frac{K'_e - K_e}{K_e} \times 100 = \frac{|0.131 - 0.132|}{0.132} \times 100 = 0.76\%$$

Small? In a system consisting of high-speed rotating masses separated by a number of stepped shafts, such as a generator coupled to a diesel engine, a series of 1 % accumulative design errors could result in disastrous failure.

1-13   ENDURANCE LIMIT, FATIGUE STRENGTH, AND FACTOR OF SAFETY

Whether or not a given stress level in a machine component will cause failure depends on several factors: some obvious and calculable—others, unfortunately, discovered after-the-fact. Foremost in the "obvious and calculable" failure category is the manner in which the load is applied: Is the load static or does it repeatedly fluctuate? More than likely, a little of each occurs. The member is subjected to the superposition of a steady load component, which produces an average stress level, and to an alternating load component (tension to tension, compression to compression, or tension to compression), which produces multivalued *fatigue stresses*. The exact mechanism of fatigue failure is not completely understood; it is assumed, however, that failure begins with a surface crack that grows larger as the load fluctuates. General observations, however, indicate that: (a) fatigue in metals depends upon the number of cycles of load change in a given stress range rather than the time rate of loading, and (b) most metals have a safe range of fluctuating stress values, called the *endurance limit*, below which failure will not occur. The endurance limit implies that the stressed component will have an infinite life—for steel members an infinite life is assumed to mean a minimum of 10 million load reversals.

Fatigue characteristics of materials are experimentally determined by repeated loading at various stress levels in either bending, tension, or torsion. Data are presented in graphic form, Fig. 1-63, in which fatigue life

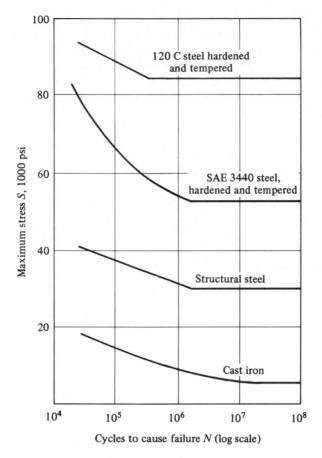

Figure 1-63 / A typical S–N diagram—cycles to failure in reversed bending.

in terms of cycles of load reversal N to cause failure is plotted against stress. The graphs, called stress–cycle diagrams or simply *S–N diagrams*, usually display stress as a function of the logarithm of the number of cycles; occasionally, both stress and number of cycles are plotted on logarithmic scales. Both types of diagrams show a sharp bend at the fatigue limit for ferrous metals; the fatigue limit for nonferrous materials is not as clearly defined.

There are several ways of approaching the problem of designing *life* into machine components. Probably the simplest is through the use of the *modified Goodman diagram*;[4] the example that follows will illustrate the method.

[4]Louis S. Clock, "Analyses of Fatigue Strength," *Machine Design*, October 15, 1959.

EXAMPLE 1-19

A solid steel machined shaft has an ultimate strength $\sigma_{ult} = 200,000$ psi and an endurance limit $\sigma_e = 64,000$ psi. The shaft is subjected to a static pull, which produces a constant average stress $\sigma_{avg} = 64,000$ psi. If the shaft is to have an infinite life (a minimum of 10 million load reversals), determine the maximum reversed stress σ_r that may be allowed in the shaft.

solution

A combination of steady and fatigue stresses act on the shaft as shown in Fig. 1-64(a). To construct the modified Goodman diagram, Fig.

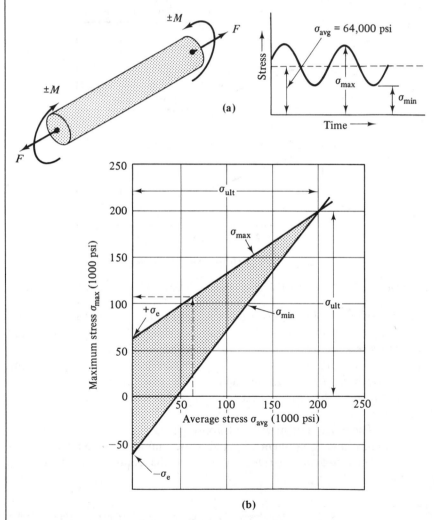

Figure 1-64 / Goodman diagram for a steel shaft (Example 1-19).

1-64(b), draw a set of coordinate axes as shown; from the given data, draw the maximum stress line from the ultimate strength coordinates to the endurance limit on the *zero average stress ordinate*. To obtain the maximum stress the shaft can sustain, construct a vertical line from the given average stress to the maximum stress line. Thus, $\sigma_{max} = 108,000$ psi for $\sigma_{avg} = 64,000$ psi (use the method of similar triangles for precise values).

$$\sigma_r = \sigma_{max} - \sigma_{avg} = 108,000 - 64,000 = 44,000 \text{ psi}$$

Deciding on a *margin of safety*, that vague relationship between service conditions and actual failure, is one of the most critical problems that the designer faces. The service conditions include such items as the type of load, how the load is applied, surface finish, stress concentration, operating temperatures, ambient temperatures, appearance, ease of repair, and the cost of failure—a broken paper clip can be tolerated; a broken linkage in a landing gear is another matter.

The usual practice in determining the design or working stress σ_d is to divide the ultimate strength σ_{ult} by a somewhat nebulous factor of safety *FS*.

$$\sigma_d = \sigma_{ult}/FS \qquad\qquad (1\text{-}38)$$

where the factor of safety *FS* may be considered a product of several independent factors *a*, *b*, *c*, etc., each relating to a particular service condition.

Most industries and government agencies have codes, standards, and specifications to which the designer must adhere. Throughout the book, reference will be made to these standards and to service conditions and the related safety factors.

QUESTIONS AND PROBLEMS

General note: Answers and/or hints to the problems of each chapter are given in the Appendix; the solutions to the design projects, however, are left to the ingenuity and creativity of the reader. Physical data that appear necessary and that are not available in the Appendix can be obtained from engineering handbooks.

P1-1   Develop an equation that will give the tensile stress in the walls of a "thin-walled" spherical pressure vessel of diameter *D* ft subjected to a pressure *p* psi.

P1-2   Draw free-body diagrams at section *a–a* for each of the machine members, loaded as illustrated. Indicate the magnitudes and directions of the forces and/or moments that act.

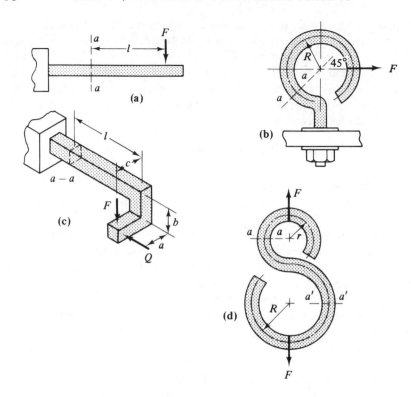

Figure P1-2

P1-3   Determine for each stress distribution pattern shown, the maximum and minimum stresses if in each instance the average stress
at the indicated section is 10,000 psi.

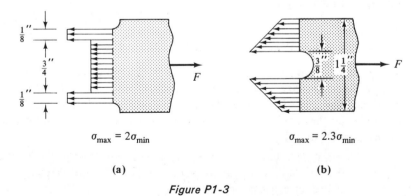

$$\sigma_{max} = 2\sigma_{min}$$

$$\sigma_{max} = 2.3\sigma_{min}$$

(a)

(b)

Figure P1-3

P1-4   The pillow block shaft assembly is subjected to belt pulls that
create forces on the slipshaft as shown. Determine the bearing

stress on the shaft, the tensile stress on plane *abcd*, and the force
in bolts *A*, *B*, *C*, and *D* if they are torqued to their limiting load.

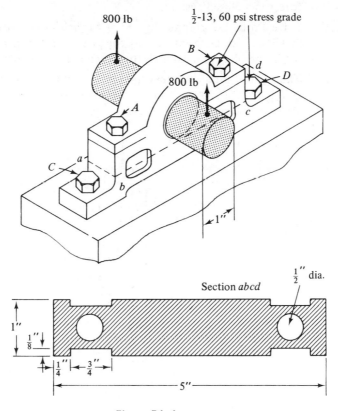

Figure P1-4

P1-5   Find the normal and shear stresses that act along plane *a–a* for
the block shown. To solve the problem, draw an appropriate free-

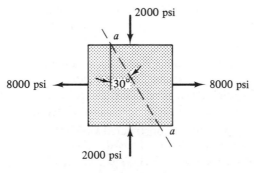

Figure P1-5

body diagram, and use the equations $\sum F_x = \sum F_y = 0$. Show, with an accurate graphical layout, that the string polygon of forces is a closed system.

P1-6   Solve Prob. 1-5 graphically using Mohr's circle; use a scale of 200 lb = 1 in. Determine, also, the principal shearing stresses acting in the block, and sketch the element at its proper orientation.

P1-7   The vertical pipe column which weighs 120 lb/ft supports a vertical load F of 5000 lb. Plot a graph that will give the compressive stress in the pipe as a function of y.

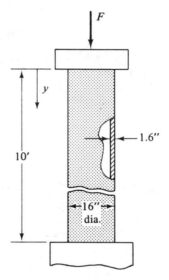

Figure P1-7

P1-8   The solid tapered steel column supports a load as shown. Plot a graph that will display the stress as a function of y. Neglect the weight of the column.

P1-9   Use Mohr's circle to determine the principal stresses (normal and shear) in each of the elements shown. Sketch the orientation in each instance.

P1-10   The element illustrated within a machine member is subjected to two states of stress, each determined experimentally through the use of strain gages. Find the principal components of stress that act on an element whose axes are inclined as shown.

P1-11   A blanking punch is constructed which will simultaneously stamp 20 standard double-edged razor blades from sheet stock having a shear failure stress of 50,000 psi. What average force is required to perform the operation? Carefully measure a blade to determine the necessary dimensions.

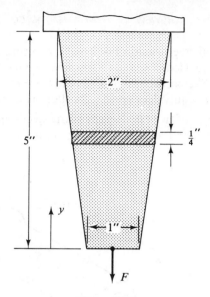

Figure P1-8

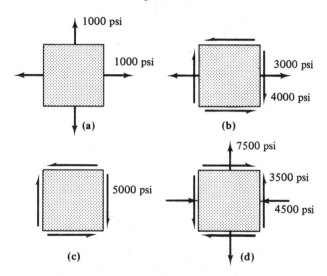

Figure P1-9

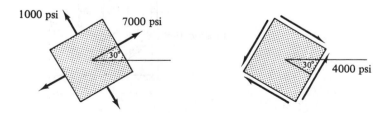

Figure P1-10

P1-12   A cylindrical steel torsion bar having a spring constant K is cut into three equal lengths, and the three segments are then connected in parallel. Sketch the system (there are several possibilities), and determine its equivalent spring constant as a function of K.

P1-13   Find the angle of twist in degrees of the steel stepped shaft shown. Neglect the junction effect in your computations.

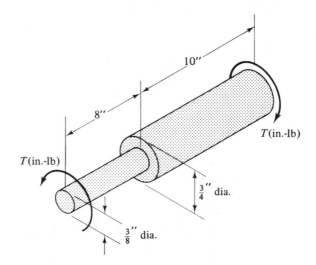

Figure P1-13

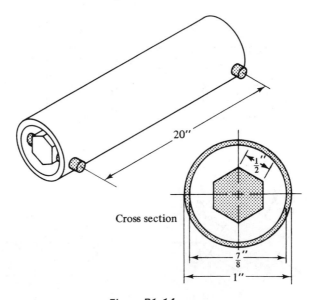

Figure P1-14

P1-14 A hexagonal brass shaft is inserted into a steel tube and securely fastened as illustrated. First, write an equation in terms of symbolic dimensions for the equivalent spring constant, and then determine the numerical value of K_e.

P1-15 The spring constant K of a system is a measure of its stiffness. How would you define $1/K$?

P1-16 Find the equivalent spring constant of the geared system shown. Each shaft has an equal constant k.

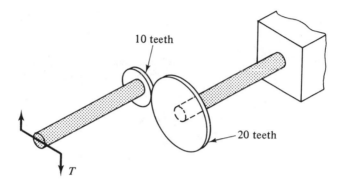

10 teeth

20 teeth

T

Figure P1-16

P1-17 A steel line shaft is driven at 630 rpm by a 100-hp motor. What are the minimum diameters, in nominal dimensions, of the three shaft segments if the shearing stress in any segment is not to exceed 20,000 psi?

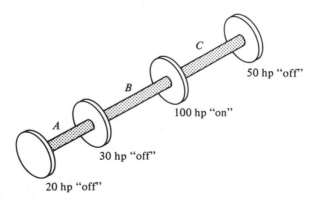

C

50 hp "off"

B

100 hp "on"

A

30 hp "off"

20 hp "off"

Figure P1-17

P1-18 Determine the equivalent spring constants of the systems shown.

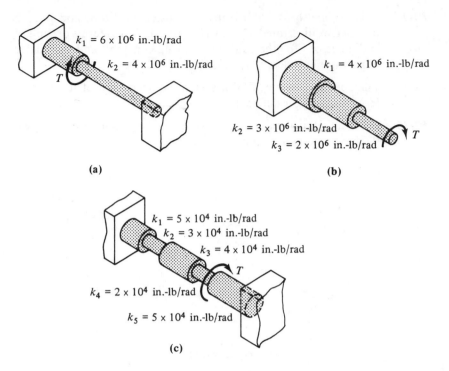

(a)   (b)

(c)

Figure P1-18

P1-19 The thin-walled hydraulic cylinder illustrated sustains an internal
pressure p of 1000 psi by force F. If the cylinder is inadvertently
twisted in its mounting by a torque of 2000 lb-in., what maximum
normal and shearing stresses are developed in the cylinder?

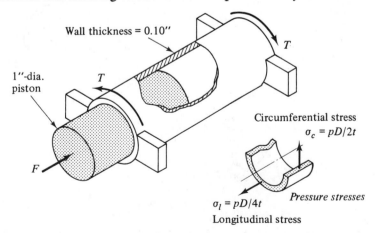

Figure P1-19

Hint: The stresses in a thin-walled cylinder caused by pressure forces *alone* are indicated in the figure.

P1-20 Point P on the two connected Brown and Sharp 20-gage steel wires drops to P', a distance $\delta = 1$ in., when a weight W is suspended as shown. Find: (a) the strain in the wire; (b) the stress in the wire; and (c) the magnitude of W.

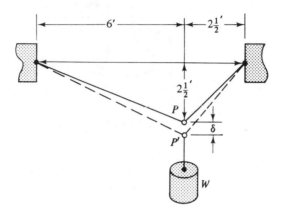

Figure P1-20

P1-21 Two similar rods, one brass and the other steel, are attached to a linkage bar as shown. Where, relative to point A, should a load F be placed if the linkage AB is to remain vertical? If $F = 5000$ lb, find the shear and bearing stress in the pins at A and B.

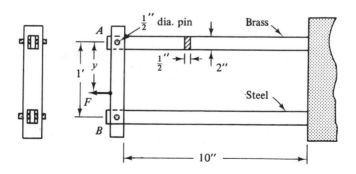

Figure P1-21

P1-22 Four identical rods of length L support the load as illustrated; the rods undergo a strain of δ. Now suppose the rods are paired off and attached to one another so that two rods each of length $2L$ support the weight. By what factor will δ change?

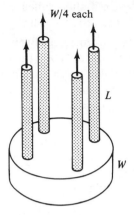

Figure P1-22

P1-23 A brass stepped shaft is subjected to the axial loads as shown. Determine: (a) the accumulative change in length of the bar; and (b) the change in dimensions w, d_1, and d_2.

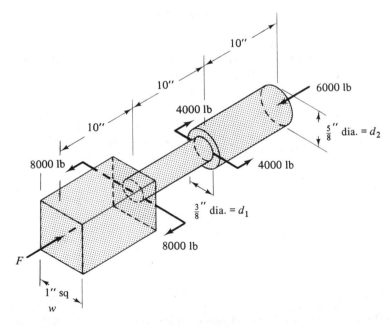

Figure P1-23

P1-24 Determine the stress in steel sections AB and BC if the right support yields 0.002 in. when the load is applied. Assume that the left support remains rigid.

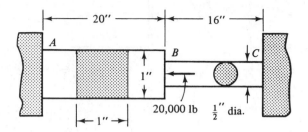

Not drawn to scale

Figure P1-24

P1-25 A $\frac{1}{4}$-in.-diameter brass rod is bent into a circular ring as illustrated. What temperature rise will cause the gap to close?

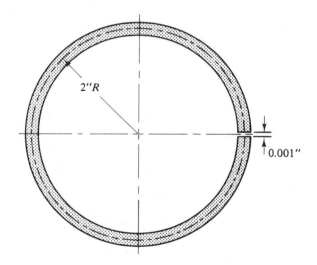

Figure P1-25

P1-26 A $\frac{1}{8}$-in.-diameter steel rod is wound on a 6-in.-diameter mandrel to form a 40-turn helix as shown. The helix is constrained at 40°F in an Invar pipe so that the diameter of the helix cannot expand. One end of the coil is securely fastened to the pipe while the other is free to move. A pointer attached to the free end is to be calibrated in Fahrenheit degrees and is to have a scale range of 40°F to 200°F. Determine the length L of the scale. Neglect frictional effects at the walls, and assume the length of the wire to be $40\pi D$.

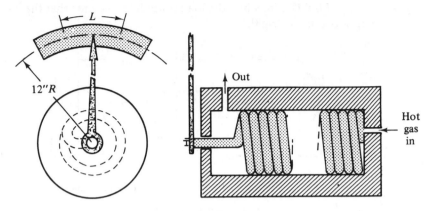

Figure P1-26

P1-27 Two square bars of equal length, one steel and the other brass, are brazed together as illustrated. They are free to expand or contract lengthwise, but are constrained to prevent bending. Determine the shear stress at the joint if the temperature increases 100°F.

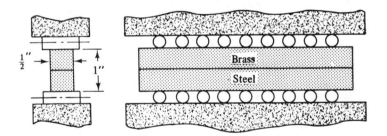

Figure P1-27

P1-28 A brass ring and a steel ring each $\frac{1}{16}$-in. thick and $\frac{1}{2}$-in. wide are press-fitted together as shown. Their difference in diameters causes a stress to develop in each of 5000 psi. Find the temperature increase necessary to cause them to just slip apart.

P1-29 In the illustration, six $\frac{1}{4}$-in.-diameter tie bolts secure the end plates to the brass cylinder. The ends of the bolts are fastened with standard $\frac{1}{4}$–28 nuts; the nuts are then tightened through one-fourth of a turn each from their "snug" position. Find the stress in both the rods and the cylinder.

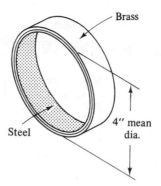

Figure P1-28

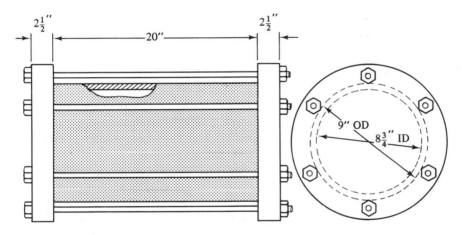

Figure P1-29

P1-30 The solid aluminum piston 6.0000″ in diameter is to "slip-fit" a heavy cylinder block with a clearance of 0.0002 in. at a maximum pressure of 1000 psi. Determine the machined, ground and then lapped, diameter of the piston. Poisson's ratio and the elastic modulus for aluminum are $\frac{1}{3}$ and 10×10^6 psi, respectively. Assume the cylinder block to be rigid.

P1-31 An adjustable temperature sensor consists of a copper bar that actuates electrical contacts through a series of gears as shown. The sensor acts as a "single-pole double-throw" switch. Determine the temperature sensitivity of the device if the contacts are adjusted as shown. Assume the gears to have zero backlash.

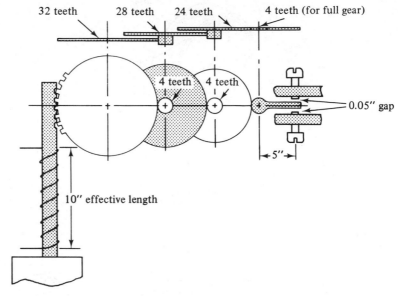

Figure P1-31

P1-32 A rectangular bar is supported as a beam as illustrated. Determine the force F if the maximum tensile and shearing stresses are not to exceed 50,000 psi and 25,000 psi, respectively.

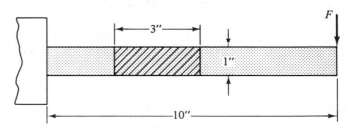

Figure P1-32

P1-33 The beam shown is subject to end moments of 2000 in.-lb. Dimension $b = 1.5$ in. Sketch the stress prisms for tension and

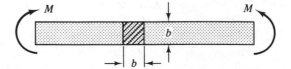

Figure P1-33

compression in the beam as in Fig. 1-29(c). Replace the two stress prisms by resultant forces and show that these resultants form a couple with a moment equal to the applied moment.

P1-34 The Tee-beam carries the load $F = 10,000$ lb as shown. What maximum tensile and compressive stresses are developed in the beam if (a) the Tee is "up"; and (b) the Tee is inverted?

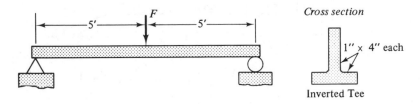

Figure P1-34

P1-35 The beam shown supports three equal loads F. Where should the loads be placed if the maximum moment in the beam is to be a minimum?

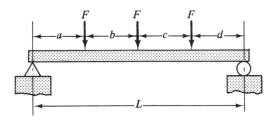

Figure P1-35

P1-36 The cross section of a box beam is shown. Determine the safe uniformly distributed load w lb per ft that can act on this section, supported as illustrated, if the permissible stresses in bending and shear are 10,000 psi and 5000 psi, respectively.

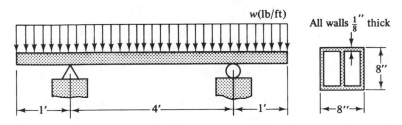

Figure P1-36

P1-37   A cylindrical bar is subjected to a pure moment of 1000 lb-in. as shown. Determine the maximum tensile, compressive, and shear stresses at sections *a–a* and at *b–b*.

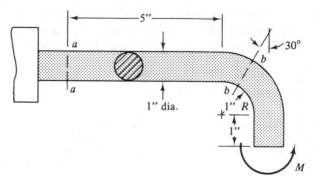

Figure P1-37

P1-38   A section of a clamp is illustrated. What force *F* can be exerted by the screw if the maximum tensile stress in the clamp is not to exceed 15,000 psi?

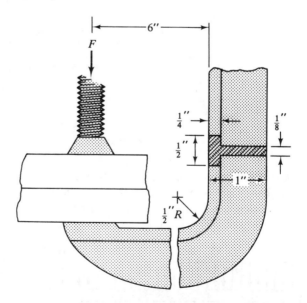

Figure P1-38

P1-39   An offset lever with end bosses is shown. Determine the maximum tensile and compressive stresses that act in the lever under the given load $F = 200$ lb.

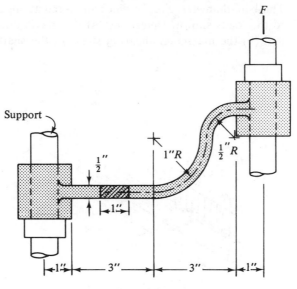

Figure P1-39

P1-40 A rectangular shaft is subjected to an eccentric load as shown. Determine the eccentricity e and the load F if the maximum tensile stress is not to exceed 10% of an allowable compressive working stress of 10,000 psi.

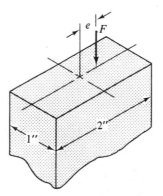

Figure P1-40

P1-41 A compressive force F lb, a torque $F/8$ in.-lb, and a bending moment $4F$ lb-in. act together on a straight circular shaft of 2-in.-diameter. Find F if the maximum tensile stress is not to exceed 25,000 psi and the maximum shearing stress is not to exceed 18,000 psi.

P1-42   The ½-in.-diameter, circular shaft anchored at one end is subjected to the loads shown. Determine (a) the maximum normal stress; and (b) the maximum shearing stress in the shaft.

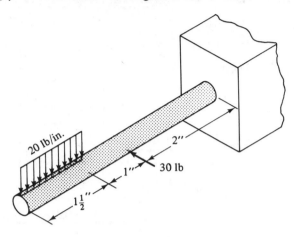

Figure P1-42

P1-43   Three pulley loads act on a 2-in.-diameter line shaft as shown. What is the maximum value of F if the bending stress is not to exceed 30,000 psi?

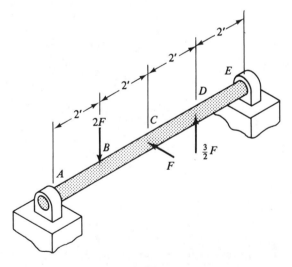

Figure P1-43

P1-44   For each of the sections shown, find the magnitude of the stress (direct, bending, or torsion—whichever applies) at section *a–a* as indicated.

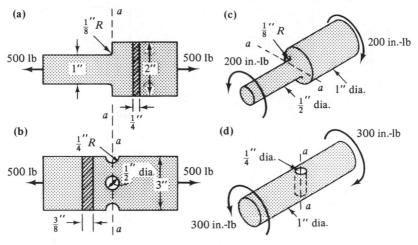

Figure P1-44

P1-45 Correct for the *junction effect*, and determine the torsional spring constant of the steel shaft illustrated.

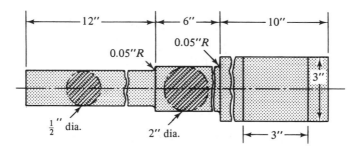

Figure P1-45

P1-46 The steel shaft shown is part of the coupling that joins a motor to a large piston pump. Determine the torsional spring constant of the coupling after correcting for the *junction effect*.

P1-47 A solid steel line shaft has an ultimate strength of 150,000 psi and an endurance limit of 50,000 psi. If the shaft is subjected to a static pull which produces a constant stress of 25,000 psi, determine the maximum reversed stress that may be developed in the shaft. The member is to sustain a minimum of 10 million cycles.

P1-48 Bearing misalignment causes the shaft of Prob. 1-46 to undergo a constant reversal of stress—tension to compression to tension— as it rotates. The shaft is also subjected to a tensile stress of 20,000 psi. The shaft material has an ultimate strength of 180,000 psi and an endurance limit of 60,000 psi. Determine the allowable

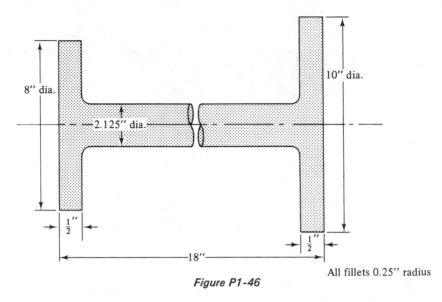

Figure P1-46

All fillets 0.25" radius

bending moment that can be sustained if the shaft is to have an infinite life.

P1-49 Convert the modulus of elasticity of steel, 30×10^6 psi, to newtons per square meter. (The result is a very large number.)

P1-50 Convert the data of the following examples in Chapter 1 to SI units and solve the examples: (a) Example 1-2; (b) Example 1-10 (note that 1°C is equal to 9/5°F).

SUGGESTED DESIGN PROJECTS

DP1-1 Universal conversion to the metric system is close to becoming a reality. Reconstruct Appendix 2 so that the *typical mechanical properties* and the *typical physical constants* will be expressed in metric units (kilograms, meters, and degrees Celsius).

DP1-2 In the development of a machine or machine component, the designer asks—and then attempts to answer—the following questions pertaining to structural considerations:

1. Does the part serve as a stationary member of the system, or does it move? How?
2. Is the load concentrated or does it move?
3. Is the load fluctuating or is it static?
4. What is the magnitude and direction of the load?
5. Is the load transmitted through sliding contact or through relative motion?

6. If the load is fluctuating, what is its frequency? At what rate is the load applied?
7. What maximum deformation caused by the load can be tolerated?
8. What maximum deformation caused by a temperature change can be tolerated?
9. What is the minimum life expectancy of the part or system?
10. What are the consequences of component or system failure?

Based on these design considerations, make, to the best of your ability, value judgments on each of the listed items for one or more of the following:

(a) A refrigerator door hinge
(b) A "bumper jack"
(c) The connecting rod for an automotive eight-cylinder gasoline engine—the car carries a 50,000 mile warranty on the engine
(d) The valve handle of a carbon dioxide fire extinguisher
(e) The propellor shaft of an inboard engine for a cabin cruiser
(f) The coupling between a motor-generator set which is used as a standby source of electricity in a hospital
(g) A spring-operated toy truck

DP1-3 A simple pendulum consisting of a steel rod and cylindrical steel "bob" is to be used as a precision timing device. The pendulum is to have a frequency of 0.500 cycles per second. Design a temperature-compensating device to be located at the "bob" which

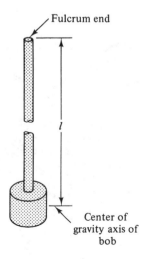

Figure DP1-3

will maintain an exact pendulum length through a temperature fluctuation of $\pm 25°F$. The frequency f of a simple pendulum as a function of length l and gravity g is

$$f = \frac{1}{2\pi}\sqrt{\frac{g}{l}}$$

DP1-4   Design a steel coupling shaft that will transmit 500 hp at 1800 rpm from a stationary diesel to a direct current generator. The angle of twist of the shaft must not exceed 2 degrees per ft, and the maximum permissible shear stress is 20,000 psi. The shaft is to be flange-coupled to both the engine and the generator. Assume the flange diameter at the bolt circle to be at least four times the diameter of the shaft, and assume the length of the shaft, which will include a half coupling at each end, to be 18-in. In starting and stopping, a thrust load of 5000 lb is exerted in the shaft.

DP1-5   A wire manufacturer wishes to test the fatigue strength in bending of cold drawn high-carbon steel wire. Sketch the essential features of a simple machine to accomplish the test. The device should be adjustable to allow for various stress levels, should have an automatic cycle counter (purchasable), and should shut itself off if and when the wire breaks.

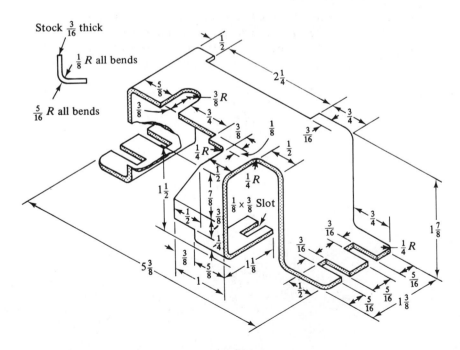

Figure DP1-6

DP1-6   The "bracket-clip" illustrated is to be stamped from a Type 431 annealed stainless steel plate. Secondary forming operations will put in the required bends. Make a layout of the "first" stamping die and determine the press force required to stamp the part.

CHAPTER 2

BEAMS AND COLUMNS: ELASTIC DEFLECTION AND BUCKLING

From the elastic curve of a beam, Euler soon progressed to his well known column formula. This appeared in a paper written in 1757 and published in the Mémoires de l'Académie de Berlin in 1759 under the title *Sur la force de colonnes*. Historically this is one of the most important contributions to the theory of Strength of Materials. Also it is unique. Here we have the equation of the elastic curve of a beam and also the column formula before the basic flexure problem had been satisfactorily solved—almost two decades before that result was to be arrived at.

HARVEY F. GIRVIN
*A Historical Appraisal
of Mechanics*

2-1   INTRODUCTION

The terms *beams* and *columns* bring to mind, by the very nature of our everyday language, large and heavy structural members—the basic components that are the skeletons, the bones, of buildings, bridges, and carports. In one sense, this is, of course, true; to the machine designer, however, beams and columns take on a totally different connotation. To the engineer, meshing gear teeth, levers and linkages, turbine rotors and fan blades, and the mainsprings of watches all are forms of beams, and automotive connecting rods, pistons, and linkage bars are types of columns.

In one sense, the analysis to determine the elastic strength and deflection proceeds through the use of the same analytical techniques whether the column or beam is measured in feet and tons or fractions of an inch and ounces, whether they support a multistory building or sustain the forces caused by variable pressures and changing temperatures. Buildings and bridges are basically subjected to static loads; machines and their components, however, move—and therein lies the difference. In the latter, deflection and buckling forces are caused by more than the mere placing of a weight here or by a push there—there are dynamic forces at play caused by fluctuating and impact loading, accelerations, centri-

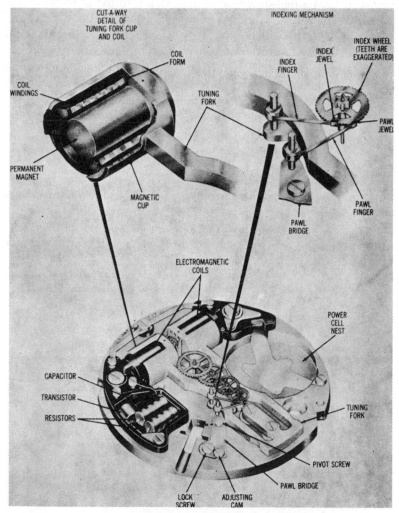

Figure 2-1 / Clock springs or curved beams? Both! (Reprinted by permission from William O. Bennett, "The Story of the Tuning Fork," American Jewelry Manufacturer, April 1970)

fugal effects of rotation, vibration, thermal expansions and contractions, and last, but not least, a teaser called the Coriolis acceleration component, which occasionally enters the total and complex picture of "the why of" elastic deformation.

2-2   BEAM DEFLECTIONS: THE MOMENT–AREA METHOD

There are at least a half-dozen-or-so approaches to the problem of computing beam deflections—some are purely mathematical and others graphical. Two approaches will be discussed in this chapter; the first is the *moment–area method*, which in reality is a graphical interpretation of Euler's mathematical equations for the elastic curve. Superposition—the addition of solutions to known beam problems—is the second approach to finding the deflection of beams under various loading arrangements.

The reciprocal of the radius of curvature ρ of a beam is defined by the relationship

$$\frac{1}{\rho} = \frac{M}{EI} \tag{2-1}$$

where M is the bending moment within the beam, E is the elastic modulus, and I is the moment of inertia with reference to the neutral plane. Segment ab of the beam shown in Fig. 2-2 is an incremental portion Δx of an arbitrary *elastic curve*—the curved form which the beam takes upon loading. From the geometry of the figure,

$$\Delta x = \rho \Delta \theta$$

which when combined with Eq. 2-1 gives

$$\Delta \theta = \frac{M \Delta x}{EI}$$

Tangents drawn to the elastic curve at a and b are separated by the same angle $\Delta \theta$ as are the normals oa and ob. By imagining the beam to consist of an infinite number of segments Δx, the slope of any arbitrary point A on the elastic curve measured relative to a second point B would be the sum of the incremental $\Delta \theta$s. Thus

$$\theta_{A/B} = \sum \Delta \theta = \frac{1}{EI} \sum_{A}^{B} M \Delta x = \frac{1}{EI} \times [\text{area}]_{AB} \tag{2-2}$$

Formally stated:

> The angle between tangents drawn at A and B on the elastic curve is equal to the area of the corresponding portion of the bending

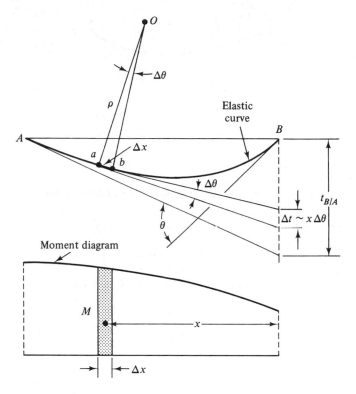

Figure 2-2 / Slope of an elastic curve.

moment diagram, divided by the *EI*, or the product of the elastic modulus and the moment of inertia taken with respect to the neutral plane.

Returning to Fig. 2-2, the term $t_{B/A}$ is defined as the *vertical deviation* of point *B* on the elastic curve measured relative to a tangent drawn at *A*. Since Δt is very nearly equal to $x\Delta\theta$, it follows that

$$t_{B/A} = \sum_{A}^{B} \Delta t = \sum_{A}^{B} x\Delta\theta = \frac{1}{EI} \sum_{A}^{B} x(M\Delta x)$$

$$= \frac{1}{EI} \times \begin{bmatrix} \text{moment of the moment diagram between } A \text{ and } B \\ \text{taken relative to a vertical line that intersects } B \end{bmatrix}$$

Again, in terms of a formal statement:

The vertical deviation of point *B* on the elastic curve that cuts a tangent drawn from point *A* is equal to the moment of the area of the bending moment diagram taken with respect to a vertical line through *B*, divided by *EI*, or the product of the elastic modulus and the moment of inertia taken relative to the neutral plane.

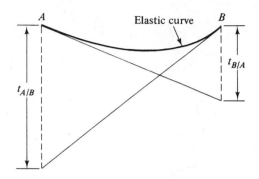

Figure 2-3 / Different deviations at two ends of a beam.

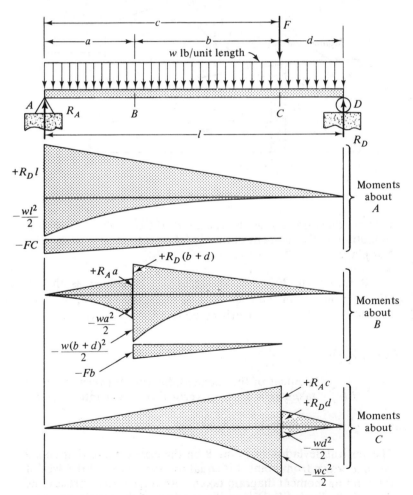

Figure 2-4 / Construction of bending moment diagrams.

It is important to note (see Fig. 2-3) that $t_{B/A}$ and $t_{A/B}$ are rarely equal, since

$$t_{B/A} = \frac{1}{EI}[\text{area}]_{AB}\bar{X}_B$$

$$(2\text{-}3)$$

and

$$t_{A/B} = \frac{1}{EI}[\text{area}]_{AB}\bar{X}_A$$

The ease of finding slopes and deflections by the moment area method depends largely on how simply the moment diagram can be drawn; the usual approach is to sketch the diagram *by parts*—by con-

TABLE 2-1

CENTROIDS OF AREAS

Section	Area	$\bar{x}_1$	$\bar{x}_2$
Rectangle 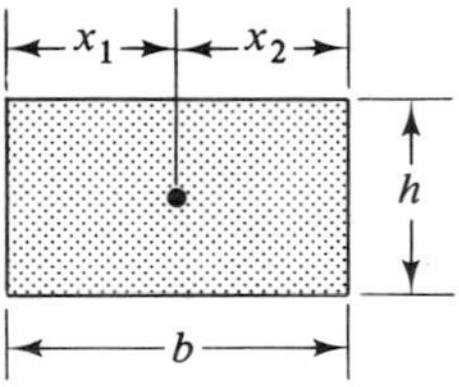	bh	$\frac{1}{2}b$	$\frac{1}{2}b$
Triangle 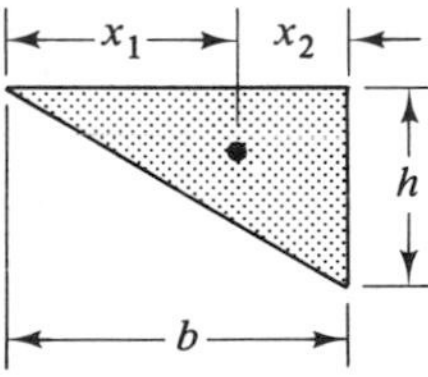	$\frac{1}{2}bh$	$\frac{2}{3}b$	$\frac{1}{3}b$
Parabola 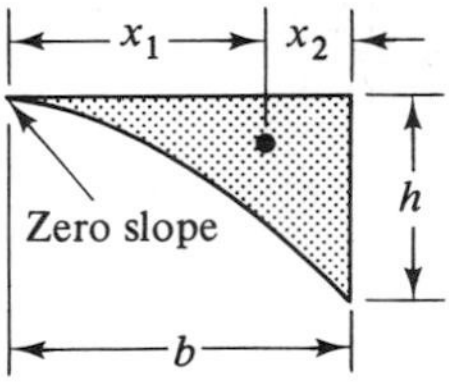	$\frac{1}{3}bh$	$\frac{3}{4}b$	$\frac{1}{4}b$
Cubic parabola 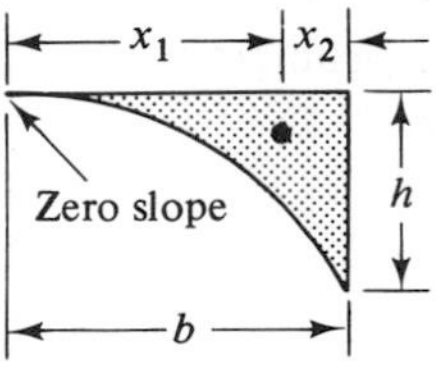	$\frac{1}{4}bh$	$\frac{4}{5}b$	$\frac{1}{5}b$

sidering the total moment diagram to be the result of the *superposition* of individual loads. Figure 2-4 illustrates how three apparently different— but exactly the same—moment diagrams can be drawn for a beam loaded as shown. The shrewd selection of a reference line will save considerable slide rule or calculator time. Table 2-1 lists the properties of areas and centroids that are frequently encountered in sketching moment diagrams by parts.

Several examples follow that will demonstrate through the use of the moment–area method (a) how slopes and deflections are computed, and (b) how the method is used to calculate statically indeterminate reactions.

EXAMPLE 2-1

Determine the maximum slope θ and deflection δ of the steel cantilevered beam shown in Fig. 2-5; develop symbolic formulas for θ and δ, and then substitute to find numerical values. The maximum bending stress in the beam must not exceed 26,000 psi.

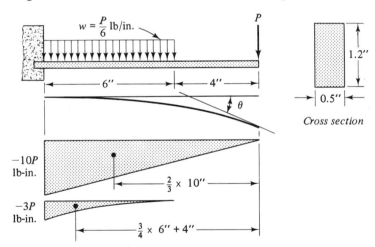

Figure 2-5 / Bending moment diagrams for a steel cantilevered beam (Example 2-1).

solution

The moment diagram is drawn by *parts* as illustrated, and the slope and deflection equations are written; the problem is simplified by the fact that the slope at the support is zero.

$$\theta = \frac{1}{EI} [\text{area}]_{A/B}$$

$$= -\frac{1}{EI}\left[\left(\frac{1}{2} \times 10 \times 10P\right) + \left(\frac{1}{3} \times 6 \times 3P\right)\right] = \frac{56P}{EI}$$

and
$$\delta = \frac{1}{EI} [\text{area}] \, \bar{X}_{A/B}$$

$$= -\frac{1}{EI} \left\{ \left[\frac{1}{2} \times 10 \times 10P \times \frac{2}{3} \times 10 \right] \right.$$

$$\left. + \left[\frac{1}{3} \times 6 \times 3P \times \left(\frac{3}{4} \times 6 + 4 \right) \right] \right\}$$

$$= -384 \frac{P}{EI}$$

Next, the magnitudes of P and I are computed:

$$\sigma = \frac{M}{Z} = \frac{M}{bh^2/6} = \frac{10P + 3P}{0.5 \times 1.2^2/6}$$

$$26,000 = \frac{13P}{0.12}$$

$$P = 240 \text{ lb}$$

$$I = \frac{1}{12} bh^3 = \frac{0.5 \times \overline{1.2}^3}{12} = 0.072 \text{ in.}^4$$

Hence
$$\theta = -56 \frac{P}{EI} = \frac{56 \times 240}{30 \times 10^6 \times 0.072} = -0.00622 \text{ rad}$$

and
$$\delta = -384 \frac{P}{EI} = \frac{384 \times 240}{30 \times 10^6 \times 0.072} = -0.0427 \text{ in.}$$

EXAMPLE 2-2

The parallel-motion, 40-lb steel connecting rod rotates at 600 rpm as shown in Fig. 2-6. Determine for the position indicated the maximum slope and deflection caused by the dynamic inertia forces.

solution

Every portion of mass in the rod is subjected to an equal downward acceleration of $r\omega^2$, where the angular velocity ω is expressed in radians per second, and r in feet. The total inertia effect F_i is

$$F_i = Ma = \frac{W}{g} r\omega^2 = \frac{40}{32.2} \times \frac{1}{2} \times \left(600 \times \frac{2\pi}{60} \right)^2$$

$$= 2450 \text{ lb}$$

This inertia force, which is evenly distributed over the entire length, constitutes a distributed load of $2450/24 = 102$ lb per in. Since the member is symmetrical, the maximum slope will occur at the ends, and the maximum deflection will occur at the center.

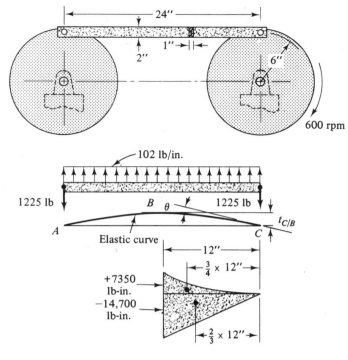

Figure 2-6 / Bending of a steel connecting rod (Example 2-2).

$$I = \frac{1}{12}bh^3 = \frac{1}{12} \times 1 \times 2^3 = \frac{2}{3} \text{ in.}^4$$

$$Z = \frac{bh^2}{6} = \frac{1 \times 2^2}{6} = \frac{2}{3} \text{ in.}^3$$

$$\theta = \frac{1}{EI}[\text{area}]_{C/B}$$

$$= \frac{1}{30 \times 10^6 \times \frac{2}{3}}\left[\left(\frac{1}{3} \times 12 \times 7350\right) - \left(\frac{1}{2} \times 12 \times 14{,}700\right)\right]$$

$$= -0.00294 \text{ rad}$$

$$\delta = \frac{1}{EI}[\text{area } \bar{X}]_{C/B}$$

$$= \frac{1}{30 \times 10^6 \times \frac{2}{3}}\left[\left(\frac{1}{3} \times 12 \times 7350 \times \frac{3}{4} \times 12\right)\right.$$

$$\left. - \left(\frac{1}{2} \times 12 \times 14{,}700 \times \frac{2}{3} \times 12\right)\right]$$

$$= -0.0221 \text{ in.}$$

Two examples follow that illustrate the use of the moment–area method applied first, to an overhanging beam—a beam that extends beyond its supports—and second, to a statically indeterminate beam—a beam containing more supports than necessary to maintain stability.

EXAMPLE 2-3

A line shaft supports the pulley forces shown. Determine the value of $EI\delta$ at point C.

solution

It is conceivable that the elastic curve could assume either the shape illustrated in Fig. 2-7(b) or Fig. 2-7(c). In one instance, by similar

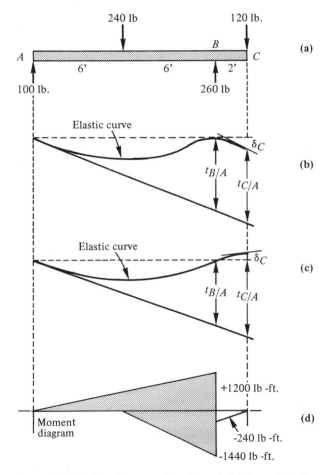

Figure 2-7 / Elastic curves for a line shaft (Example 2-3).

triangles

$$\frac{\delta_C + t_{C/A}}{14} = \frac{t_{B/A}}{12} \qquad \text{[See Fig. 2-7(b)]}$$

$$\delta_C = \tfrac{7}{6} \times t_{B/A} - t_{C/A}$$

and in the other

$$\frac{t_{C/A} - \delta_C}{14} = \frac{t_{B/A}}{12} \qquad \text{[See Fig. 2-7(c)]}$$

$$\delta_C = -\tfrac{7}{6} \times t_{B/A} + t_{B/A}$$

The equations differ only in algebraic sign; either formula could be used. An incorrect "educated guess" will produce a negative answer; the magnitude of the solution, however, will be correct. Therefore, assume that δ_C will be downward as shown in Fig. 2-7(b).

$$EI\delta_C = \tfrac{7}{6} \times [\text{area } \bar{X}]_{B/A} - [\text{area } \bar{X}]_{C/A}$$

$$EI\delta_C = \tfrac{7}{6}\{[\tfrac{1}{2} \times 12 \times 1200 \times \tfrac{12}{3}] + [\tfrac{1}{2} \times 6 \times (-1440) \times \tfrac{6}{3}]\}$$

$$- \{[\tfrac{1}{2} \times 12 \times 1200 \times (\tfrac{12}{3} + 2)]$$

$$+ [\tfrac{1}{2} \times 6 \times (-1440) \times (\tfrac{6}{3} + 2)]$$

$$+ [\tfrac{1}{2} \times 2 \times (-240) \times \tfrac{2}{3} \times 2]\}$$

$$= -2080 \text{ lb-ft}^3$$

A negative answer! The "educated guess" is incorrect, hence point *C* lies above the horizontal axis of the beam, and the deflection is $2080/EI$ ft.

EXAMPLE 2-4

The perfectly "restrained" machine member, Fig. 2-8, is subjected to the load shown. Find the reactions at *A* and *B*, and then sketch the composite transverse shear and moment diagrams.

solution

Ends *A* and *B* share a common horizontal tangent. It, therefore, follows that

$$t_{A/B} = t_{B/A} = 0$$

and

$$\theta_{A/B} = \theta_{B/A} = 0$$

These four relationships plus the two allowable formulas of static equilibrium provide more than enough simultaneous algebraic equa-

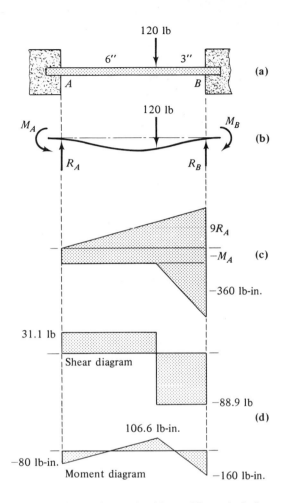

Figure 2-8 / A restrained beam (Example 2-4).

tions than needed to find the four unknowns R_A, R_B, M_A, and M_B. The simplest approach would be to use the following four equations:

$$\theta_{A/B} = \frac{1}{EI} [\text{area}]_{A/B} = 0 \qquad (a)$$

$$t_{A/B} = \frac{1}{EI} [\text{area } \bar{X}]_{A/B} = 0 \qquad (b)$$

$$\sum F_y = 0 \qquad (c)$$

$$\sum M = 0 \qquad (d)$$

Numerical substitutions are made:

$$[\text{area}]_{A/B} = 0 \qquad (1)$$

$$(\tfrac{1}{2} \times 9 \times 9R_A) - (9M_A) - (\tfrac{1}{2} \times 9 \times 360) = 0$$

$$9R_A - 2M_A = 360$$

$$[\text{area } \bar{X}]_{A/B} = 0 \qquad (2)$$

$$(\tfrac{1}{2} \times 9 \times 9R_A \times \tfrac{1}{3} \times 9) - (9 \times M_A \times \tfrac{1}{2} \times 9)$$
$$- (\tfrac{1}{2} \times 3 \times 360 \times \tfrac{1}{3} \times 3) = 0$$

$$9R_A - 3M_A = 40$$

Simultaneous solution of (1) and (2) gives

$$R_A = 31.1 \text{ lb}$$

and

$$M_A = 80 \text{ lb-in.}$$

Next, the equations of static equilibrium are applied to the free-body:

$$\bar{Z}F_y = 0$$

$$R_B = 120 - R_A = 120 - 31.1 = 88.9 \text{ lb}$$

$$\bar{Z}M_B = 0$$

$$M_B = 9R_A - M_A - 360 = 9 \times 31.1 - 80 - 360 = -160 \text{ lb-in.}$$

The transverse shear and moment diagrams can now be drawn, as in Fig. 2-8(d).

2-3   BEAM DEFLECTIONS: THE SUPERPOSITION METHOD

The easiest and most direct approach to the general beam problem—support reactions, slopes, and deflections—is through the use of *superposition*. Table 2-2, which describes 18 of the most frequently encountered loading conditions, is an invaluable designer's tool. In effect, the table describes an almost endless variety of beam loadings when one considers that two or more of the various arrangements can be superimposed upon one another. The number of possible variations is over 10^{16}—a noble sum.

TABLE 2-2

BEAMS OF UNIFORM CROSS SECTIONS, LOADED TRANSVERSELY: SHEARS, MOMENTS, AND DEFLECTIONS

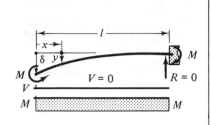

$$y = \frac{M(l - x)^2}{2EI}$$

$$\delta = \frac{Ml^2}{2EI}$$

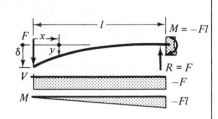

$$y = \frac{F}{6EI}(2l^3 - 3l^2x + x^3)$$

$$\delta = \frac{F^3}{3EI}$$

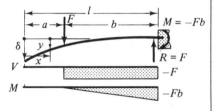

$$y = \frac{Fb^2}{6EI}(3l - 3x - b) \text{ for } x < a$$

$$y = \frac{F(l - x)^2}{6EI}(3b - l + x) \text{ for } x > a$$

$$\delta = \frac{Fb^2}{6EI}(3l - b)$$

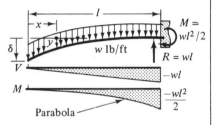

$$y = \frac{w}{24EI}(x^4 - 4l^3x + 3l^4)$$

$$\delta = \frac{wl^4}{8EI}$$

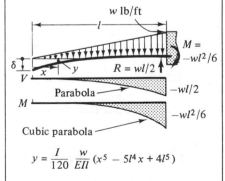

$$y = \frac{l}{120}\frac{w}{EIl}(x^5 - 5l^4x + 4l^5)$$

$$\delta = \frac{1}{30}\frac{wl^4}{EI}$$

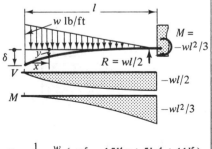

$$y = \frac{1}{120}\frac{w}{EIl}(-x^5 - 15l^4x + 5lx^4 + 11l^5)$$

$$\delta = \frac{11}{120}\frac{wl^4}{EI}$$

TABLE 2-2 (Cont.)

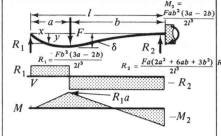

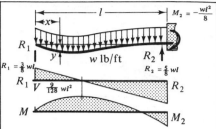

$$y = \frac{Fb^2}{12EIl^3}(3al^2x - 2lx^3 - ax^4) \text{ for } x < a$$

$$y = \frac{Fa(l-x)^2}{12EIl^3}(3l^2x - a^2x - 2a^2l) \text{ for } x > a$$

$$\delta = \frac{Fa^2b^3}{12EIl^3}(3l + a) \text{ at point of load}$$

$$y = \frac{w}{48EI}(l^3x - 3lx^+ + 2x^4)$$

$$\delta = \frac{wl^4}{185EI} \text{ at } x = 0.422l$$

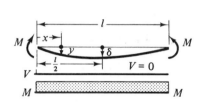

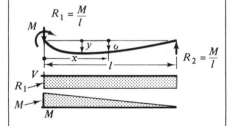

$$y = \frac{M}{2EI}(lx - x^2)$$

$$\delta = \frac{Ml^2}{8EI} \text{ at } \frac{l}{2}$$

$$y = \frac{M}{6lEI}(2l^2x - 3lx^2 + x^3)$$

$$\delta = \frac{Ml^2}{9\sqrt{3}EI} \text{ at } l(1 - \frac{1}{\sqrt{3}})$$

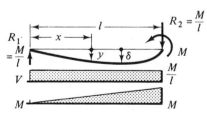

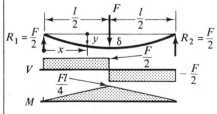

$$y = \frac{M}{6lEI}(l^2x - x^3)$$

$$\delta = \frac{Ml^2}{9\sqrt{3}EI} \text{ at } x = \frac{l}{\sqrt{3}}$$

$$y = \frac{F}{48EI}(3l^2x - 4x^3) \text{ for } x < \frac{l}{2}$$

$$\delta = \frac{Fl^3}{48EI} \text{ at } x = \frac{l}{2}$$

TABLE 2-2 (Cont.)

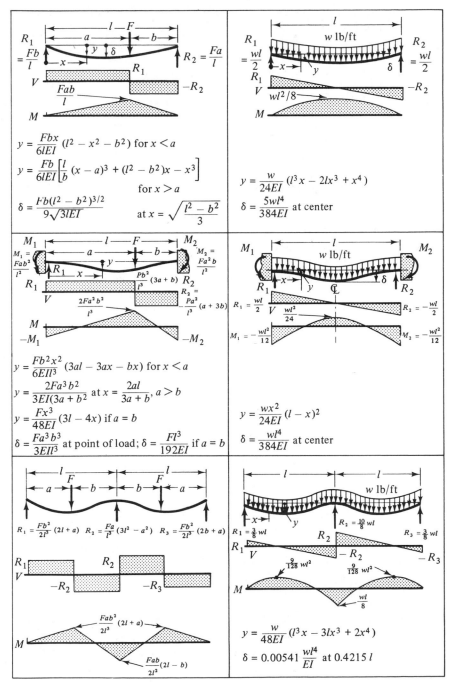

$$y = \frac{Fbx}{6lEI}(l^2 - x^2 - b^2) \text{ for } x < a$$

$$y = \frac{Fb}{6lEI}\left[\frac{l}{b}(x-a)^3 + (l^2 - b^2)x - x^3\right]$$
$$\text{for } x > a$$

$$\delta = \frac{Fb(l^2 - b^2)^{3/2}}{9\sqrt{3}lEI} \qquad \text{at } x = \sqrt{\frac{l^2 - b^2}{3}}$$

$$y = \frac{w}{24EI}(l^3 x - 2lx^3 + x^4)$$

$$\delta = \frac{5wl^4}{384EI} \text{ at center}$$

$$y = \frac{Fb^2x^2}{6EIl^3}(3al - 3ax - bx) \text{ for } x < a$$

$$y = \frac{2Fa^3b^2}{3EI(3a+b^2)} \text{ at } x = \frac{2al}{3a+b}, a > b$$

$$y = \frac{Fx^3}{48EI}(3l - 4x) \text{ if } a = b$$

$$\delta = \frac{Fa^3b^3}{3EIl^3} \text{ at point of load; } \delta = \frac{Fl^3}{192EI} \text{ if } a = b$$

$$y = \frac{wx^2}{24EI}(l - x)^2$$

$$\delta = \frac{wl^4}{384EI} \text{ at center}$$

$$y = \frac{w}{48EI}(l^3 x - 3lx^3 + 2x^4)$$

$$\delta = 0.00541 \frac{wl^4}{EI} \text{ at } 0.4215\, l$$

123

EXAMPLE 2-5

Find the midspan value of the deflection δ of the restrained beam shown in Fig. 2-9.

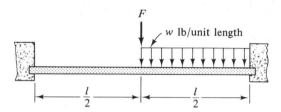

Figure 2-9 / Loading diagram of a restrained beam (Example 2-5).

solution

Case 15 of Table 2-2 contains the solution to that portion of the problem concerned with the concentrated load. Solution of the half-distributed load, which does not appear in the table, can be obtained from Case 16. Imagine the beam of Case 16 to consist of two beams: a left-hand portion with a distributed load over the left half of the span and a right-hand portion carrying an equal distributed load over the right-hand section of the beam. These two imaginary beams have equal deflections and when superimposed upon one another, they would be equivalent to Case 16.

$$\delta = \delta_{\text{Case 15}} - \left(\frac{1}{2} \times \delta_{\text{Case 16}}\right)$$

$$= \frac{Fl^3}{192EI} + \left(\frac{1}{2} \times \frac{wl^4}{384EI}\right)$$

$$= \frac{5Fl^3 + wl^4}{768EI}$$

EXAMPLE 2-6

The eight-bladed steel *impeller* shown in Fig. 2-10 is brought to a speed of 1800 rpm from rest in 1 second. Determine: (a) the approximate deflection in any one blade relative to the hub; and (b) the approximate bending stress created at the junction of the hub and the blade. Steel weighs 0.282 lb per cu in.

solution

(a) The angular acceleration α is computed, and the incremental elements of mass in the impeller when multiplied by their respective accelerations react as a force, opposing the direction of acceleration

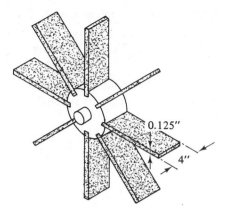

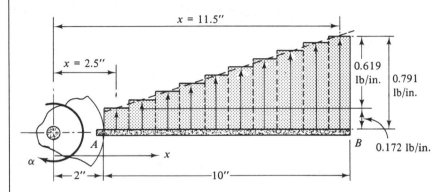

Inertia force approximation

Figure 2-10 / Bending load in an impeller blade (Example 2-6).

and distributed as shown. An *approximate* solution is obtained by selecting elements of mass Δm that are each 1 in. long and that weigh

$$\Delta w = 0.125 \times 4 \times 0.282 = 0.141 \text{ lb/in.}$$

Angular acceleration:

$$\alpha = \frac{\omega_f - \omega_0}{t} = \frac{1800 - 0}{1} \times \frac{2\pi}{60} = 60\,\pi \text{ rad/sec}^2$$

Inertia force per inch at $x = 2.5$ in.:

$$F_{2.5} = ma = \frac{\Delta w r \alpha}{g} = \frac{0.141}{32.2} \times \frac{2.5}{12} \times 60\pi = 0.172 \text{ lb/in.}$$

Inertia force per inch at $x = 11.5$ in.:

$$F_{11.5} = \frac{0.141}{32.2} \times \frac{11.5}{12} \times 60\pi = 0.791 \text{ lb/in.}$$

If the step-function *force diagram* is approximated by a straight line, the loading consists of the superposition of Case 4 and Case 6 of Table 2-2.

$$\delta = \delta_4 + \delta_6 = \frac{wl^4}{8EI} + \left(\frac{11}{60} \times \frac{w'l^4}{EI}\right) = \frac{l^4}{EI}\left(\frac{w}{8} + \frac{11}{120}w'\right)$$

where $I = \frac{1}{12}bh^3 = \frac{1}{12} \times 4 \times 0.125^3 = 651 \times 10^{-6} \text{ in.}^4$

Thus, $\delta = \dfrac{10^4}{30 \times 10^6 \times 651 \times 10^{-6}} \left(\dfrac{0.172}{8} + \dfrac{11 \times 0.619}{120}\right) = 0.04 \text{ in.}$

Moment at A:

$$M = M_1 + M_2 = \frac{1}{2}wl^2 + \frac{w'l^2}{3} = \frac{0.172 \times 10^2}{2} + \frac{0.619 \times 10^2}{3}$$

$$= 29.2 \text{ lb-in.}$$

(b)   Stress at A:

$$\sigma = \frac{M}{Z} = \frac{M}{bh^2/6}$$

$$= \frac{29.2}{4 \times 0.125^2/6} = 2800 \text{ psi}$$

2-4   DEFLECTION OF CURVED-END CANTILEVERS

Fundamental design equations and charts[1] for an easy determination of vertical and horizontal deflections in curved-end cantilevers under various load conditions are presented in this section. Derivations for these equations and charts are based on the approach by Castigliano: the mechanical energy which is stored in an elastically stressed system can be recovered upon the removal of external loads. The charts give both the vertical and horizontal deflections as a function of the product Fr^3/EI and a constant C as is illustrated in Figs. 2-11, 2-12, 2-13, and 2-14. An example which illustrates the use of the charts follows.

---

[1] Alexander Blake, "Graphs Simplify Deflection Calculations for Curved-End Cantilevers," *Machine Design*, July 23, 1959.

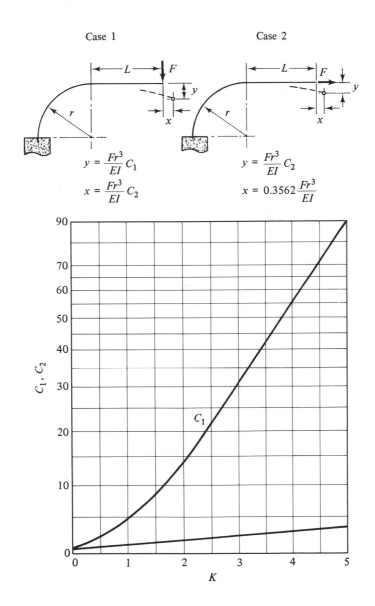

Case 1   Case 2

$$y = \frac{Fr^3}{EI} C_1$$

$$x = \frac{Fr^3}{EI} C_2$$

$$y = \frac{Fr^3}{EI} C_2$$

$$x = 0.3562 \frac{Fr^3}{EI}$$

Figure 2-11 / *Charts and equations for determination of deflections in curved-end cantilevers: quarter curve—load applied on straight section.*

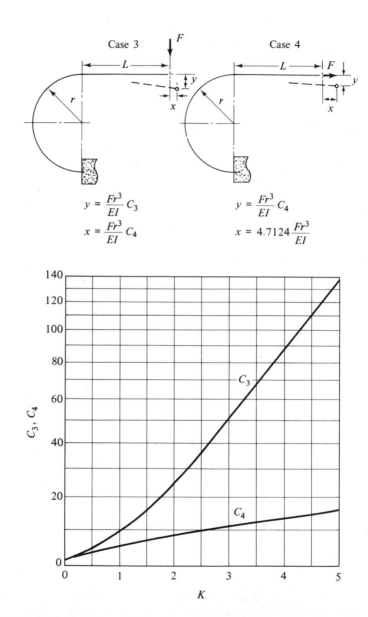

Figure 2-12 / Charts and equations for determination of deflections in curved-end cantilevers: half curve—load applied on straight section.

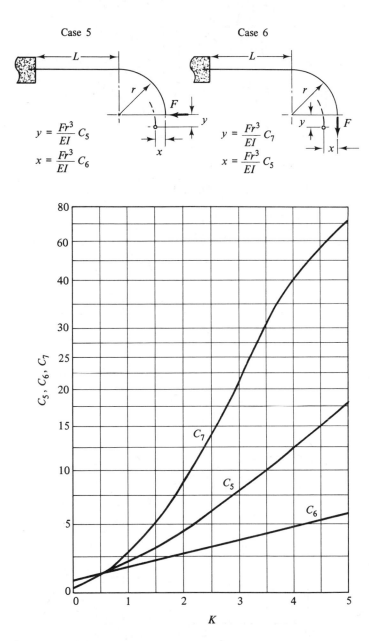

Figure 2-13 / Charts and equations for determination of deflections in curved-end cantilevers: quarter curve—load applied at curved end.

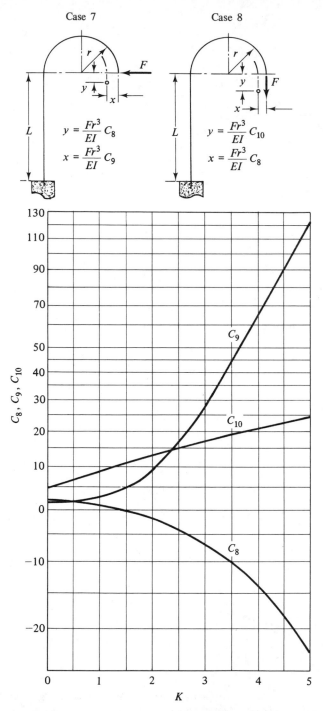

Figure 2-14 / Charts and equations for determination of deflections in curved-end cantilevers: half curve—load applied at curved end.

EXAMPLE 2-7

A tubular steel bar is shaped as shown in Fig. 2-15. The bar carries a vertical load of $F = 400$ lb. Determine the magnitude and direction of the end displacement.

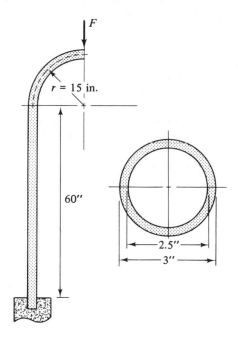

Figure 2-15   /   Tubular steel bar under end load (Example 2-7).

solution

Case 5 applies.

$$x = \frac{Fr^3}{EI} \times C_6$$

for $L/r = 60/15 = 4$, $C_6 = 4.8$. The moment of inertia for the tubular section is 2.06 in.^{4}, therefore

$$x = \frac{400 \times \overline{15}^3}{30 \times 10^6 \times 2.06} \times 4.8 = 0.105 \text{ in.}$$

$$y = \frac{Fr^3}{EI} \times C_5 = \frac{400 \times \overline{15}^3}{30 \times 10^6 \times 2.06} \times 12.5 = 0.273 \text{ in.}$$

The resultant displacement is

$$\delta = \sqrt{0.105^2 + 0.273^2} = 0.290 \text{ in.}$$

2-5   DEFLECTION OF STEPPED SHAFTS

There are several ways to determine the deflection at a particular point in a shaft of nonuniform diameter. One method makes use of the theoretical elastic energy equation:

$$EF\delta = \Sigma \frac{MM'\Delta x}{I} \tag{2-4}$$

where F represents an auxiliary load, usually assumed to be one lb or so, placed on the shaft, Fig. 2-16(a); where the deflection δ is desired; where

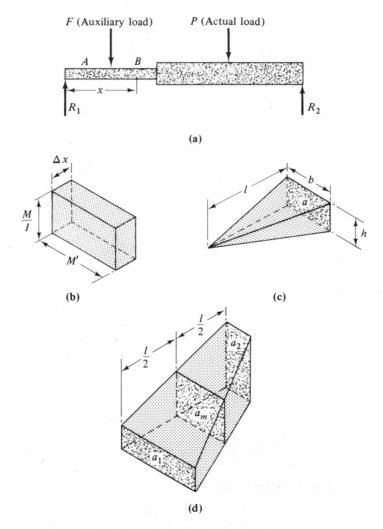

(a)

(b)   (c)

(d)

Figure 2-16 / Deflection of a stepped shaft.

M is the bending moment in terms of x for any general point B; and where M' is the bending moment at B caused by the auxiliary small load F. E and I are the modulus of elasticity and moment of inertia of the section, respectively. The summation proceeds from one end of the shaft to the other. Fortunately, Eq. (2-4) has a geometric interpretation.[2] The summation $\sum (MM'\Delta x/I)$ can be considered to be a solid, Fig. 2-16(b), with the three dimensions M/I, M', and Δx. If the moment diagrams for M and M' consist of segments of straight lines, the solid may be considered to have a finite length, and the volumes may be found by solid geometry. Since odd shapes will invariably appear, it is well to know that the volume of a pyramid, Fig. 2-16(c), is equal to

$$V = \tfrac{1}{3} \times l \times \text{area of base} \qquad (2\text{-}5)$$

and the volume of a prismoidal solid, Fig. 2-16(d), is

$$V = \frac{l}{6}(a_1 + 4a_m + a_2) \qquad (2\text{-}6)$$

where a_1 and a_2 are areas of the end surfaces, l is the distance between a_1 and a_2, and a_m is the area of the cross section midway between a_1 and a_2.

EXAMPLE 2-8

Find the deflection at point A of the circular steel shaft loaded as shown in Fig. 2-17. Assume the moments of inertia of the smaller and larger sections to be 6 in.[4] and 10 in.[4], respectively.

solution

The three diagrams M, M/I, and M' are drawn as illustrated. An auxiliary load of 5 lb is selected to keep the arithmetic simple—the magnitude of this small load is purely arbitrary.

The total volume of the prism is

$$v_1 = \tfrac{1}{3} \times 12 \times 2000 \times 36 = 288,000 \text{ lb/in.}$$
$$v_2 = \tfrac{8}{6}[(1200 \times 36) + (4 \times 1600 \times 28) + (2000 \times 20)]$$
$$= 350,000 \text{ lb/in.}$$
$$v_3 = \tfrac{1}{3} \times 10 \times 2000 \times 20 = 133,000 \text{ lb/in.}$$
$$E\delta = V = 288,000 + 350,000 + 133,000 = 771,000 \text{ lb/in.}$$

Therefore
$$\delta = \frac{771,000}{30 \times 10^6} = 0.026 \text{ in.}$$

---

[2]M. F. Spotts, *Design of Machine Elements*, 3rd ed. (Englewood Cliffs, N.J.: Prentice-Hall, Inc., 1961) pp. 128–31.

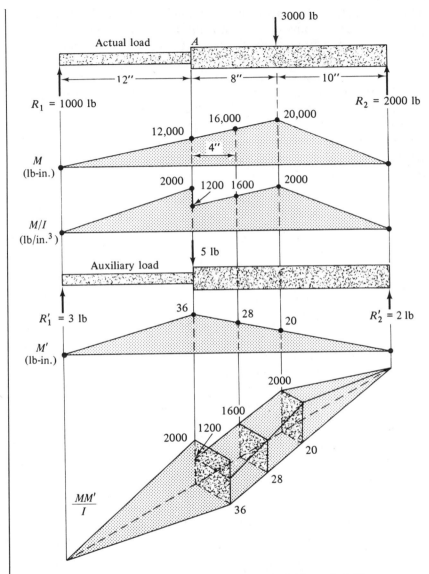

Figure 2-17 / Deflection of a stepped shaft (Example 2-8).

2-6   BEAMS AS SPRINGS

As was shown in Chapter 1, elastic materials and machine components that have a *linear force–deflection* relationship can be viewed as mechanical springs. A *series system*, one in which equal forces act in each element, has an equivalent constant $K_e = F/\Delta$ of

$$K_e = 1/\sum \frac{1}{K_n} \qquad (2\text{-}7)$$

Parallel systems are those whose components deflect equally. The equivalent spring constant $K_e = F/\Delta$ of this type of arrangement is

$$K_e = \sum K_n \qquad (2\text{-}8)$$

Two examples follow that demonstrate the virtues of the *equivalent spring concept.*

EXAMPLE 2-9

Find the deflection at point B of the pinned system of cantilever beams shown in Fig. 2-18. Both beams are steel, and $F = 200$ lb.

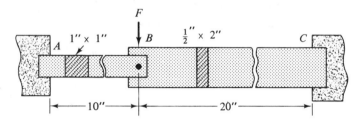

Figure 2-18 | Pinned cantilever beams (Example 2-9).

solution

Since both beams deflect equally, the equivalent constant of the combination is

$$K_e = K_{AB} + K_{BC}$$

where $\quad K_{AB} = \dfrac{3EI}{l^3} = \dfrac{3 \times 30 \times 10^6}{10^3} \times \dfrac{1}{12} \times 1 \times 1^3$

$$= 7.5 \times 10^3 \text{ lb/in.}$$

and $\quad K_{BC} = \dfrac{3EI}{l^3} = \dfrac{3 \times 30 \times 10^6}{20^3} \times \dfrac{1}{12} \times \dfrac{1}{2} \times 2^3$

$$= 3.75 \times 10^3 \text{ lb/in.}$$

Hence $\quad K_e = (7.5 \times 10^3) + (3.75 \times 10^3) = 11.25 \times 10^3$ lb/in.

Deflection: $\quad \delta = \dfrac{F}{K_e} = \dfrac{200}{11.25 \times 10^3} = 0.0178$ in.

EXAMPLE 2-10

Determine the deflection at the point of application of the load $F = 200$ lb as shown in Fig. 2-19. The vertical rod is bronze, and the beam is steel.

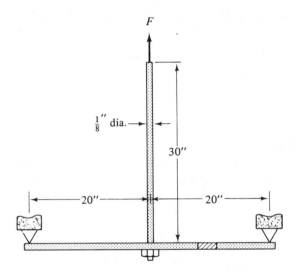

Figure 2-19 / A series system of rods (Example 2-10).

solution

The combination represents a series system—the force is transmitted through the rod to the beam.

For the rod:

$$K_R = \frac{EA}{l} = \frac{15 \times 10^6 \times \pi \times \overline{0.125^2}}{30 \times 4} = 6.13 \times 10^3 \text{ lb/in.}$$

For the beam:

$$K_B = \frac{48EI}{l^3} = \frac{48 \times 30 \times 10^6}{40^3} \times \frac{1}{12} \times 2 \times 1^3 = 3.75 \times 10^3 \text{ lb/in.}$$

For the system:

$$K_e = \frac{1}{\dfrac{1}{K_R} + \dfrac{1}{K_B}} = \frac{10^3}{\dfrac{1}{6.13} + \dfrac{1}{3.75}} = 2.33 \times 10^3 \text{ lb/in.}$$

Deflection:

$$\delta = \frac{F}{K_e} = \frac{200}{2.33 \times 10^3} = 0.0858 \text{ in.}$$

2-7 BUCKLING OF COLUMNS

Members that fail by a total *collapse* under the action of an axial compressive load are loosely defined as columns. Little concern is given to degree of buckling or collapsing—the state is one of *instability*; the member is assumed to have failed. From a design point of view, a column is a long, slender member whose crippling load is not a direct function of an allowable stress nor is it found by the simple division of force by area.

The Euler column formula defines the critical buckling force F_{cr} in terms of the elastic and geometric properties of the material.

$$F_{cr} = C\frac{\pi^2 EI}{l^2} \qquad (2\text{-}9)$$

where C is a constant defined by the end restraints and bracing conditions as shown in Fig. 2-20, E is the elastic modulus, and l is the length. The equation has its practical limitations; by dividing both sides of the equality

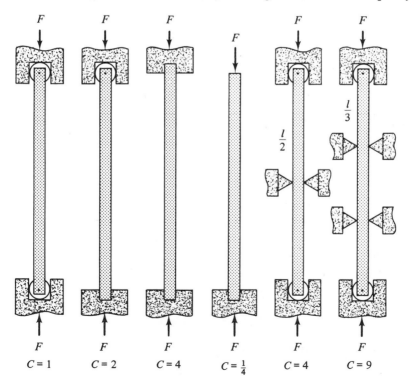

Figure 2-20 / End restraints for columns.

by the cross-sectional area, the buckling stress is found to be

$$\sigma_{cr} = \frac{F_{cr}}{A} = C\frac{\pi^2 EI}{l^2 A} \qquad (2\text{-}10)$$

Since I/A is the square of the radius of gyration k of the section, the critical stress can be written as

$$\sigma_{cr} = C\frac{\pi^2 E}{(l/k)^2} \qquad (2\text{-}11)$$

Or, in terms of the critical load F_{cr}

$$F_{cr} = \sigma_{cr}A = \frac{\pi^2 EA}{(l/k)^2} \qquad (2\text{-}12)$$

where the term l/k is called the *slenderness ratio*. Mathematically, this ratio can assume any value greater than zero. From the practical standpoint, however, severe restrictions have to be placed on the term. For a rough approximation, consider Case 1 $(C = 1)$ applied to steel with a yield point stress of $\sigma_{yp} = 30{,}000$ psi

$$\frac{l}{k} = \frac{\pi^2 \times 30 \times 10^6}{30{,}000} \approx 100$$

This is a minimum value; any magnitude of $l/k < 100$ will stress the column beyond the yield point.

Columns whose slenderness ratio falls below the Euler limits can be analyzed by a variety of other formulas. One such widely used relationship is the Johnson equation:

$$F_{cr} = \sigma_{yp}A\left[1 - \frac{\sigma_{yp}(l/k)^2}{4C\pi^2 E}\right] \qquad (2\text{-}13)$$

It is interesting to note that when l/k approaches zero, $F_{cr} = \sigma_{yp}A$; the column becomes a *post*—a member in pure compression.

A more precise limiting minimum value of the slenderness ratio for Euler's formula can be found by setting Eqs. (2-12) and (2-13) equal, assuming that for large l/k, the "one" in the Johnson formula becomes insignificant.

$$C\frac{\pi^2 EA}{(l/k)^2} = \frac{\sigma_{yp}^2 A(l/k)^2}{4C\pi^2 E}$$

Solving for the critical l/k gives

$$\left(\frac{l}{k}\right)_{cr} = \sqrt{\frac{2C\pi^2 E}{\sigma_{yp}}} \qquad (2\text{-}14)$$

TABLE 2-3

PROPERTIES OF CROSS SECTIONS

Cross section	Moment of inertia	Section modulus	Radius of gyration
Rectangle (dimensions $b \times h$)	$I = \dfrac{bh^3}{12}$	$Z = \dfrac{bh^2}{6}$	$k = \dfrac{h}{\sqrt{12}} = 0.289h$
Hollow square (h_o, h_i)	$\dfrac{h_o^4 - h_i^4}{12}$	$\dfrac{h_o^4 - h_i^4}{6h_o}$	$\dfrac{h_o^2 - h_i^2}{12}$
Triangle (base b, height h, centroid $h/3$)	$\dfrac{bh^3}{36}$	$\dfrac{bh^2}{24}$	$\dfrac{b}{\sqrt{18}} = 0.236h$
Hexagon (radius r)	$\dfrac{5\sqrt{3}}{16}r^4 = 0.541r^4$	$\dfrac{5r^3}{8}$	$\sqrt{\dfrac{5}{24}}\,r = 0.456r$
Octagon (radius r)	$\dfrac{1 + 2\sqrt{2}}{6}r^4$ $= 0.638r^4$	$0.691r^3$	$0.475r$
Circle (diameter d)	$\dfrac{\pi d^4}{64}$	$\dfrac{\pi d^3}{32}$	$\dfrac{d}{4}$
Hollow circle (d_o, d_i)	$\dfrac{\pi}{64}(d_o^4 - d_i^4)$	$\dfrac{\pi}{32}\dfrac{(d_o^4 - h_i^4)}{d_o}$	$\dfrac{\sqrt{d_o^2 - d_i^2}}{4}$

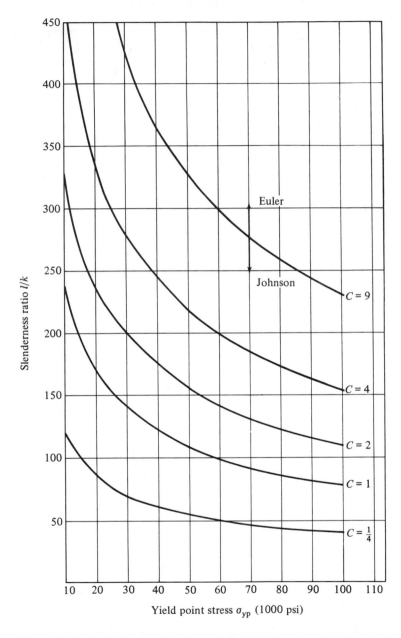

Figure 2-21 / Critical values for column slenderness ratio for steel ($E = 30 \times 10^6$) as a function of yield point stress and the end-restraint and bracing constant C. For any C, values of $1/k$ falling above the curves are Euler columns; if $1/k$ falls below the curve, columns are solved by the Johnson equation.

Euler's equation is valid when

$$\frac{l}{k} > \sqrt{\frac{2C\pi^2 E}{\sigma_{yp}}}$$

and the Johnson equation is used when

$$\frac{l}{k} < \sqrt{\frac{2C\pi^2 E}{\sigma_{yp}}}$$

Figure 2-21 is a plot of critical values of l/k for steel ($E = 30 \times 10^6$) columns as a function of yield point stress and the end-restraint and bracing constant C.

EXAMPLE 2-11

A ball and socket-ended steel column, 8-in. long, is braced at the midpoint as shown in Fig. 2-22. Determine the working load F_d based on a factor of safety of 2.5 and a yield point stress of 60,000 psi.

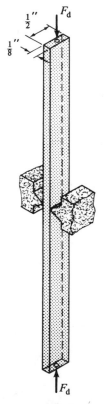

Figure 2-22 / Braced column (Example 2-11).

solution

Two columns are actually involved—the member could fail in either the free position or in the braced position; both configurations must be checked.

Radius of gyration:

$$k_x = \sqrt{\frac{I}{A}} = \sqrt{\frac{1}{12}\frac{bh^3}{bh}} = \sqrt{\frac{h^2}{12}}$$

$$= \sqrt{\frac{0.5^2}{12}} = 0.144 \text{ in.}$$

$$k_y = \sqrt{\frac{0.125^2}{12}} = 0.036 \text{ in.}$$

Area: $A = bh = 0.125 \times 0.5 = 0.0625 \text{ in.}^2$

For the unbraced configuration ($C = 1$):

$$\frac{l}{k} = \frac{8}{0.144} = 55.6; \qquad 55.6 < 100$$

Hence, the Johnson equation is valid.

$$F_{cr} = \sigma_{yp}A\left[1 - \frac{\sigma_{yp}(l/k)^2}{4C\pi^2 E}\right]$$

$$= 60{,}000 \times 0.0625\left[1 - \frac{60{,}000 \times \overline{55.6^2}}{4 \times 1 \times \pi^2 \times 30 \times 10^6}\right]$$

$$= 3160 \text{ lb}$$

For the braced configuration ($C = 4$):

$$\frac{l}{k} = \frac{8}{0.036} = 222; \qquad 222 > 200$$

Hence, the Euler equation is valid.

$$F_{cr} = 4\frac{\pi^2 EA}{(l/k)^2} = 4 \times \frac{\pi^2 \times 30 \times 10^6 \times 0.0625}{222^2}$$

$$= 1500 \text{ lb}$$

The design load F_d, therefore, must be based on the braced configuration:

$$F_d = \frac{F_{cr}}{2.5} = \frac{1500}{2.5} = 600 \text{ lb}$$

2-8 DESIGN OF LONG COLUMNS LOADED BETWEEN SUPPORTS

When long hinged-ended columns sustain concentric loads applied at points other than the ends, Fig. 2-23, Euler's formula becomes

$$F_{cr} = \frac{\pi^2 EI}{l_r^2} \qquad (2\text{-}15)$$

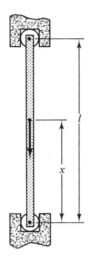

Figure 2-23 | Column with a middle load.

where l_r is called the *reduced length*, the length of a hinged column which has the same critical buckling load as the column loaded between supports.

Values of l_r may be obtained from the graph[3,4] of Fig. 2-24, which relates the ratios x/l and l_r/l. The former is obtained from known data and the latter from the graph.

For validity, Euler's equation must give a critical stress less than the yield stress, which means that a failure will occur through buckling rather than through a combination of buckling and compression. This point is easily checked by simply computing the value of the critical stress at the buckling load:

$$\sigma = \frac{F_{cr}}{A}$$

---

[3] James Dow, "Columns Loaded Between Supports," *Machine Design*, February 1961.

[4] Irving J. Levinson, *Mechanics of Materials*, 2nd ed. (Englewood Cliffs, N.J.: Prentice-Hall, Inc. 1970).

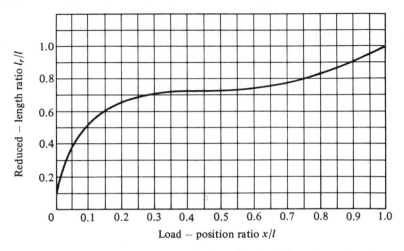

Figure 2-24 / Graph for determination of reduced length as a function of load position for Euler columns loaded between supports.

EXAMPLE 2-12

The control linkage shown in Fig. 2-25 consists of a high-strength steel rod and a pivot arm. The load is transmitted to the rod through pin A. Determine: (a) the critical value of F based on the buckling strength of the rod; and (b) the stress in the rod at this critical load. $\sigma_{yp} = 60,000$ psi.

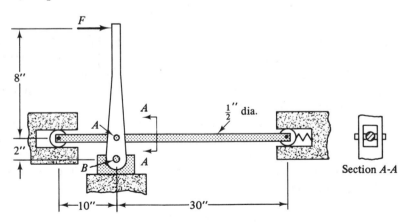

Figure 2-25 / A control linkage (Example 2-12).

solution

(a) The ratio of $x/l = 30/40 = 0.75$; l_r/l from the graph of Fig. 2-24 is 0.8; hence,

$$l_r = 0.8 \times 40 = 32 \text{ in.}$$

Pin A can tolerate a critical force of

$$F_{cr} = \frac{\pi^2 E I}{l_r^2}$$

where $\qquad I = \dfrac{\pi d^4}{64} = \dfrac{\pi \times \overline{0.5^4}}{64} = 0.00307 \text{ in.}^4$

Thus $\qquad F_A = F_{cr} = \dfrac{\pi^2 \times 30 \times 10^6 \times 0.00307}{32^2} = 887 \text{ lb}$

Stress, at the critical load, is

$$\sigma_{cr} = \frac{F_{cr}}{A} = \frac{887}{\pi \times 0.25^2} = 4520 \text{ psi}$$

which is well below the yield stress of 60,000 psi.

Moments taken about the pin B will give the critical force F:

$$10F = 2 \times 887$$
$$F = 177 \text{ lb}$$

2-9 ECCENTRICALLY LOADED COLUMNS: CONSERVATIVE APPROACH

Both the Euler and Johnson equations apply to columns that are initially straight and centrally loaded—an idealized set of conditions. Virtually every column has an inherent built-in eccentricity—"perfectly straight"

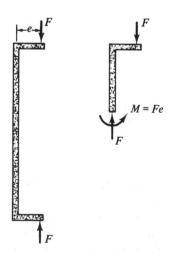

Figure 2-26 / Direct and bending loads.

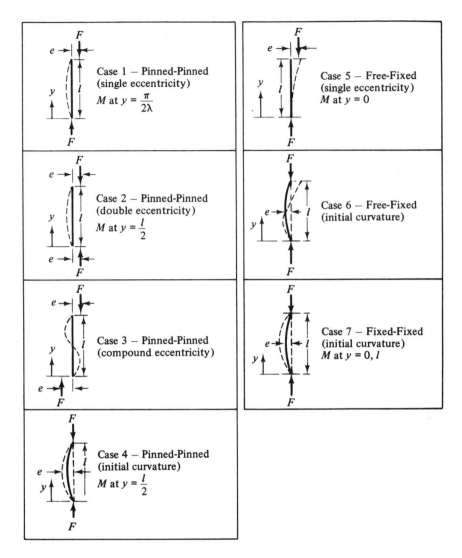

*Figure 2-27 / Classification of types of column eccentricities. Location
y of the maximum moment is given for each case.*

simply cannot be achieved nor can a load be situated on the exact centroidal axis. This built-in eccentricity demands high factors of safety.

An axial load applied eccentrically, or applied to a curved column,
will generate compressive as well as bending stresses. The result is an increased deflection with an increase in load—in buckling, no deflection
occurs until the state of instability is reached. The combination of direct

and bending loads is illustrated in Fig. 2-26, where the combined maximum stress

$$\sigma_{max} = \frac{F}{A} + \frac{Fe}{Z} \qquad (2\text{-}16)$$

is obtained through either the Euler or Johnson equations, depending, of course, on both the magnitude of the slenderness ratio and end restraints.

A graphical[5] approach to the problem of eccentricity, which is based on elastic beam theory, provides a rather conservative—conservatism is occasionally a necessary ingredient in design—method of analysis. Seven "cases" of column eccentricities are illustrated in Fig. 2-27. The following steps are used to determine the total stress on an eccentrically loaded or an initially curved column:

1 / Determine the load parameter $\lambda = l\sqrt{F/EI}$.

2 / From the appropriate chart, Fig. 2-28, determine $(M/Fe)_{max}$.

3 / Multiply $(M/Fe)_{max}$ by Fe to obtain the maximum bending moment M.

4 / Determine the stress from Eq. (2-16) $\sigma_{max} = F/A + M/Z$.

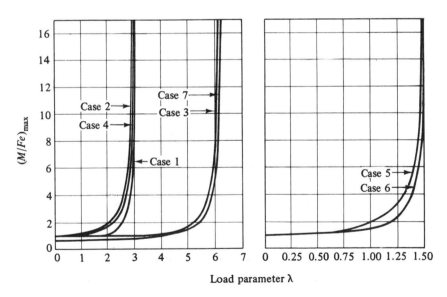

Figure 2-28 / Maximum moment for the seven cases of column eccentricities described in Figure 2-27.

[5]Carl F. Zorowski, "Eccentrically Loaded Columns," *Machine Design*, May 1962.

EXAMPLE 2-13

Determine the maximum stress in the pin-ended, doubly eccentric steel column shown in Fig. 2-29. $\sigma_{yp} = 60,000$ psi.

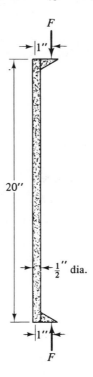

Figure 2-29 / Eccentrically loaded steel column (Example 2-13).

solution

Section properties are first computed.

Area:

$$A = \pi d^2/4 = \pi \times \overline{0.5^2}/4 = 0.196 \text{ in.}^2$$

Moment of inertia:

$$I = \pi d^4/64 = \pi \times \overline{0.5^2}/64 = 0.003 \text{ in.}^4$$

Section modulus:

$$Z = \pi d^3/32 = \pi \times \overline{0.5^3}/32 = 0.0123 \text{ in.}^3$$

Radius of gyration:

$$k = d/4 = 0.5/4 = 0.125 \text{ in.}$$

Slenderness ratio:

$$l/k = 20/0.125 = 160 \qquad \text{(an Euler column)}$$

Load parameter:

$$\lambda = l\sqrt{\frac{F}{EI}} = 20\sqrt{\frac{200}{30 \times 10^6 \times 0.003}} = 0.944$$

From Fig. 2-28, for a Case 2 column:

$$\frac{M}{Fe} = 1.2; \qquad M = 1.2 \times 200 \times 1 = 240 \text{ lb-in.}$$

Combined stress, Eq. (2-16):

$$\sigma = \frac{F}{A} + \frac{M}{Z} = \frac{200}{0.196} + \frac{240}{0.0123} = 20,530 \text{ psi}$$

For an interesting comparison, compute the critical Euler load and stress for a centrally loaded, pin-ended column ($C = 1$).

$$F_{cr} = C\sqrt{\frac{\pi^2 EI}{l^2}} = 1 \times \sqrt{\frac{\pi^2 \times 30 \times 10^6 \times 0.003}{20^2}} = 2220 \text{ lb}$$

$$\sigma_{cr} = \frac{F_{cr}}{A} = \frac{2200}{0.196} = 11,200 \text{ psi}$$

The eccentrically loaded column roughly supports only $\frac{1}{10}$ the load.

QUESTIONS AND PROBLEMS

P2-1
to
P2-6
Draw the shear and moment diagrams "by parts" for each of the beams illustrated and at the sections indicated. The areas should consist of only rectangles, triangles, parabolas, and cubic-parabolas.

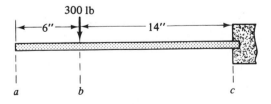

300 lb

├─6"──┤├────14"────┤

a   b   c

Figure P2-1

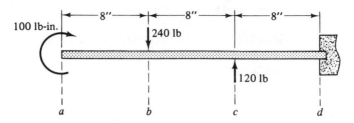

Figure P2-2

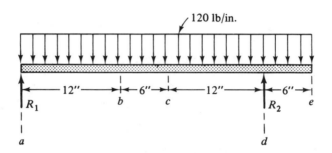

Figure P2-3

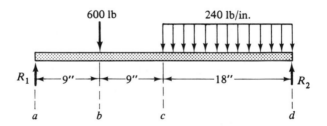

Figure P2-4

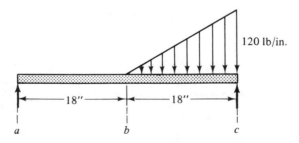

Figure P2-5

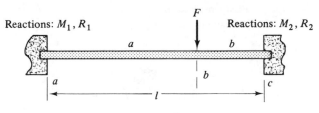

Figure P2-6

P2-7   Determine the slope and deflection at the free end of the beam
to   shown by the moment–area method. Check the answers for deflec-
P2-10 tions in each instance by the appropriate formulas given in Table
           2-2.

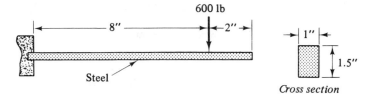

Figure P2-7

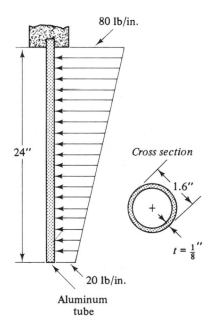

Figure P2-8

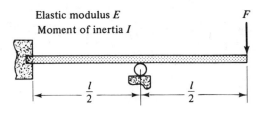

Elastic modulus E
Moment of inertia I

F

$\dfrac{l}{2}$ $\dfrac{l}{2}$

Figure P2-9

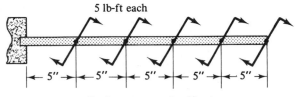

5 lb-ft each

← 5″ →|← 5″ →|← 5″ →|← 5″ →|← 5″ →

Steel — stress grade: 60 psi

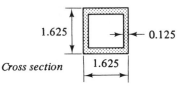

1.625   →|← 0.125

Cross section   1.625

Figure P2-10

P2-11 A simple beam having a span of 36 in. supports a concentrated load at midspan of F lb and a distributed load over the entire span of $F/18$ lb per in. The beam is a rectangular steel tube, 2-in. wide and 3-in. high, with a wall thickness of $\frac{1}{8}$ in. Find F under the following conditions:

> Bending stress: 20,000 psi max
> Shearing stress: 15,000 psi max
> Deflection: span/100 max
> Factor of safety: 2.5

P2-12 Both deflection and stress are criteria involved in beam design. Describe situations where:
(a)   limiting stress is more critical than deflection; and
(b)   where deflection is more critical than stress.

P2-13 Find the equivalent spring constant in symbolic terms F, l, E, and I of the double beam arrangement illustrated. Determine F if the deflection at the center is 0.05 in. for $l = 30$ in.

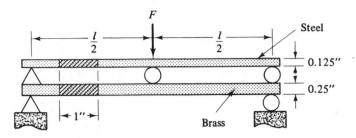

Figure P2-13

P2-14 A double-beam steel shock-absorbing spring system is shown. Find the force *F* required to cause beam *B* to deflect 0.20 in. Prior to application of the load, the clearance *c* between beams *A* and *B* is 0.10 in.

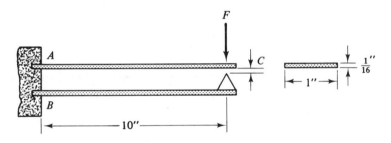

Figure P2-14

P2-15 Strap iron $2 \times \frac{1}{4}$ in. in section is used for hoops around a 30-ft diameter water tank made of wood staves. For a straight strap bent to this diameter, what is the stress in the strap?

P2-16 Belt loads act as shown on the solid steel shaft which is rotating at a constant angular velocity. Determine the required diameter of the shaft for the following set of design limitations:

> Bending stress: 40,000 psi max
> Shear stress: 20,000 psi max
> Deflection: 0.01 in. max

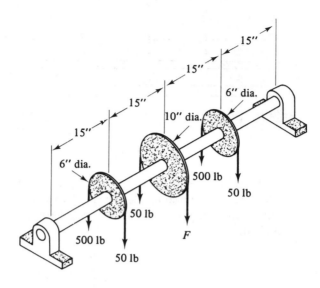

Figure P2-16

P2-17   A flat steel spring employed in a mechanical device is precurved as
shown. Determine the decrease in distance between free ends for
a force $F = 5$ lb.

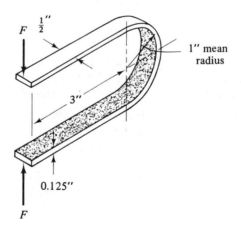

Figure P2-17

P2-18   A rigidly supported curved preformed steel beam is illustrated. Determine F if the vertical and horizontal deflections are not to exceed 0.20 and 0.05 in., respectively.

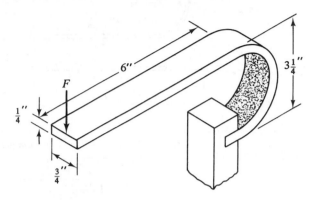

Figure P2-18

P2-19   The aluminum cantilever shown is acted on by two forces $F_1 = 10$ lb, and $F_2 = 20$ lb. Determine the horizontal and vertical displacement of the free end. The beam cross section is circular and has a diameter $d = \frac{1}{4}$ in. $E = 10 \times 10^6$.

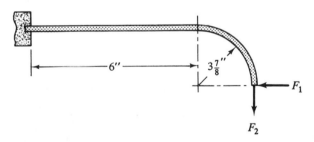

Figure P2-19

P2-20   Find the energy stored in the system if the cam shown is rotated 90° counterclockwise. The beams are preformed steel and the cross sections are indicated.

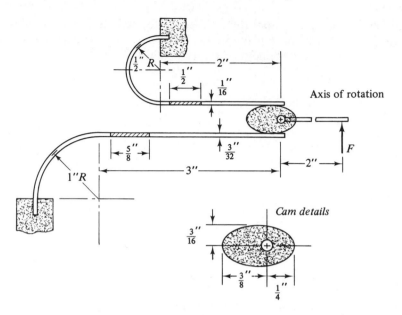

Figure P2-20

P2-21 The column cross sections shown are straight and centrally loaded.
to   Identify each, with respect to the indicated axes, as being a
P2-26 "Euler" or a "Johnson" column, and find the working load and
           working stress.

Length = 20″, $C = \frac{1}{4}$
Material: Steel, σ_{yp} = 60,000 psi

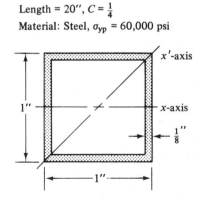

Figure P2-21

Length = 30″, C = 1
Material: Aluminum, σ_{yp} = 20,000 psi

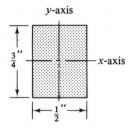

Figure P2-22

Length = 100″, C = 2
Material: Steel, σ_{yp} = 30,000 psi

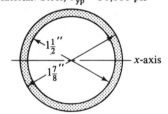

Figure P2-23

Length = 25″, C = 1
Material: Brass, σ_{yp} = 25,000 psi

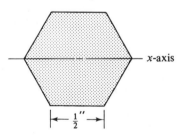

Figure P2-24

Length = 40", $C = \frac{1}{4}$
Material: Steel, σ_{yp} = 30,000 psi

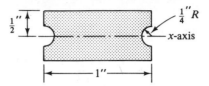

Figure P2-25

Length = 40", $C = \frac{1}{4}$
Material: Steel, σ_{yp} = 70,000 psi

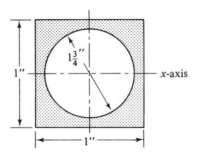

Figure P2-26

P2-27 A pin-ended steel column 36 in. long has a slenderness ratio of 150. The designer can select any material from the following list. A percentage dollar per pound premium must be paid, however, as the yield point stress increases. Select the most economical material.

TYPE (psi)	PERCENTAGE OF BASE COST
30,000	100
40,000	110
50,000	117
60,000	125
70,000	150

P2-28 An initially stress-free and straight steel member is rigidly held between nonyielding supports as shown. Determine the increase in temperature that will cause instability. $l = 20$ in., $d = \frac{1}{4}$ in.

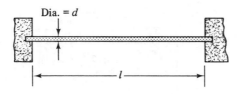

Figure P2-28

P2-29 Compare the load-carrying capacity of two pin-ended steel columns: one is initially straight, and the other is curved with an eccentricity of $\frac{1}{8}$ in. Both columns are 20 in. long; each has a diameter of $\frac{1}{4}$ in. and a yield point stress of 70,000 psi.

P2-30 Determine the permissible initial curvature for a fixed-ended aluminum tube with the following dimensions: length = 45 in.; nominal diameter = 1 in.; wall thickness = $\frac{1}{16}$ in.; elastic modulus = 10×10^6 psi. The curvature is such that the allowable load F is at least 80% of F_{cr} for a maximum compressive stress of 40,000 psi.

SUGGESTED DESIGN PROJECTS

DP2-1 Prepare a graph, similar to that of Fig. 2-21, that will show the Euler–Johnson limits for the following stress grades of aluminum alloys: σ_{yp} = 20,000 psi, 25,000 psi, 30,000 psi, 35,000 psi, and 40,000 psi. In each instance, $E = 10 \times 10^6$ psi.

DP2-2 A delicate optical instrument weighing 350 lb with its 24-in. diameter base is to be supported by three vertical telescoping legs. The height of the instrument must be adjustable from a minimum of 12 in. to a maximum of 36 in. above a base plane. Design the legs using an aluminum alloy. Weight, stability, and cost are important factors.

DP2-3 An outline of the essential features of a bench-mounted industrial precision drill press is shown. The "head" weighs 450 lb, and the location of its center of gravity is indicated. During a boring operation, a thrust load of 250 lb is developed in the spindle. The adjustable table weighs 150 lb, and its maximum height is illustrated. Design the tubular steel column. Select an appropriate grade of steel, and account for an expected eccentricity of 0.0005 in. in its manufacture. Cost is a factor, since 5000 machines are to be produced.

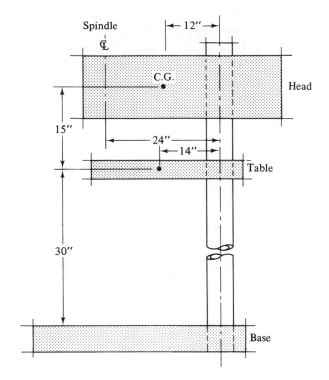

Figure DP2-3

DP2-4 Design a shaft to support the loads shown and in accord with the following criteria:

Maximum deflection at A: 0.0015 in.
Maximum deflection at B: 0.0035 in.
Maximum bending stress: 30,000 psi

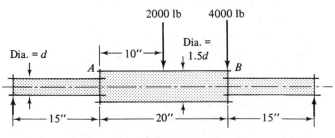

Figure DP2-4

CHAPTER 3

CAMS AND ECCENTRICS

"Better slow than never, Jack," replied I. Experience has even taught and mechanical observations have established as a principle, that what is gained in speed is lost in strength: the purpose of the crow is not to enable us to raise anything rapidly, but to raise what is exceedingly heavy; and the heavier the thing we would move, the slower the mechanical operation.

JOHANN WYSS
The Swiss Family Robinson

3-1 INTRODUCTION

Controlling the ignition of fuel in each of the cylinders in your car—four, six, or eight—is a small cam attached to the distributer shaft. With as many lobes as the engine has cylinders, the cam actuates the ignition at an astounding rate of forty or fifty times a second. Opening and closing the intake and exhaust valves—again on your car—is a shaft aligned with cams, two for each cylinder, that open and close the intake and exhaust valves through a linkage system called the *valve train*.

You awake in the morning by an alarm, tripped by a cam or an eccentric. Typewriters, sewing machines, can openers, and spacecraft all contain cams and eccentrics of one sort or another—cams that either transform one type of motion into another or that transform force into torque or torque into force.

Figure 3-1 / Camshaft for 1971 Pontiac V-8. (Courtesy of Pontiac Motor Division, General Motors Corporation)

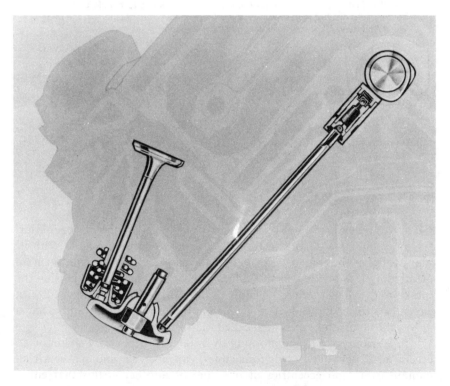

Figure 3-2 / Valve train. (Courtesy of Pontiac Motor Division, General Motors Corporation)

Use of cams and eccentrics for industrial purposes had its beginnings in antiquity. An early automatic metal shears, Fig. 3-3,[1] invented by Polhem and used in Sweden in the 1700s, made the country the prime producer of nails. Actuating the metal shears was a pegged "jumping" cam. As can be seen in the drawing, the machine was built of both wood and metal; it is also quite obvious that the operator had to keep his mind on his work.

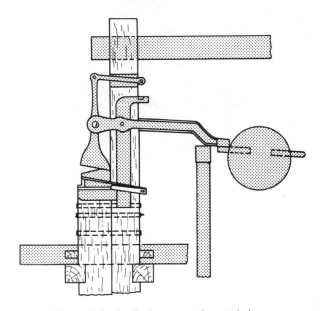

Figure 3-3 / Early automatic metal shears.

3-2   CAM SYSTEMS

Categorization of cams can be based on the two essential ingredients: the follower and the cam itself. In the case of the former, variables include contact, motion, position, and constraints; for the cam, shape and motion are definable qualities. Figure 3-4 describes seven types of follower contact and two types each of follower motion, position, and constraint. Briefly, each can be described as follows:

  a / *Point or knife-edge translating follower.* Mainly used with slow-speed cams where accuracy is not an essential factor. Construction

---

[1]Abbott Payson Usher, *A History of Mechanical Inventions* (Cambridge, Mass.: Harvard University Press, 1954), pp. 376–77.

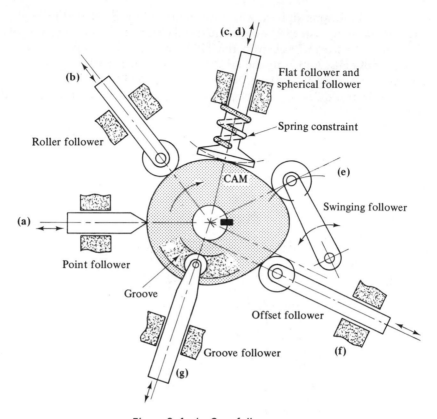

Figure 3-4 / Cam followers.

is simple but the edge quickly wears. A gravity constraint is illustrated.

b / *Translating roller follower.* Functions at moderate to low speeds; malfunctions through slipping at high speeds. The follower can jam if the cam is too steep.

c / *Flat* and *spherical translating follower.* Operates at moderate to
d / high speeds and can be used with steep cam profiles. Also described in the figure is a spring constraint.

e / *Swinging or oscillating follower.* Rotational cam motion is converted into swinging follower motion.

f / *Offset translating follower.* Offset is used to reduce the possibility of jamming in a steep cam.

g / *Translating groove follower.* Roller follows path in groove of cam; the groove affords a positive constraint.

Figure 3-5 illustrates four of an almost endless variety of *yoke* cams. These four *positive constraint systems* are shown because they represent some interesting cases of transformation of rotational motion into reciprocating motion.

a / Uniform harmonic follower motion is produced by the captive rotating pin. Accelerations are a maximum at top and bottom dead center; velocities are a maximum when the driving pin is on the horizontal centerline.

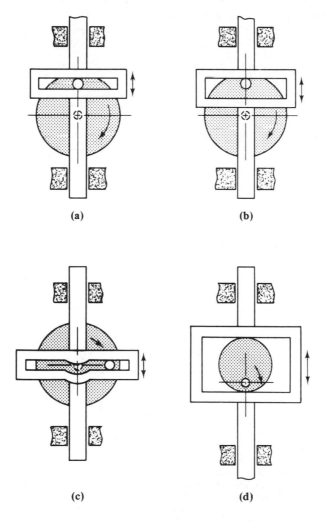

(a)   (b)

(c)   (d)

Figure 3-5 / Yoke cams.

b / Since the yoke is larger than the driving pin, the follower recip-
rocates with four intermittent stops.

c / Depression in the yoke causes the follower to momentarily stop
or dwell at bottom dead center; otherwise the follower moves with
harmonic motion. The yoke could be designed to cause the follower to dwell at top dead center.

d / Reciprocating motion is derived from an eccentric circular cam.

Two unusual arrangements are illustrated in Fig. 3-6; shown first
is a two-dimensional system in which the displacement, velocity, and
acceleration of point *P* is a function of an *x*-cam and a *y*-cam. The second

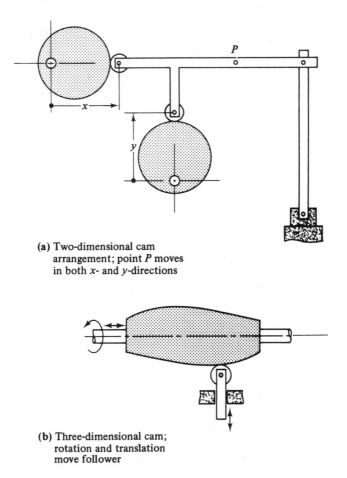

(a) Two-dimensional cam
arrangement; point *P* moves
in both *x*- and *y*-directions

(b) Three-dimensional cam;
rotation and translation
move follower

Figure 3-6 / Two-dimensional and three-dimensional cam systems.

illustration is representative of a three-dimensional system—the barrel cam both rotates about an axis of revolution and translates along this same axis.

3-3 TERMINOLOGY

Several different names are used by the industry to designate similar shapes and types of cams. For the sake of both clarification and a firsthand view of what is being manufactured, Fig. 3-7 illustrates several cams. Their designations, as labeled in Fig. 3-7, are:

a / Plate cam; flat cam; radial cam.

b / Double profile cam; conjugate follower cam; complementary cam; double disk cam.

c / Ribbed barrel cam; double-end cam.

d / Grooved barrel cam; plain cylindrical cam; drum cam.

e / End cam; bell cam.

f / Face cam; face-groove cam; plate-groove cam; box cam; closed-track cam.

g / Conjugate face cam—gear can act as driver or driven member; double-purpose face cam.

h / Double-cut barrel cam.

3-4 TYPES OF FOLLOWER MOTION

Displacement, velocity, acceleration, and jerk are the important factors in the design and the final layout of a cam. Fortunately or unfortunately, as the case may be, all are mathematically related. What might prove to be an ideal lift-time relationship between displacement and velocity would also prove to have extremely poor acceleration characteristics—accelerations could approach an infinite magnitude and cause the follower, even if spring constrained, to leave the cam surface. Searching for perfection and hunting for the "lost chord" are just about equal in terms of achievement.

Four basic follower movements and their variations are considered. Each movement is based on geometry, and several cams are drawn using the approach as an aid to understanding. Actual layout for manufacturing purposes requires a great deal of precision—a goal that cannot be attained by even the most expert draftsman.

a

b

c

d

e

f

g

h

Figure 3-7 / Types of cams. (Courtesy of Ferguson Machine Company)

168

CONSTANT VELOCITY

Figure 3-8 is a plot of follower lift as a function of increments of cam angle θ. The abscissa is divided into 12 units—a purely arbitrary number; each unit could represent 1°, 5°, 10°, or even 30° of cam rotation. In the last instance, the follower would abruptly drop from its total lift or rise to its *base circle* or zero position. The ordinate of the plot represents the position of the follower at the respective cam angle. Since the velocity is constant, the displacements $y_n - y_{n-1}$ will always be constant. Because the velocity of the follower is constant, the acceleration during the lift is zero—no inertia forces are involved. However, at the beginning and at the end of the rise, the acceleration has a theoretical magnitude of infinity. The beginning and end of the curve are, therefore, usually modified to reduce the accelerations to manageable quantities.

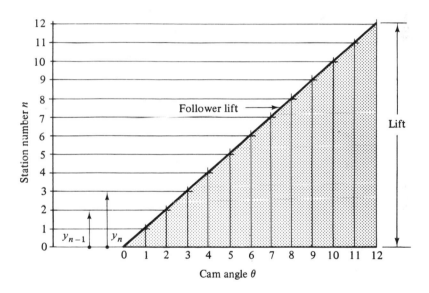

Figure 3-8 / Constant velocity cam lift.

CONSTANT ACCELERATION—PARABOLIC MOTION

The displacement diagram is constructed as illustrated in Fig. 3-9. An even number of angle divisions are selected for the abscissa. Ordinate AC is divided into segments proportional to $1:3:5:7:7:5:3:1$ if eight angle divisions are used; for twelve angle divisions, the proportionality would be $1:3:5:7:9:11:11:9:7:5:3:1$. These relative lengths are carefully drawn on any convenient line AB and projected by similar triangles to the vertical line AC. Intersections of the particular stations with verticals of the

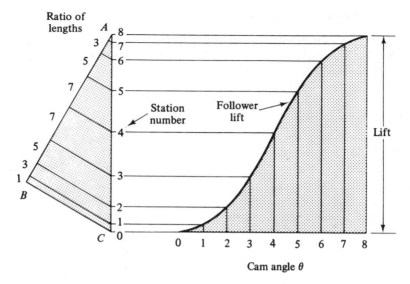

Figure 3-9 / Constant acceleration cam lift.

corresponding cam angle give a curve that depicts follower lift as a function of cam rotation. Parabolic motion gives a follow lift with the least acceleration of any motion; it is far from ideal, however, and is recommended for only low- or moderate-speed cam rotation.

HARMONIC MOTION

Construction of this type of follower motion begins with the drawing of a semicircle, Fig. 3-10, on the ordinate or lift axis of the displacement dia-

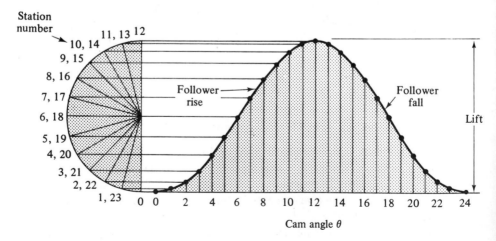

Figure 3-10 / Harmonic motion.

gram. The semicircle is then divided into as many equal parts as are chosen for the cam angle axis; in the illustration, both a harmonic rise and a fall are shown. Description of the construction of the follower motion is illustrated in the figure. As with the previous types of motion, harmonic displacements are recommended for low- or moderate-speeds.

CYCLOIDAL MOTION

Of the four motions described, cycloidal has the singular distinction of having zero acceleration at the beginning and at the end of the lift cycle as well as at the beginning and end of the follower fall. Also, it is the most difficult to graphically describe and construct. A lift axis *AB*, Fig. 3-11, is drawn that has a length equal to the circumference of circle *C*. The cycloidal curve is the path that a point *P* on the periphery of the circle generates as the circle rolls upward. To draw the cycloid, divide the circle into as many equal increments as are selected for the cam angle; the centerline of the circle is projected upward and a partial circle drawn at

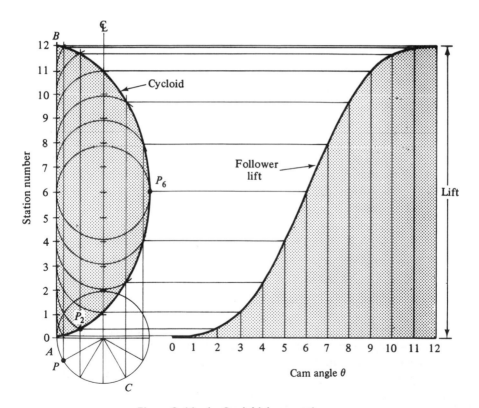

Figure 3-11 / Cycloidal cam action.

each intersection of this centerline with a station line. The construction proceeds as indicated, and the follower lift path is generated. Because of its zero acceleration characteristics, cycloidal motion can be used at high speeds.

3-5   CHARACTERISTICS OF FOLLOWER MOTION

Velocity, acceleration, and jerk are each time functions of displacement and of one another:

$$\text{Velocity} = \text{time rate of change of displacement}; \ \Delta s/\Delta t$$
$$= \text{slope of the displacement curve}$$

$$\text{Acceleration} = \text{time rate of change of velocity}; \ \Delta v/\Delta t$$
$$= \text{slope of the velocity curve}$$

$$\text{Jerk} = \text{time rate of change of acceleration}; \ \Delta a/\Delta t$$
$$= \text{slope of the acceleration curve}$$

These relationships are displayed in Fig. 3-12. The advantages and disadvantages of the uniform, parabolic, harmonic, and cycloidal motion become readily apparent:

1 / *Uniform motion:* infinite accelerations (and, therefore, infinite inertia forces) at beginning and end of lift.

2 / *Parabolic motion:* acceleration is undefined at the point of maximum velocity. Jerk, the rate of change of acceleration, is infinite at this point.

3 / *Harmonic motion:* the slope of the acceleration curve is zero at both the beginning and end of the displacement curve. Jerk, therefore, becomes an infinite quantity—at high speeds the follower would momentarily leave the surface of the cam.

4 / *Cycloidal motion:* of the four types of motion, this is the only one in which the acceleration and jerk have finite values during the complete cycle.

3-6   GRAPHICAL LAYOUT OF CAMS

While there are more precise ways of arriving at an appropriate cam profile—and these will be discussed—the graphical analysis adds both appreciation and understanding. Critical dimensions are usually given along with a series of events, a schedule, that are to occur as the cam rotates. As an example, consider the design of a cam profile that will conform to the following set of conditions:

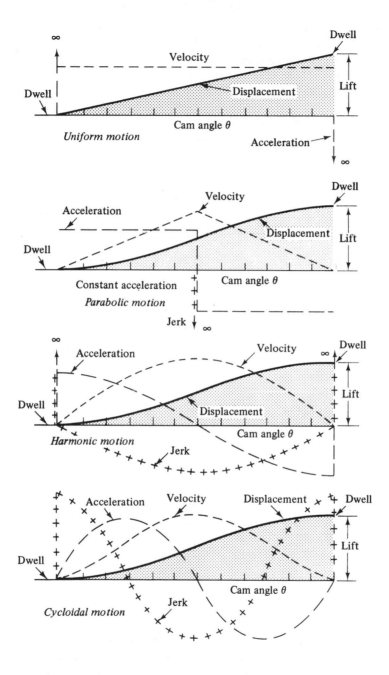

Figure 3-12 / Relationships among velocity, acceleration, and jerk.

Base circle: 1.5-in.-dia (this is the smallest circle that can be drawn
         through the cam surface)
Cam rotation: Counterclockwise
Follower style: Knife edge; on center
Follower schedule:
         Rise: 0.75 inches with harmonic motion in 120° of cam rotation
         Dwell: 30° of cam rotation
         Fall: 0.75 inches with harmonic motion in 180° of cam rotation
         Dwell: 30° of cam rotation

Construction of the cam profile is described in Fig. 3-13. The base
circle is first drawn to scale along with a horizontal *angle line* having a
length πd equal to the circumference of the base circle. For followers other
than knife-edge, a longer line is used, and the rationale for this length
determination is discussed in the next section.

The angle line, which represents 360° of cam rotation, is proportionally divided according to the schedule. Harmonic motion is required
for the *rise* and *fall*; so a circle is drawn with its diameter as the ordinate at
the 0° mark on the angle line. For convenience of description—not for
accuracy—the circumference of the circle is divided into six equal increments. On the angle line, each increment of length represents 20° of cam
rotation during the rise; during the fall, each increment represents 30° of
rotation—the spacing during the fall is, therefore, proportionally greater.

The cam, which is to rotate counterclockwise, is held fixed; the
projections of follower displacement are transferred to radial *station lines*
in a clockwise direction (opposite to the cam rotation) as shown. Finally,
the cam profile is drawn as the best smooth curve through the fourteen
points of intersection of lift with the radial station lines.

Inflection points, which in this case occur at stations 3 and 10,
identify the maximum *steepness* of the cam profile. The magnitude of this
steepness is a measure of the tendency of the follower to *jam* as the cam
rotates. The *jamming angle* is called the *pressure angle* ϕ—it is a maximum
at station 3, and, on the cam profile, it is measured between the follower
axis and a perpendicular projected from a tangent to the profile as
illustrated.

3-7 *CAM TABLES*

To lay out the cam of Fig. 3-13 with mathematical precision rather than
graphical, use is made of precalculated cam tables—in this instance Table
3-2. Contained in the table is the proportional rise—or working backward,
the proportional fall—in terms of 120-angle stations θ. During the rise,
which occurs in 120° of cam rotation, each station θ given in the table is

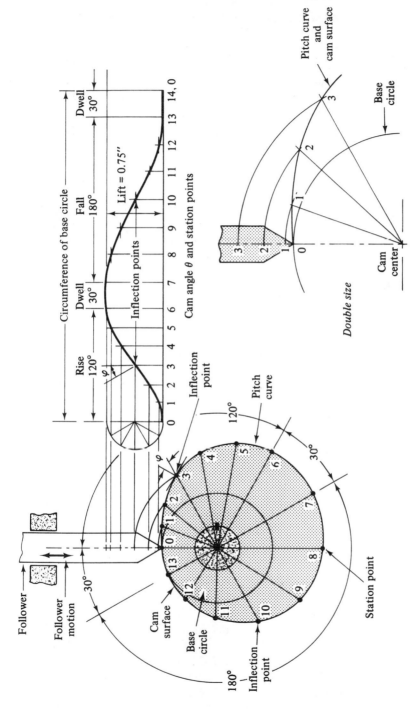

Figure 3-13 | Graphical construction of a cam profile.

175

TABLE 3-1
CAM DISPLACEMENTS FOR UNIFORM ACCELERATION

θ	y	θ	y	θ	y
1	0.000139	41	0.233471	81	0.788745
2	0.000556	42	0.244999	82	0.799439
3	0.001250	43	0.256804	83	0.809856
4	0.002222	44	0.268887	84	0.819995
5	0.003472	45	0.281248	85	0.829856
6	0.005000	46	0.293887	86	0.839439
7	0.006806	47	0.306804	87	0.848744
8	0.008889	48	0.319998	88	0.857772
9	0.011250	49	0.333470	89	0.866522
10	0.013889	50	0.347220	90	0.874994
11	0.016806	51	0.361248	91	0.883188
12	0.020000	52	0.375553	92	0.891105
13	0.023472	53	0.390136	93	0.898744
14	0.027222	54	0.404997	94	0.906105
15	0.031250	55	0.420136	95	0.913188
16	0.035556	56	0.435553	96	0.919994
17	0.040139	57	0.451247	97	0.926522
18	0.045000	58	0.467219	98	0.932772
19	0.050139	59	0.483469	99	0.938744
20	0.055556	60	0.500000	100	0.944438
21	0.061250	61	0.516525	101	0.949855
22	0.067222	62	0.532776	102	0.954994
23	0.073472	63	0.548747	103	0.959855
24	0.080000	64	0.564441	104	0.964438
25	0.086806	65	0.579858	105	0.968744
26	0.093889	66	0.594997	106	0.972772
27	0.101250	67	0.609858	107	0.976522
28	0.108889	68	0.624441	108	0.979994
29	0.116806	69	0.638746	109	0.983188
30	0.125000	70	0.652774	110	0.986105
31	0.133472	71	0.666524	111	0.988744
32	0.142222	72	0.679996	112	0.991105
33	0.151250	73	0.693190	113	0.993188
34	0.160555	74	0.706107	114	0.994994
35	0.170138	75	0.718746	115	0.996522
36	0.180000	76	0.731107	116	0.997772
37	0.190138	77	0.743190	117	0.998744
38	0.200555	78	0.754995	118	0.999438
39	0.211249	79	0.766523	119	0.999855
40	0.222221	80	0.777773	120	1.000000

TABLE 3-2
CAM DISPLACEMENTS FOR HARMONIC MOTION

θ	y	θ	y	θ	y
1	0.000171	41	0.261420	81	0.761249
2	0.000685	42	0.273004	82	0.772319
3	0.001541	43	0.284744	83	0.783203
4	0.002739	44	0.296631	84	0.793892
5	0.004277	45	0.308658	85	0.804380
6	0.006155	46	0.320816	86	0.814660
7	0.008372	47	0.333096	87	0.824724
8	0.010926	48	0.345491	88	0.834565
9	0.013815	49	0.357992	89	0.844177
10	0.017037	50	0.370590	90	0.853553
11	0.020590	51	0.383277	91	0.862687
12	0.024471	52	0.396044	92	0.871572
13	0.028679	53	0.408882	93	0.880202
14	0.033209	54	0.421782	94	0.888572
15	0.038060	55	0.434736	95	0.896676
16	0.043227	56	0.447735	96	0.904508
17	0.048707	57	0.460770	97	0.912063
18	0.054496	58	0.473832	98	0.919335
19	0.060591	59	0.486911	99	0.926320
20	0.066987	60	0.500000	100	0.933012
21	0.073679	61	0.513088	101	0.939408
22	0.080664	62	0.526167	102	0.945503
23	0.087936	63	0.539229	103	0.951292
24	0.095491	64	0.552264	104	0.956772
25	0.103323	65	0.565263	105	0.961939
26	0.111427	66	0.578217	106	0.966790
27	0.119797	67	0.591117	107	0.971320
28	0.128427	68	0.603955	108	0.975528
29	0.137312	69	0.616722	109	0.979409
30	0.146446	70	0.629409	110	0.982962
31	0.155822	71	0.642007	111	0.986184
32	0.165434	72	0.654508	112	0.989073
33	0.175275	73	0.666903	113	0.991627
34	0.185339	74	0.679183	114	0.993844
35	0.195619	75	0.691341	115	0.995722
36	0.206107	76	0.703368	116	0.997260
37	0.216796	77	0.715255	117	0.998458
38	0.227680	78	0.726995	118	0.999314
39	0.238750	79	0.738579	119	0.999828
40	0.250000	80	0.750000	120	1.000000

TABLE 3-3
CAM DISPLACEMENTS FOR CYCLOIDAL MOTION

θ	y	θ	y	θ	y
1	0.000003	41	0.208188	81	0.816808
2	0.000030	42	0.221240	82	0.828728
3	0.000102	43	0.234646	83	0.840250
4	0.000243	44	0.248391	84	0.851365
5	0.000474	45	0.262460	85	0.862065
6	0.000818	46	0.276837	86	0.872343
7	0.001297	47	0.291507	87	0.882195
8	0.001932	48	0.306451	88	0.891616
9	0.002745	49	0.321651	89	0.900603
10	0.003755	50	0.337089	90	0.909154
11	0.004984	51	0.352745	91	0.917270
12	0.006451	52	0.368599	92	0.924949
13	0.008173	53	0.384630	93	0.932195
14	0.010171	54	0.400818	94	0.939010
15	0.012460	55	0.417141	95	0.945398
16	0.015058	56	0.433576	96	0.951365
17	0.017980	57	0.450102	97	0.956917
18	0.021240	58	0.466697	98	0.962061
19	0.024854	59	0.483337	99	0.966808
20	0.028834	60	0.500000	100	0.971165
21	0.033191	61	0.516662	101	0.975145
22	0.037938	62	0.533302	102	0.978759
23	0.043082	63	0.549897	103	0.982019
24	0.048634	64	0.566423	104	0.984941
25	0.054601	65	0.582859	105	0.987539
26	0.060989	66	0.599181	106	0.989828
27	0.067804	67	0.615369	107	0.991826
28	0.075050	68	0.631400	108	0.993548
29	0.082729	69	0.647254	109	0.995015
30	0.090845	70	0.662910	110	0.996244
31	0.099396	71	0.678348	111	0.997254
32	0.108383	72	0.693548	112	0.998067
33	0.117804	73	0.708492	113	0.998702
34	0.127656	74	0.723162	114	0.999181
35	0.137934	75	0.737539	115	0.999525
36	0.148634	76	0.751608	116	0.999756
37	0.159749	77	0.765353	117	0.999897
38	0.171271	78	0.778759	118	0.999969
39	0.183191	79	0.791811	119	0.999996
40	0.195501	80	0.804498	120	1.000000

equivalent to 120/120, or 1°. During the fall, each table station θ represents 180/120 or $1\frac{1}{2}$°. The value y of the table gives the linear displacement based on a total rise or fall of one inch. For the cam of Fig. 3-13, each value of y would have to be multiplied by 0.75. The number of stations used depends upon the required accuracy. Suppose for the intended purposes of the cam that a layout point must be established every 10° during the rise and every 15° during the fall. The following schedule Table 3-4, derived from Table 3-2, will give precise follower position with six-place accuracy.

TABLE 3-4

CAM LAYOUT SCHEDULE

Rise: 120° with harmonic motion
Dwell: 30°
Return: 180° with harmonic motion
Dwell: 30°
Base Circle: 1.5-in. diameter
Lift: 0.75 in.

CAM ANGLE (deg)	TABLE VALUES		LIFE: 0.75y (in.)	CAM ANGLE (deg)	TABLE VALUES		LIFE: 0.75y (in.)
	θ	y			θ	y	
0	0	0	0				
10	10	0.017037	0.012778	165	110	0.982962	0.737222
20	20	0.066987	0.050240	180	100	0.933012	0.699759
30	30	0.146446	0.109835	195	90	0.853553	0.640165
40	40	0.250000	0.187500	210	80	0.750000	0.562500
50	50	0.370590	0.277943	225	70	0.629409	0.472057
60	60	0.500000	0.375000	240	60	0.500000	0.375000
70	70	0.629409	0.472057	255	50	0.370590	0.277943
80	80	0.750000	0.562500	270	40	0.250000	0.187500
90	90	0.853553	0.640165	285	30	0.146446	0.109835
100	100	0.933012	0.699759	300	20	0.066987	0.050240
110	110	0.982962	0.737222	315	10	0.017037	0.012778
120	120	1.000000	0.750000	330	0	0	0
120– 150	Dwell		0.750000	330– 360	Dwell		0

3-8 ROLLER, OFFSET, FLAT, AND OSCILLATING FOLLOWERS; GRAPHICAL LAYOUTS

For the sake of comparison in the examples that follow, the schedule described in Sec. 3-6 will be used to graphically derive the various cam profiles.

Base circle: 1.5-in.-dia
Cam rotation: Counterclockwise
Follower style: Roller, offset, flat, and oscillating
   Rise: 0.75 in. with harmonic motion in 120° of cam
      rotation
   Dwell: 30° of cam rotation
   Fall: 0.75 in. with harmonic motion in 180° of cam
      rotation
   Dwell: 30° of cam rotation

ROLLER FOLLOWER

Figure 3-14 illustrates the construction. In this instance, to maintain a reasonable pressure angle (which should not exceed 30°), the circumference of the *prime circle* is used as the length of the cam angle axis. Analogous to the *pitch circle* for a knife-edge follower, the *prime circle* is the smallest circle that can be drawn from the cam center to the roller center. The line generated by the roller center as the cam rotates is called the *pitch curve*.

OFFSET ROLLER FOLLOWER

Development of the cam surface is shown in Fig. 3-15. An *offset circle* having a radius equal to the magnitude of the offset is first drawn. Radial lines, or rays, and then tangents to these rays are next constructed for each station point. These tangents represent the axis of the follower. The roller is partially drawn at each station point, as illustrated, and the best smooth curve tangent to the roller position at each station defines the cam surface.

FLAT FOLLOWER

Construction, Fig. 3-16, begins as for the knife-edge follower. Straight lines representing the flat face are drawn perpendicular to each radial station line. The cam profile is tangent to each of the follower lines.

OSCILLATING FOLLOWER

A circle is drawn from the cam center through the follower's center of oscillation as shown in Fig. 3-17. Radial lines representing station points are constructed, and the oscillator is positioned at each point as illustrated. As with the previous examples, the cam surface is the best smooth curve tangent to each roller position.

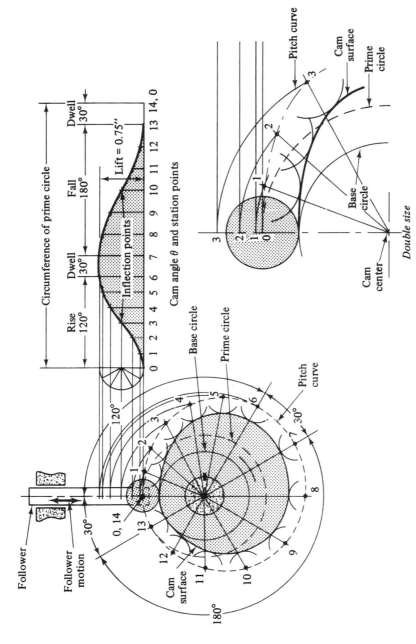

Figure 3-14 / Cam profile for a roller follower.

181

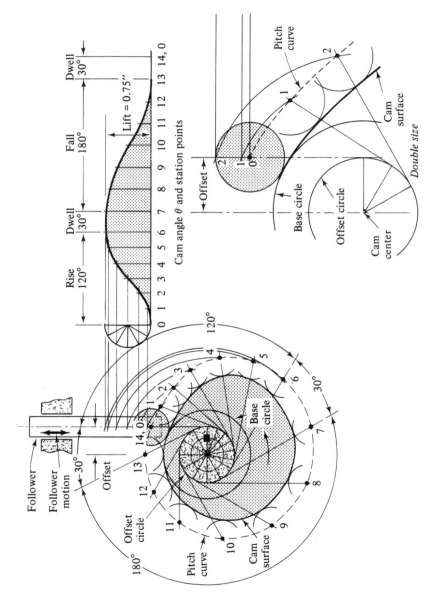

Figure 3-15 / Cam construction for an offset roller follower.

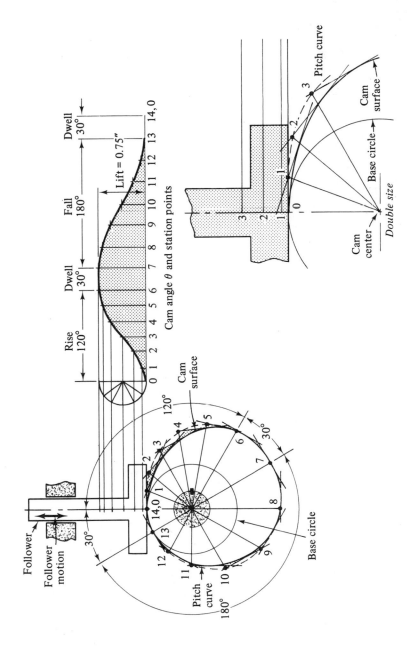

Figure 3-16 / Cam construction for a flat face follower.

183

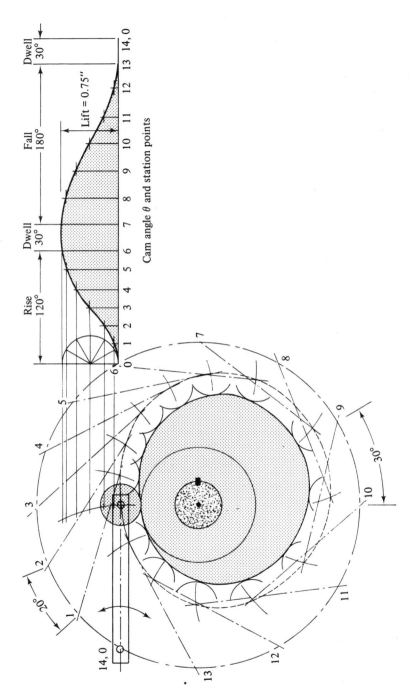

Dwell 30°

Lift = 0.75"

Fall 180°

Dwell 30°

Rise 120°

Cam angle θ and station points

0 1 2 3 4 5 6 7 8 9 10 11 12 13 14, 0

20°

30°

14, 0

Figure 3-17 | *Cam construction for an oscillating follower.*

3-9   ECCENTRICS

Figure 3-18(a) describes a circular cam rotating counterclockwise about an eccentric center *O*. The advantage of a zero pressure angle is readily apparent since the cam profile is always tangent to the follower surface. Although a flat follower is illustrated, any follower system including the oscillating type can be adapted to a rotating eccentric cam, and the advantage of the zero pressure angle would still be present.

Momentary stops at both top and bottom dead center, along with harmonic follower motion, are unique characteristics of eccentrics. Displacement analysis is similar to that used with the yoke cams (a) and (d) illustrated in Fig. 3-5. A circle with radius equal to the eccentricity *e* is drawn through *O* as shown in Fig. 3-18(b); for the sake of illustration, the circumference of this circle is divided into twelve equal segments or stations. Arcs equal to the cam radius *R* are drawn from each station, and horizontal tangents to these arcs are extended to construct the *displacement-cam angle diagram* as shown in Fig. 3-18(c).

Variations of the circular eccentric are many; the bell-crank locking lever shown in Fig. 3-19 and the fixed-pin slotted arm eccentric described in Fig. 3-20 are two such variations. In the latter, the top of the arm *A* has both horizontal and vertical motion, and the path traced is an ellipse.

3-10   JAMMING EFFECTS; PRESSURE ANGLE LIMITATIONS

Pressure angles, as was discussed, attain their maximum values at inflection points in the curves that describe follower displacement as a function of cam angle. Figure 3-21 describes two cams with the same lift *l*—the only difference in the two cams is in the diameters of their respective base circles. In the first instance, Fig. 3-21(a), the cam will jam or lock—the cam is too steep, which is a simple way of saying that the pressure angle is too large. Increasing the diameter of the base circle, Fig. 3-21(b), decreases the pressure angle. To eliminate the problem and keep the pressure angle below 30° (a practical maximum), why not always specify large-diameter base circles? Space limitations, too, must be considered.

3-11   GENEVA MOVEMENTS

Two basic elements, a continuously rotating arm with its roller and an intermittently revolving slotted wheel, comprise a Geneva movement. The wheel may have three or more slots, with the action progressing from

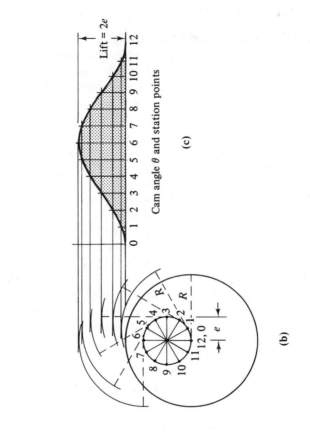

Lift = 2e

Cam angle θ and station points

0 1 2 3 4 5 6 7 8 9 10 11 12

(c)

(b)

R

R

e

7 6 5 4 3 2 1

8

9

10

11 12, 0

(a)

O

Eccentricity

e

R

Figure 3-18 / An eccentric.

186

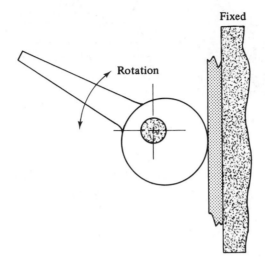

Figure 3-19 / *A locking lever.*

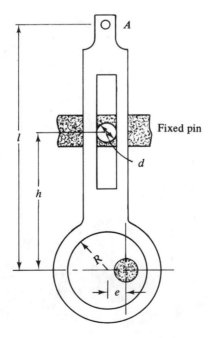

Figure 3-20 / *A fixed-pin slotted arm eccentric.*

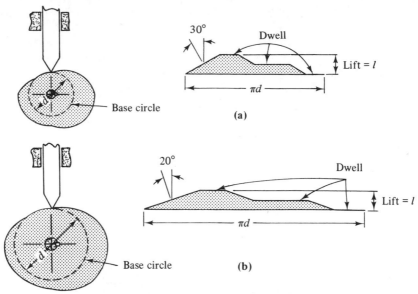

(a)

(b)

Figure 3-21 / Jamming effects.

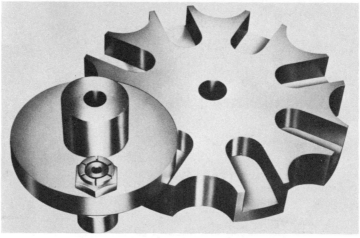

Figure 3-22 / Geneva movements. (Courtesy of Geneva Mechanisms Corporation)

abrupt to smooth as the number of slots increase. The device is used when rotational motion must be transformed into intermittent motion. Of interest to the design engineer, along with physical size, is the indexing rate. Figure 3-24 describes the critical design dimensions of an eight-point Geneva drive. The driver rotates at a constant speed causing the cam follower to engage in a slot in the follower or Geneva wheel, which in this case rotates through an angle of $360/8 = 45°$. When the follower leaves the slot, the driver hub engages with the matching surface of the Geneva lying between the slots—the driver continues to rotate until once again it engages with a slot in the follower.

Figure 3-23 / Geneva movements. (Courtesy of Geneva Mechanisms Corporation)

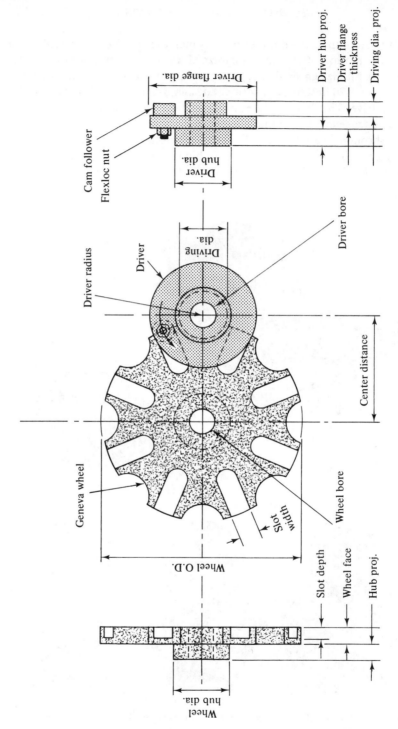

Cam follower
Flexloc nut
Driver flange dia.
Driver hub dia.
Driver hub proj.
Driver flange thickness
Driving dia. proj.

Driver radius
Driver
Driving dia.
Driver bore

Center distance

Geneva wheel
Wheel bore
Slot width
Wheel O.D.

Slot depth
Wheel face
Hub proj.
Wheel hub dia.

Figure 3-24 / Geneva drive.

190

QUESTIONS AND PROBLEMS

Design radial cams according to the schedules given in Problems 3-1, 3-2, 3-3, and 3-4. Positive displacements are those that increase the cam radius. In each instance, determine the maximum pressure angle.

P3-1   *Diameter of base circle:* 2 in.
Follower style: Knife-edge
Cam rotation: Clockwise

SCHEDULE

CAM ANGLE (deg)	FOLLOWER LIFT (in.)	FOLLOWER MOTION
0 to 90	0 to +0.75	Constant velocity
90 to 180	+0.75 to +0.75	Dwell
180 to 330	+0.75 to 0	Constant velocity
330 to 360	0 to 0	Dwell

P3-2   *Diameter of base circle:* 2.5 in.
Follower style: 0.5-in.-dia. roller
Cam rotation: Counterclockwise

SCHEDULE

CAM ANGLE (deg)	FOLLOWER LIFT (in.)	FOLLOWER MOTION
0 to 180	0 to +1	Constant acceleration
180 to 210	+1 to +1	Dwell
210 to 330	+1 to 0	Constant deceleration
330 to 360	0 to 0	Dwell

P3-3   *Diameter of base circle:* 2.25 in.
Follower style: Flat face
Follower width: Minimum necessary plus a 0.25-in. overhang on left and right edges
Cam rotation: clockwise

SCHEDULE

CAM ANGLE (deg)	FOLLOWER LIFT (in.)	FOLLOWER MOTION
0 to 90	0 to +1	Constant acceleration
90 to 180	+1 to +1	Dwell
180 to 330	+1 to 0	Harmonic
330 to 360	0 to 0	Dwell

P3-4 *Diameter of base circle:* 5 cm
Follower style: Offset roller; see Fig. P3-4
Cam rotation: Clockwise

SCHEDULE

CAM ANGLE (deg)	FOLLOWER LIFT (cm)	FOLLOWER MOTION
0 to 180	0 to +2	Harmonic
180 to 360	+2 to 0	Harmonic

Note: Plot the path of point *P*.

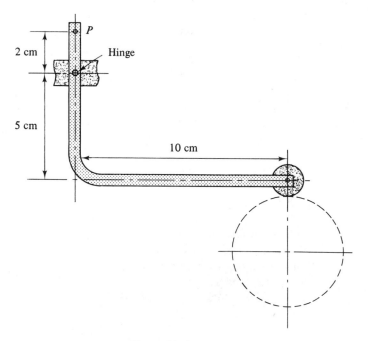

Figure P3-4

P3-5   The oscillating flat-face follower illustrated is to rise through an
arc of 30° with cycloidal motion during 150° of clockwise cam
rotation, dwell for 30°, and return with constant acceleration.
Construct the cam profile, and determine the dimensions of the
flat face. Allow ¼-in. clearance at each end. Determine the appropriate direction of rotation.

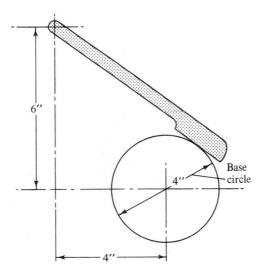

Figure P3-5

P3-6   Determine the displacement diagram and the maximum pressure
angle for the cam illustrated.

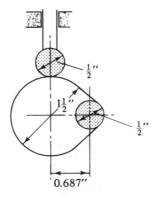

Figure P3-6

P3-7   Plot the displacement diagram for the yoke cam shown.

Radial measurements of
cam surface:

0– 1	30°	0.675″
0– 2	60°	0.835
0– 3	90°	1.01
0– 4	120°	1.16
0– 5	150°	1.285
0– 6	180°	1.375
0– 7	210°	1.225
0– 8	240°	1.05
0– 9	270°	0.90
0–10	300°	0.75
0–11	330°	0.66
0–12	0°	0.625

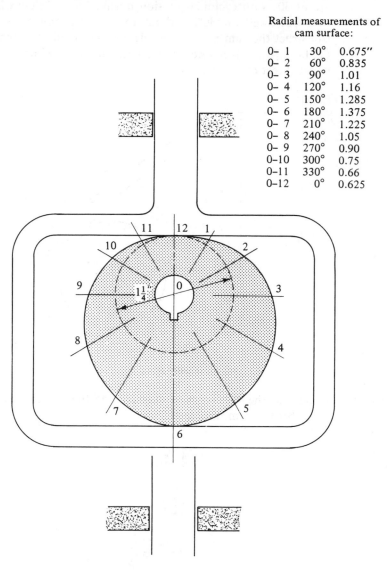

Figure P3-7

P3-8   Draw the displacement diagram for the yoke cam illustrated, and
determine the period of the dwell. The cam speed is 400 rpm.

Radial measurements of
cam surface:

0– 1	30°	1.50″
0– 2	60°	1.60
0– 3	90°	1.91
0– 4	120°	2.42
0– 5	150°	2.75
0– 6	180°	2.75
0– 7	210°	2.75
0– 8	240°	2.42
0– 9	270°	1.91
0–10	300°	1.60
0–11	330°	1.50
0–12	0°	1.50

Dwell 60°, position 11 to 1
position 5 to 7

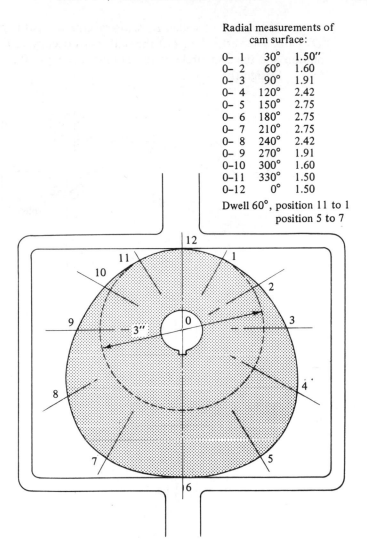

Figure P3-8

P3-9 A two-dimensional cam layout is shown. Roller *C* is mounted on a horizontal slide that carries a vertical slide for roller *D*. Roller *C* is a cam follower driven by a cam on camshaft *A*. Roller *D* is a cam follower driven by a cam on camshaft *B*. The two camshafts are geared together and rotated at the same speed. Camshaft *B* moves horizontally in synchronism with roller *C*.

The center of roller D is to traverse the path shown, a "D" path. The vertical part of the path must occupy 120° of cam rotation and the semicircle of the D must occupy 240°.

Design the two cams.

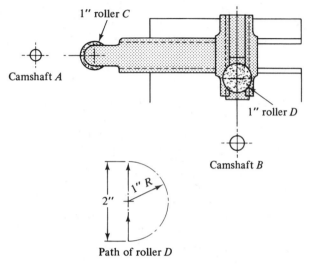

Figure P3-9

P3-10 Plot the path of point A, Fig. 3-20. The following are the critical dimensions:

$$R = 3 \text{ in.} \quad h = 6 \text{ in.}$$
$$e = \tfrac{1}{2} \text{ in.} \quad l = 14 \text{ in.}$$
$$d = \tfrac{1}{2} \text{ in.}$$

SUGGESTED DESIGN PROJECTS

DP3-1 Design a yoke cam similar to that shown in Fig. 3-5(d). The follower is to make 40 complete strokes per min with simple harmonic motion. The total displacement is to be 1 in.

DP3-2 Design a double cam system similar to that shown in Fig. 3-6(a). During one complete cycle, the path of point P is to be a square 1 in. on edge.

CHAPTER 4

SPRINGS

DETROIT—Imagine it. The perfect car. It is easy to maneuver and park. It's practically silent, yet has all the pep and power of any of today's big cars. And it costs a mere fraction of what current cars cost to operate.

What's more, no pollution of any kind comes out of its tail pipe. Indeed, it doesn't even have a tail pipe.

What powers this new miracle car? Electric batteries? A steam engine? An exotic chemical fuel cell? Wrong on all counts, friends. It's a wind-up car.

In fact, there seems to be only one over-riding problem: the automobile makers think the whole idea is nuts.

THE WALL STREET JOURNAL
March 15, 1971

4-1 INTRODUCTION

Modern spring theory began with Hooke[1]—in 1678 in his *De Potentia Restivtutivâ* he states: "Take a quantity of even-drawn wire, either Steel, Iron or Brass, and coyl it on an even Cylinder into a Helix of what number of turns you please, then turn the ends of the Wire into loops, by one of which suspend this coyl upon a nail and by the other restrain the weight that you would have to extend it, and hanging on several weights observe

---

[1] Robert Hooke (1635–1703), a genius with many talents, made scientific contributions in biology, geology, astronomy, mechanics, and physics. He invented the spring balance, an anchor escapement for watches, and the wheel barometer.

197

exactly to what length its own weight doth extend it, and, you shall find
that if one ounce, or one pound, or one certain weight doth lengthen it
one line or one inch or one certain length, then two ounces, or two pounds,
or two weights will extend it two lines, two inches or two lengths and
three ounces, pounds or weights, three lines, inches or lengths and so
forwards." Displacement is proportional to load!

4-2   HOOKE'S IDEALIZED HELICAL SPRING

By superimposing on a loaded helical spring the two types of shear stress
that act—direct and torsional—theoretical formulas for stress, deflection,
and spring constant can be obtained, which supply credence to the observations of Robert Hooke. The spring shown in Fig. 4-1(a) is subjected to
a load F. A free-body diagram, Fig. 4-1(b), shows that two loads must
act at the *cut* to maintain equilibrium, a direct load F and a torsional load
$F \times R$. It is assumed that the spring has been developed, as Hooke described, by winding a helical coil of a solid circular wire of diameter d
on a mandril of diameter D_M. The mean radius R of the coil is

$$R = \frac{D_M + d}{2} \qquad\qquad (4\text{-}1)$$

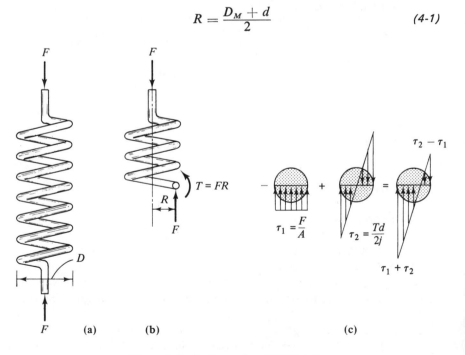

Figure 4-1   /   Forces in a helical spring.

The direct stress τ_1 and the torsional stress τ_2 are superimposed, Fig. 4-1(c), to give an equation for the maximum stress τ_{max}

$$\tau_{max} = \tau_1 + \tau_2 = \frac{F}{A} + \frac{Td}{2J}$$

$$\tau_{max} = \frac{4F}{\pi d^2} + \frac{16FR}{\pi d^3} = \frac{16FR}{\pi d^3}\left(1 + \frac{d}{4R}\right) \qquad (4\text{-}2)$$

It can be seen that the greatest stress at the cut occurs at the innermost portion of the coil. When the wire diameter is small compared to the mean radius, the term $d/4R$ can be neglected—the stress in the spring is mainly a function of the torsional load FR. It is interesting to note that the stress given by Eq. (4-2) is dependent on load and geometry—the elastic properties of the wire are not involved; not so, however, with deflection.

The extension of compression under load of Hooke's idealized helical spring is principally caused by twisting of the wire rather than by direct shear. To obtain a relationship for deflection, imagine the spring, which consists of n coils, to be a straight shaft $2\pi Rn$ long, as shown in Fig. 4-2. The deflection δ at a distance R from the center of the wire in terms of the angle of twist θ is approximately

$$\delta = R\theta$$

and since $\theta = TL/GJ$,

$$\delta = R \times \frac{TL}{GJ} = R \times \frac{FR \times 2\pi Rn}{G \times \pi d^4/32}$$

$$\delta = 64\frac{FR^3 n}{Gd^4} \qquad (4\text{-}3)$$

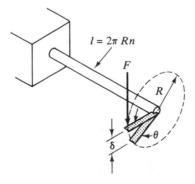

$l = 2\pi Rn$

Figure 4-2 | Deflection of a helical spring.

The *spring rate* or *spring constant* k can now be defined in terms of force and deflection.

$$k = \frac{F}{\delta} = \frac{Gd^4}{64R^3n} \qquad\qquad (4\text{-}4)$$

4-3   SERIES, PARALLEL, AND MIXED SYSTEMS

A *series spring system* is one in which an equal force acts in each spring: the total deflection is the sum of the individual deflections. For the system shown in Fig. 4-3, the total deflection δ is

$$\delta = \delta_1 + \delta_2 + \delta_3$$

where

$$\delta = \frac{F}{k_1} + \frac{F}{k_2} + \frac{F}{k_3}$$

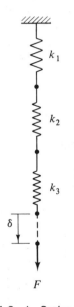

Figure 4-3 / Series system.

Dividing both sides of the equation by F gives

$$\frac{\delta}{F} = \frac{1}{k_1} + \frac{1}{k_2} + \frac{1}{k_3}$$

Since δ/F is the reciprocal of an equivalent spring whose constant is k_e, it follows that

$$k_e = \cfrac{1}{\cfrac{1}{k_1} + \cfrac{1}{k_2} + \cfrac{1}{k_3}} \qquad (4\text{-}5)$$

In a *parallel system*, the springs deform equally, as they each share a portion of the load. The equivalent constant in this case is the sum of the individual constants. To prove this statement, consider the system shown in Fig. 4-4. If the bar is to remain horizontal, each spring will support a share of the load.

$$F = F_1 + F_2 + F_3 = k_1\delta + k_2\delta + k_3\delta$$

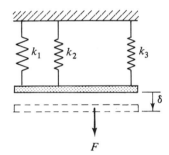

Figure 4-4   /   Parallel system.

Dividing this equation by the deflection δ gives an expression for the equivalent parallel spring constant.

$$\frac{F}{\delta} = k_e = k_1 + k_2 + k_3 \qquad (4\text{-}6)$$

EXAMPLE 4-1

Two helical springs, nested one inside the other as shown in Fig. 4-5(a), support a load $F = 200$ lb. The critical dimensions of each spring are as follows:

	OUTER SPRING	INNER SPRING
Mean diameter	5 in.	3 in.
Wire diameter	0.6 in.	0.4 in.
Active turns	25	15
Free length	15 in.	15 in.
Material	Steel	Steel

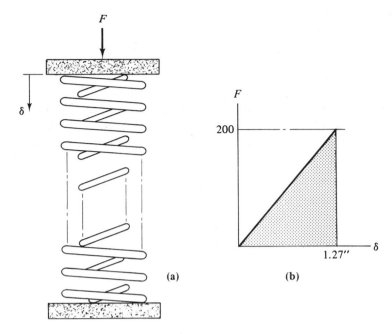

Figure 4-5 / Nested springs (Example 4-1).

Determine: (a) the spring rate of the system; (b) the deflection of the system; (c) the maximum stress in each spring; and (d) the energy stored in each spring.

solution
(a) Spring rates are found by substitution of numerical data into Eq. (4-4).
      Outer spring:

$$k_o + \frac{Gd^4}{64R^3n} = \frac{12 \times 10^6 \times \overline{0.6^4}}{64 \times \overline{2.5^3} \times 25} = 62.2 \text{ lb/in.}$$

      Inner spring:

$$k_i = \frac{12 \times 10^6 \times \overline{0.4^4}}{64 \times \overline{1.5^3} \times 15} = 94.8 \text{ lb/in.}$$

The outer and inner springs are in parallel, hence

$$k_e = k_o + k_i = 62.2 + 94.8 = 157 \text{ lb/in.}$$

(b) Deflection based on the equivalent spring constant of the system is next computed:

$$\delta = \frac{F}{k_e} = \frac{200}{157} = 1.274 \text{ in.}$$

To find the shearing stresses, the individual forces F_o and F_i must be computed.

$$F_o = k_o \delta = 62.2 \times 1.274 = 79.2 \text{ lb}$$
$$F_i = k_i \delta = 94.8 \times 1.274 = 120.8 \text{ lb}$$

Check! $$F_o + F_i = 79.2 + 120.8 = 200 \text{ lb}$$

(c) Next, the stresses are found using Eq. (4-2).
Outer spring:

$$\tau_{max} = \frac{16FR}{\pi d^3}\left(1 + \frac{d}{4R}\right)$$

$$= \frac{16 \times 79.2 \times 2.5}{3.14 \times 0.6^3}\left(1 + \frac{0.6}{4 \times 215}\right) = 4950 \text{ psi}$$

Inner spring:

$$\tau_{max} = \frac{16 \times 120.8 \times 1.5}{3.14 \times 0.4^3}\left(1 + \frac{0.4}{4 \times 1.5}\right) = 15{,}400 \text{ psi}$$

(d) The energy stored in the springs is equivalent to the external work required to compress the system by amount δ. This work is equal to the shaded area of the force–displacement diagram, Fig. 4-5(b),

$$\text{Stored energy} = \tfrac{1}{2} \times F_{max} \times \delta_{max} = \tfrac{1}{2} \times 200 \times 1.27 = 127 \text{ lb-in.}$$

EXAMPLE 4-2

A 100-ton barge traveling at 1 mph strikes the spring assembly shown (top view) in Fig. 4-6(a). Although the springs have equal constants k, their free lengths differ as illustrated. Find k if the maximum deflection of the spring assembly is not to exceed 10 in.

solution

The kinetic energy $\tfrac{1}{2}Mv^2$ of the barge is transferred into stored energy (work) in the springs. This stored energy is equal to the sum of the

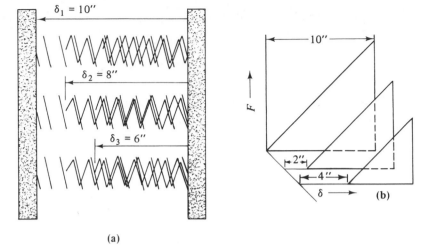

(a)

(b)

Figure 4-6 / Energy absorption of a spring system (Example 4-2).

areas of the force–deflection graphs, as plotted in Fig. 4-6(b),

$$\text{Kinetic energy} = \text{Work done}$$

$$|\tfrac{1}{2}M(v_f^2 - v_o^2)| = \sum \text{area}; \quad \text{where } v_f = 0$$

In units of foot-pounds with k in lb/in.

$$\frac{1}{2} \times \frac{100 \times 2000}{32.2} \times \left(1 \times \frac{22}{15}\right)^2 = 3\left[\left(\frac{1}{2} \times \frac{6}{12} \times 6k\right)\right.$$
$$+ \left(\frac{1}{2} \times \frac{8}{12} \times 8k\right)$$
$$\left. + \left(\frac{1}{2} \times \frac{10}{12} \times 10k\right)\right]$$
$$k = 267 \text{ lb/in.}$$

EXAMPLE 4-3

Figure 4-7(a) shows a *mixed system*, a configuration that combines series and parallel springs. Find: (a) the equivalent spring constant; and (b) the energy stored in the system (the potential energy) for a total stretch of 2 in. from a released position. Assume that the supports and tie-bars remain horizontal.

solution

(a)   The parallel combination of springs k_3 and k_4 is in series with k_2; the equivalent constant of the combination is in parallel with k_1.

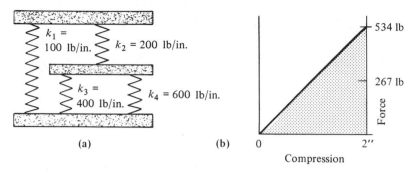

Figure 4-7 / Series and parallel (mixed) system (Example 4-3).

Expressed in a single symbolic equation, the equivalent constant k_e is

$$k_e = k_1 + \cfrac{1}{\cfrac{1}{k_2} + \cfrac{1}{k_3 + k_4}}$$

Substitution of numerical data in units of lb/in. gives

$$k_e = 100 + \cfrac{1}{\cfrac{1}{200} + \cfrac{1}{400 + 600}} = 267 \text{ lb/in.}$$

(b)   Stored or potential energy is equivalent to the work required to stretch the system; this equivalence is graphically illustrated in Fig. 4-7(b). The shaded triangular area represents this potential energy *PE*.

$$PE = \text{area of work diagram}$$
$$= \tfrac{1}{2} \times \text{base} \times \text{height}$$
$$= \tfrac{1}{2} \times 2 \text{ in.} \times 2 \text{ in.} \times 267 \text{ lb/in.} = 534 \text{ lb-in.}$$

4-4   HELICAL COMPRESSION AND EXTENSION SPRINGS

Figure 4-8 describes some of the more common types of helical springs in terms of geometric configuration and use. The spring rate k for the constant-diameter constant-pitch helical springs manufactured from round wire is given by

$$k_e = \frac{F}{\delta} = \frac{Gd^4}{64R^3n} = \frac{Gd^4}{8D^3n}$$

where F/δ = load/deflection (lb/in.)
  G = shear modulus of elasticity (psi)

Formulas and Reference

Type	Configuration	Action	Load/Deflection	Stress/Load
Round and rectangular wire	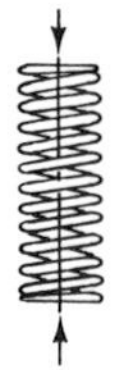	Push — wide load and deflection range — constant rate.	$P = \dfrac{fGd^4}{8D^3 N}$	$S = \dfrac{8\,PD}{\pi d^3}\,K_w$
Conical	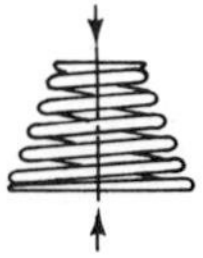 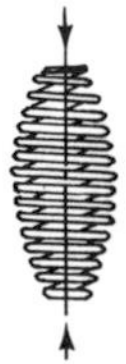 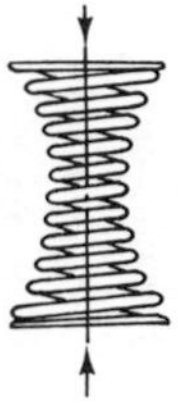 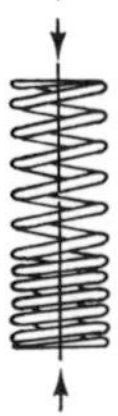 Barrel	Push — wide load and deflection range. Conical spring can be made with minimum solid height and with constant or increasing rate. Barrel, hourglass, and variable-pitch springs used to minimize resonant surging and vibration	Calculate as helical compression spring of uniform diameter, using average mean diameter of active coils. This applies only until first active coil "bottoms" or touches next coil. The spring is recalculated as each coil deflects until it becomes inactive.	
Hourglass	Variable-pitch			

Figure 4-8 / Types of helical springs. [Reprinted by permission from Design Handbook—Springs, Custom Metal Parts (Bristol, Conn.: Associated Spring Corporation, 1967), 6.]

d = wire diameter (in.)
R = mean radius of the spring (in.)
D = mean diameter of the spring (in.)
n = number of active coils

Spring rates for the conical, barrel, and hourglass springs are calculated as helical compression springs of uniform diameter using, however, the average mean diameter of the active coils. The analysis of variable pitch springs can proceed coil-by-coil as illustrated in the example that follows.

EXAMPLE 4-4

Phosphor-bronze ($G = 6.3 \times 10^6$ psi) is wound as a variable-pitch constant-diameter spring as shown in Fig. 4-9. The outside diameter of the spring is 0.500 in., and the wire diameter is 0.030 in. Prior to the application of a "pushing force," assume the spring to have 5 active coils. Determine the force required to just close the spring.

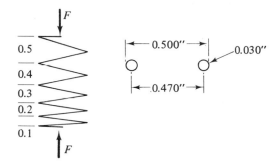

Figure 4-9 / Variable-pitch spring of constant diameter (Example 4-4).

solution
The compressive force F, given by Eq. (4-4), can be expressed in terms of the mean radius R or the mean diameter D:

$$F = \frac{\delta G d^4}{64 R^3 n} = \frac{\delta G d^4}{8 D^3 n}$$

The deflection δ and the number of active coils n will change as the spring closes. Factoring δ/n will give

$$F = \frac{G d^4}{8 D^3 n} \times \frac{\delta}{n} = \frac{6.3 \times 10^6 \times \overline{0.030}^4}{8 \times \overline{0.47}^3} \times \frac{\delta}{n}$$

$$= 6.14 \frac{\delta}{n}$$

The total force must, therefore, equal the sum of the forces required to close each coil. It must be remembered, of course, that as one coil closes, they all tend to close; in other words, when the bottom coil closes, the separation between all coils diminishes by 0.1 inches. As the next coil closes, the separation of each of the remaining coils again diminishes by 0.1 in. and so on.

$$F = 6.14\left(\frac{0.1}{5} + \frac{0.1}{4} + \frac{0.1}{3} + \frac{0.1}{2} + \frac{0.1}{1}\right) = 1.40\,\text{lb}$$

4-5 HELICAL COMPRESSION SPRINGS: STRESS CONCENTRATION FACTORS AND DESIGN CONSIDERATIONS

While bending stresses and direct shear stresses are present in a compressed (or stretched) helical spring, their magnitudes are negligible compared to torsional stresses. However, the ever-present stress concentration considerations—the deviations of the "real" from the "theoretical"—must be accounted for in actual design, particularly when dynamic loads are involved. The fundamental design considerations[2] and equations for springs made from round wire are as follows:

1 | Spring constant:

$$k = \frac{\text{force}}{\text{deflection}} = \frac{Gd^4}{64R^3n}\,\text{psi}$$

where $G =$ shear modulus of elasticity (psi)
$d =$ wire diameter (in.)
$R =$ mean radius of the spring (in.)
$n =$ number of active coils

2 | Spring index: $c = 2R/d =$ ratio of mean diameter of the spring to the diameter of the wire. The acceptable index range is 4 to 12. Under 4 the spring may be difficult to make and too highly stressed. Over 12, the spring will be flimsy; the coils may slip over one another if loaded to solid.

3 | Shear stress:

$$\tau = \frac{16FR}{\pi d^3}K_w\,\text{psi}$$

---

[2] *Design Handbook* (Bristol, Connecticut: Associated Spring Corporation, 1967)

where $F =$ applied force (lb)

$K_w =$ Wahl's correction factor

$$= \frac{4c-1}{4c-4} + \frac{0.615}{c} \text{ or as graphically presented in}$$

Fig. 4-10

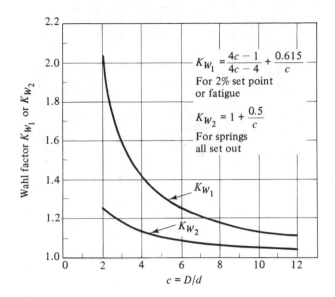

Figure 4-10 / Wahl's correction factor. [Reprinted by permission from Design Handbook—Springs, Custom Metal Parts (*Bristol, Conn.: Associated Spring Corporation, 1967*), 10.]

4 | *Solid height:* Length of the spring with all coils closed.

5 | *Free height:* Overall length of the spring in free or unloaded position.

6 | *Direction of coiling:* May be left-hand or right-hand (similar to screw threads). If springs are nested, they should be coiled in opposite directions. If a spring works over a screw thread, it should be coiled opposite to the thread direction.

7 | *Diametral clearance:* Space that must be allowed to permit the free functioning of a spring that operates in a hole or over a rod. Adequate clearance is provided by an allowance of 10% of the spring diameter.

8 | *Coil clearance:* The clearance between L_2 and $L_{S'}$, Fig. 4-11, should be about 10% of L_2 to avoid the nonlinear region that occurs when the spring approaches a solid member.

9 | *Buckling:* When the free length of a compression spring is about

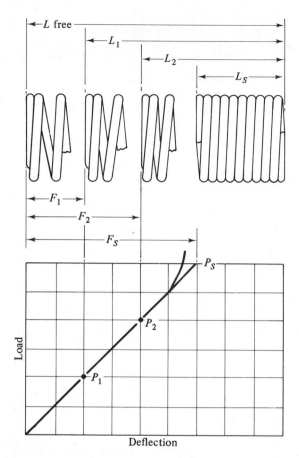

Figure 4-11 / Coil clearance. [*Reprinted by permission from* Design Handbook—Springs, Custom Metal Parts (*Bristol, Conn.: Associated Spring Corporation, 1967*), *13.*]

four times greater than its mean diameter, the spring becomes unstable. If the spring is guided over a rod or tube, buckling can be minimized.

10 / *End conditions:* Geometry of spring ends, Fig. 4-12, affects the number of active coils, the free length, and the solid length, as indicated in the figure.

4-6 HELICAL TORSION SPRINGS

Since the torque about the axis of a helix, Fig. 4-13, acts as a bending moment on every section of the wire, a fair prediction of average stress

Spring ends

for helical springs may be either plain, plain ground, squared, or squared and ground as shown below. This results in a decrease of the number of active coils and affects the free length and solid length of the spring as shown below.

Type of ends	Total coils	Solid length	Free length
Plain	n	$(n + 1)d$	$np + d$
Plain ground	n	nd	np
Squared	$n + 2$	$(n + 3)d$	$np + 3d$
Squared and ground	$n + 2$	$(n + 2)d$	$np + 2d$

p = pitch, n = number of active coils, d = wire diameter

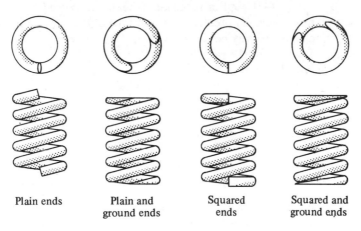

| Plain ends | Plain and ground ends | Squared ends | Squared and ground ends |

Figure 4-12 / Spring ends. [Reprinted by permission from Hall, Holowenda, and Laughlin, Theory and Problems of Machine Design *(New York: Schaum Publishing Company, 1961), 194.]*

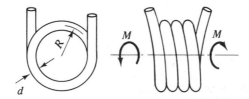

Figure 4-13 / Torsion in a helical spring.

and angular deformation can be obtained from the beam equations

$$\sigma = Mc/I \qquad (4\text{-}7)$$

and

$$\theta = \frac{Ml}{EI} \qquad (4\text{-}8)$$

where σ = stress (psi)
M = applied moment (in.-lb)
c = $d/2$ for round wire (in.)
  = $h/2$ for rectangular wire (in.)
I = moment of inertia about the bending axis (in.^{4})
  = $\pi d^4/64$ for round wire
  = $bh^3/12$ for rectangular wire
θ = angle of twist (radians)
l = length of the coiled wire
  = $\pi D \times$ number of turns
E = modulus of elasticity (psi)

Since Eq. (4-7) applies to straight beams, a correction factor is necessary to more precisely predict the magnitude of the stress, particularly in a spring made of closely coiled heavy wire. This correction can be expressed as a function of the ratio R/d.

For *round wire* at the inner edge:

$$\sigma = \frac{Mc}{I}\left(\frac{8R/d - 1}{8R/d - 4}\right) \qquad\qquad (4\text{-}9)$$

For *rectangular wire* at the inner edge:

$$\sigma = \frac{Mc}{I}\left(\frac{6R/h - 1}{6R/h - 3}\right) \qquad\qquad (4\text{-}10)$$

where R is the mean radius of the coil, and h is the depth of the wire section perpendicular to the axis of the coil.

The spring constant or spring rate k_t for torsional springs is defined as the ratio of torque to angle of twist; appropriate units for k_t are in.-lb per radian, in.-lb per degree, or in.-lb per revolution.

EXAMPLE 4-5

A helical torsion spring subjected to static loading, Fig. 4-14, is to have the following properties:

Wire diameter: 0.063 in.
Material: music wire, ultimate tensile strength = 300,000 psi
Rotation: 360°
Torque at maximum rotation: 3.5 in.-lb

Find: (a) required number of coils n; (b) approximate tightly wound length; (c) weight of 1000 such springs; (d) stress in wire when fully loaded; (e) factor of safety (based on minimum tensile strength) in fully loaded position; and (f) stored energy in fully loaded position.

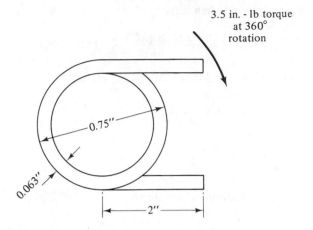

Figure 4-14 / Helical torsion spring (Example 4-5).

solution

(a) Required number of coils:

$$\theta = \frac{Ml}{EI} \quad \text{[Eq. (4-8)]}$$

$$2\pi = \frac{3.5l}{30 \times 10^6 \times \pi \times 0.063^4/64}$$

$$l = 41.6 \text{ in.}$$

Number of coils based on mean diameter of coil:

$$n = \frac{l}{\pi D} = \frac{41.6}{\pi(0.75 - 0.063)} = 19.3 \text{ coils}$$

Total length of wire L:

$$L = l + \text{length of ends} = 41.6 + 2 + 2 = 45.6 \text{ in.}$$

(b) Approximate tightly wound body length:
At section $A–A$

$$\text{body length} = n \times \text{wire diameter}$$
$$= 19 \times 0.063 = 1.2 \text{ in.}$$

(c) Weight of 1000 springs: See Table 4-3, p. 234.

$$W_{1000} = \frac{1.062}{100 \times 12} \text{ lb/in.} \times 41.6 \text{ in.} \times 1000 = 36.8 \text{ lb}$$

(d)   Stress in wire when fully loaded:

$$\sigma = \frac{Mc}{I} \times \frac{8R/d - 1}{8R/d - 4} \quad \text{[Eq. (4-9)]}$$

where $8R/d = 8 \times 0.344/0.063 = 43.7$. Hence,

$$\sigma = \frac{3.5 \times 0.063/2}{\pi \times 0.063^4/64} \times \frac{43.7 - 1}{43.7 - 4}$$

$$= 153,000 \text{ psi}$$

(e)   Factor of safety *FS*:

Ultimate tensile strength $= 300,000$ psi

$$FS = \frac{300,000}{153,000} = 2$$

(f)   Stored energy *PE* in fully loaded position:

$$PE = \frac{M\theta}{2} = \frac{3.5 \times 2\pi}{2} = 11 \text{ lb-in.}$$

4-7   LEAF SPRINGS

Beam theory and its accompanying equations for stress, deflection, and the spring constant present the basis for leaf spring design. While this theory was covered in Chapter 2, Fig. 4-15 gives design values—deflection and stress—for three of the more commonly encountered leaf springs.

EXAMPLE 4-6

A trapezoidal leaf spring and auxiliary mechanism are shown in Fig. 4-16. Find: (a) the force P required to cause a maximum end deflection δ of 0.32 in.; (b) the maximum stress developed in the spring; and (c) the factor of safety of the design. The spring is made from 24-gage (American Standard) steel strip that has been heat-treated and tempered to a Rockwell hardness of $R_c = 50$.

solution

(a)   Load P [Fig. 4-15(c) and Table 4-3]:

$$b/b_o = \tfrac{1}{2}; \quad Q = 1.15$$

$$P = \frac{E b_o h^3 \delta}{4 l^3 Q} = \frac{30 \times 10^6 \times 1 \times \overline{0.025}^3 \times 0.32}{4 \times \overline{2.5}^3 \times 1.15}$$

$$= 2.09 \text{ lb}$$

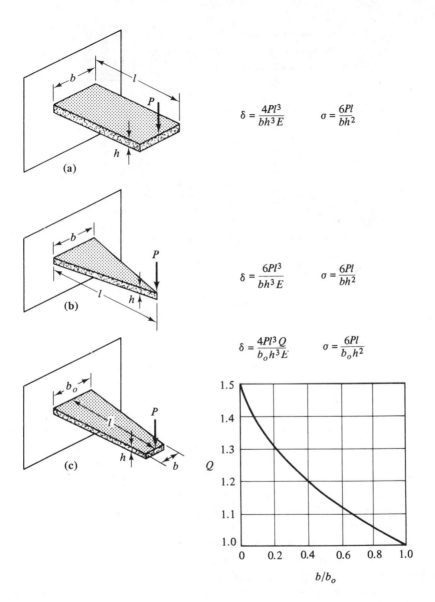

$$\delta = \frac{4Pl^3}{bh^3 E} \qquad \sigma = \frac{6Pl}{bh^2}$$

$$\delta = \frac{6Pl^3}{bh^3 E} \qquad \sigma = \frac{6Pl}{bh^2}$$

$$\delta = \frac{4Pl^3 Q}{b_o h^3 E} \qquad \sigma = \frac{6Pl}{b_o h^2}$$

Figure 4-15 / Formulas for leaf springs.

(b) Maximum stress:

$$\sigma = \frac{6Pl}{b_o h^2} = \frac{6 \times 2.09 \times 2.5}{1 \times 0.025^2} = 50{,}200 \text{ psi}$$

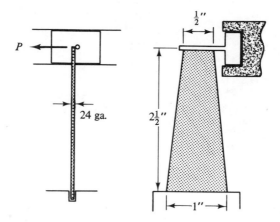

Figure 4-16 / A trapezoidal leaf spring (Example 4-6).

(c) Factor of safety:

$$FS = \frac{\text{allowable stress}}{\text{working stress}} = \frac{270{,}000}{50{,}200} = 5.4$$

4-8   CONSTANT-FORCE EXTENSION SPRINGS

Load versus the deflection relationship, whether it be in a spring, beam, or solid mass of material, has fascinated scientists and engineers for centuries. In fact, most basic designs usually begin with the assumption that load and deflection are reasonably proportional—a provision imposed by nature. There are times, however, when the designer would like to "beat" nature and provide something other than the linear relationship. Since he has little control over elastic properties, the designer has only one choice—variation of geometry. By astute manipulation of configuration, physical behavior can be directed to suit a given need—the constant-force spring is a classic example. So different are the properties of this mechanical element, that many "spring-thinking" engineers would rather it be called a motor or a work-delivering mechanism.

The device, a Neg'ator® spring[3], is a prestressed strip of flat spring stock which coils around a bushing or around successive layers of itself, as shown in Fig. 4-17. When this spring is deflected by pulling on the outer end of the coil, Fig. 4-18, the following occurs:

---

[3]Neg'ator® spring manufactured by The Hunter Spring Co., Hatfield, Pennsylvania.

Figure 4-17 / Neg'ator® spring. (Courtesy of Hunter Spring Company, A Division of Ametek, Inc.)

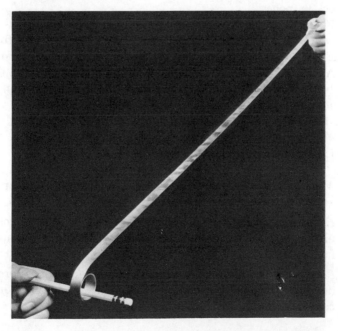

Figure 4-18 / Spring deflection. (Courtesy of Hunter Spring Company, A Division of Ametek, Inc.)

1 /   A constant resisting force P with a line of action through 0 develops. The force P does not increase with deflection.

2 /   The change from a curved section to a straight section takes place in the transition zone.

3 /   Once through the transition zone, the material appreciably flattens in the direction of the pull.

4 /   The straightened material stores energy, but adds nothing further to the force.

5 /   There is a tendency for the material to recoil around itself.

6 /   Friction is limited to that occurring at the spindle bearing.

A graphical comparison between a normal extension and a typical constant-force spring is shown in Fig. 4-19.

The force in pounds developed by a constant-force extension spring is given by

$$P = Qbt \qquad (4\text{-}11)$$

where Q is a factor based on effective working stress and modulus of elasticity, b is the width of the spring stock in inches, and t is the thickness of the stock in inches.

Figure 4-20 gives values of Q for two materials, high-carbon spring steel (SAE 1095) and stainless steel (Types 301/302) as a function of fatigue life. It is readily seen that Q diminishes as life expectancy increases. An infinite variety of springs could be obtained through the use of Eq. (4-11). Practically speaking, however, the designer is confronted with economic considerations—the use of standard thicknesses and widths; Table (4-1)[4] represents some precalculated spring designs based on Eq. (4-11).

Design procedures involve the following steps:

1 /   *Material selection.* Cost, appearance, and application are variables to consider.

2 /   *Estimation of fatigue life.* Overdesign can be just as costly as underdesign—experience and reliability testing are important factors.

3 /   *Load determination.* The first two steps will locate the appropriate design table, which then will give alternate spring selections that can be used.

---

[4]Courtesy of The Hunter Spring Co., Hatfield, Pennsylvania.

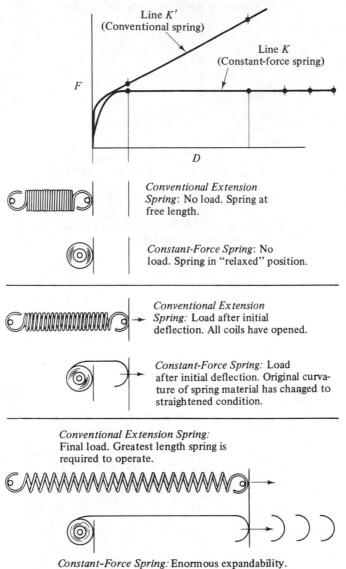

Line K'
(Conventional spring)

Line K
(Constant-force spring)

F

D

Conventional Extension Spring: No load. Spring at free length.

Constant-Force Spring: No load. Spring in "relaxed" position.

Conventional Extension Spring: Load after initial deflection. All coils have opened.

Constant-Force Spring: Load after initial deflection. Original curvature of spring material has changed to straightened condition.

Conventional Extension Spring: Final load. Greatest length spring is required to operate.

Constant-Force Spring: Enormous expandability. Final load is limited only by length of spring material. At all extensions, load remains same.

Figure 4-19 / Constant-force springs compared with conventional extension springs. (Courtesy of Hunter Spring Company, A Division of Ametek, Inc., from The Hunter Neg'ator® and Other Spring Devices, Bulletin 310-67)

TABLE 4-1

CONSTANT FORCE SPRINGS—NEG'ATOR®
EXTENSION SPRING DESIGN DATA[1]

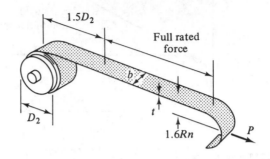

Symbols used in NEG'ATOR® Design Tables.

t — material thickness (in.)
b — material width (in.)
$2Rn$ — natural coil diameter (in.)
D_2 — recommended drum
      diameter (in.)
P — load (lbs.)

		b →	B ⅛	C ³⁄₁₆	D ¼	E ⁵⁄₁₆	F ⅜	G ½	J ⅝	K ¾	P 1	R 1¼	S 1½	U 2	V 2½	W 3	Y 4
t	$2Rn$	D_2						LOAD P									

chart A fatigue life—4,000 cycles   high carbon spring steel

t	$2Rn$	D_2	B	C	D	E	F	G	J	K	P	R	S	U	V	W	Y
.002	.174	.209	.130	.195	.260	.325	.390										
.003	.261	.313	.195	.292	.390	.487	.585	.780	.975								
.004	.348	.418		.390	.520	.650	.780	1.04	1.30	1.56	2.08						
.005	.435	.523			.650	.813	.975	1.30	1.63	1.95	2.60						
.006	.522	.627			.780	.975	1.17	1.56	1.95	2.34	3.12	3.90					
.007	.609	.730				1.14	1.36	1.82	2.28	2.73	3.64	4.55	5.46				
.008	.696	.835					1.56	2.08	2.60	3.12	4.16	5.20	6.24				
.010	.870	1.04					1.95	2.60	3.25	3.90	5.20	6.50	7.80	10.4			
.012	1.04	1.25						3.12	3.90	4.68	6.24	7.80	9.36	12.5			
.014	1.22	1.46							4.55	5.46	7.28	9.10	10.9	14.6	18.2		
.015	1.31	1.57								5.85	7.80	9.75	11.7	15.6	19.5	23.4	
.016	1.39	1.67									8.32	10.4	12.5	16.6	20.8	25.0	
.018	1.57	1.89									9.36	11.7	14.0	18.7	23.4	28.1	
.020	1.74	2.09									10.4	13.0	15.6	20.8	26.0	31.2	41.6
.022	1.91	2.29										14.3	17.2	22.9	28.6	34.3	45.8
.025	2.18	2.62											19.5	26.0	32.5	39.0	52.0
.028	2.43	2.92											21.8	29.1	36.4	43.7	58.2
.032	2.78	3.34												33.3	41.6	49.9	66.5

[1] *Courtesy of Hunter Spring Company, A Division of Ametek, Inc.—Bulletin 310–67.*

TABLE 4-1 (Cont.)

		b →	B 1/8	C 3/16	D 1/4	E 5/16	F 3/8	G 1/2	J 5/8	K 3/4	P 1	R 1¼	S 1½	U 2	V 2½	W 3	Y 4
t	2Rn	D₂	LOAD P														

chart B fatigue life—5,000 cycles   high carbon spring steel

t	2Rn	D₂	B	C	D	E	F	G	J	K	P	R	S	U	V	W	Y
.002	.200	.240	.104	.156	.208	.260	.312										
.003	.300	.360	.156	.234	.312	.390	.469	.623	.782								
.004	.400	.480		.312	.415	.520	.623	.830	1.04	1.25	1.66						
.005	.500	.600			.520	.650	.780	1.04	1.30	1.56	2.08						
.006	.600	.720			.623	.780	.936	1.25	1.56	1.87	2.50	3.12					
.007	.700	.840				.910	1.09	1.46	1.82	2.18	2.91	3.64	4.37				
.008	.800	.960					1.25	1.67	2.08	2.50	3.33	4.16	5.00				
.010	1.00	1.20					1.56	2.08	2.60	3.12	4.16	5.20	6.24	8.32			
.012	1.20	1.44						2.50	3.12	3.74	5.00	6.24	7.49	9.99			
.014	1.40	1.68							3.64	4.36	5.82	7.28	8.72	11.6	14.5		
.015	1.50	1.80								4.68	6.24	7.80	9.36	12.5	15.6	18.7	
.016	1.60	1.92									6.66	8.32	9.99	13.3	16.6	20.0	
.018	1.80	2.16									7.50	9.36	11.2	15.0	18.7	22.4	
.020	2.00	2.40									8.32	10.4	12.5	16.6	20.8	24.9	33.3
.022	2.20	2.64										11.4	13.7	18.3	22.9	27.4	36.6
.025	2.50	3.00											15.6	20.8	26.0	31.2	41.6
.028	2.80	3.38											17.4	23.2	29.0	34.8	46.4
.032	3.20	3.84												26.6	33.2	39.7	53.2

chart C fatigue life—10,000 cycles   high carbon spring steel

t	2Rn	D₂	B	C	D	E	F	G	J	K	P	R	S	U	V	W	Y
.002	.260	.312	.0675	.101	.135	.169	.202										
.003	.390	.468	101	.152	.202	.253	.304	.405	.506								
.004	.520	.625		.202	.270	.337	.405	.539	.675	.819	.108						
.005	.649	.779			.337	.422	.506	.675	.845	1.01	1.35						
.006	.779	.935			.405	.506	.607	.809	1.01	1.21	1.62	2.02					
.007	.910	1.07				.592	.708	.945	1.17	1.42	1.89	2.36	2.84				
.008	1.04	1.25					.810	1.08	1.35	1.62	2.16	2.70	3.24				
.010	1.30	1.56					1.01	1.35	1.69	2.02	2.70	3.37	4.05	5.40			
.012	1.56	1.87						1.62	2.02	2.43	3.24	4.05	4.85	6.48			
.014	1.82	2.19							2.37	2.84	3.78	4.73	5.67	7.56	9.45		
.015	1.95	2.34								3.04	4.04	5.05	6.06	8.10	10.1	12.1	
.016	2.08	2.50									4.32	5.39	6.48	8.64	10.8	12.9	
.018	2.34	2.81									4.86	6.06	7.28	9.70	12.1	14.5	
.020	2.60	3.12									5.04	6.75	8.10	10.8	13.5	16.2	21.6
.022	2.85	3.42										7.40	8.93	11.9	14.8	17.8	23.7
.025	3.25	3.90											10.1	13.5	16.9	20.2	27.0
.028	3.64	4.34											11.3	15.1	18.9	22.7	30.2
.032	4.15	4.96												17.3	22.2	26.7	35.6

chart D fatigue life—20,000 cycles   high carbon spring steel

t	2Rn	D₂	B	C	D	E	F	G	J	K	P	R	S	U	V	W	Y
.002	.336	.403	.0421	.0631	.0840	.106	.126										
.003	.504	.605	.0631	.0971	.126	.158	.189	.252	.316								
.004	.672	.805		.126	.168	.211	.252	.336	.420	.504	.672						
.005	.840	1.01			.211	.264	.316	.420	.525	.631	.840						
.006	1.01	1.21			.252	.316	.378	.504	.631	.756	1.01	1.26					
.007	1.18	1.42				.367	.441	.589	.735	.883	1.18	1.47	1.76				
.008	1.34	1.61					.504	.673	.840	1.01	1.34	1.68	2.02				
.010	1.68	2.02					.631	.840	1.05	1.26	1.68	2.10	2.52	3.36			
.012	2.02	2.42						1.01	1.26	1.51	2.02	2.52	3.02	4.03			
.014	2.35	2.82							1.47	1.76	2.35	2.94	3.53	4.70	5.88		
.015	2.52	3.02								1.89	2.52	3.15	3.78	5.04	6.30	7.56	
.016	2.69	3.23									2.69	3.36	4.03	5.37	6.72	8.06	
.018	3.02	3.62									3.02	3.78	4.54	6.05	7.56	9.07	
.020	3.36	3.91									3.36	4.20	5.04	6.72	8.40	10.1	13.4
.022	3.70	4.44										4.62	5.54	7.39	9.24	11.1	14.8
.025	4.20	5.04											6.30	8.40	10.5	12.6	16.8
.028	4.71	5.64											7.05	9.41	11.8	14.1	18.8
.032	5.37	6.44												10.8	13.4	16.1	21.5

TABLE 4-1 (Cont.)

		b →	B ⅛	C 3⁄16	D ¼	E 5⁄16	F ⅜	G ½	J ⅝	K ¾	P 1	R 1¼	S 1½	U 2	V 2½	W 3	Y 4
t	2Rn	D₂									LOAD P						

chart E fatigue life—35,000 cycles   high carbon spring steel

t	2Rn	D₂	B	C	D	E	F	G	J	K	P	R	S	U	V	W	Y
.002	.400	.480	.0308	.0462	.0614	.0769	.0923										
.003	.600	.720	.0462	.0693	.0921	.115	.138	.185	.231								
.004	.800	.960		.0924	.123	.154	.185	.246	.307	.369	.493						
.005	1.00	1.20			.154	.192	.231	.308	.384	.461	.616						
.006	1.20	1.44			.184	.231	.277	.369	.461	.554	.738	.923					
.007	1.40	1.68				.269	.323	.431	.538	.646	.862	1.07	1.29				
.008	1.60	1.92					.369	.492	.615	.738	.984	1.23	1.48				
.010	2.00	2.40					.461	.615	.769	.923	1.23	1.54	1.85	2.46			
.012	2.40	2.88						.739	.923	1.11	1.48	1.84	2.21	2.95			
.014	2.80	3.36							1.07	1.29	1.72	2.15	2.58	3.44	4.31		
.015	3.00	3.60								1.38	1.85	2.31	2.77	3.69	4.61	5.54	
.016	3.20	3.84									1.97	2.46	2.95	3.94	4.92	5.90	
.018	3.60	4.32									2.22	2.77	3.32	4.43	5.54	6.64	
.020	4.00	4.80									2.46	3.08	3.69	4.92	6.15	7.38	9.84
.022	4.40	5.28										3.38	4.06	5.41	6.77	8.12	10.8
.025	5.00	6.00											4.61	6.15	7.69	9.23	12.3
.028	5.60	6.72											5.14	6.89	8.61	10.3	13.8
.032	6.40	7.68												7.88	9.84	11.8	15.7

chart F fatigue life—100,000 cycles   high carbon spring steel

t	2Rn	D₂	B	C	D	E	F	G	J	K	P	R	S	U	V	W	Y
.002	.444	.533	.0253	.0379	.0505	.0631	.0758										
.003	.667	.800	.0379	.0568	.0758	.0947	.114	.152	.189								
.004	.889	1.07		.0758	.101	.126	.152	.202	.253	.303	.404						
.005	1.11	1.33			.126	.158	.189	.253	.316	.379	.505						
.006	1.33	1.60			.152	.189	.227	.303	.379	.455	.606	.758					
.007	1.56	1.87				.221	.265	.354	.441	.530	.707	.884	1.06				
.008	1.78	2.14					.303	.404	.505	.606	.808	1.01	1.21				
.010	2.22	2.66					.379	.505	.631	.758	1.01	1.26	1.52	2.02			
.012	2.67	3.20						.606	.758	.909	1.21	1.52	1.82	2.42			
.014	3.11	3.73							.884	1.06	1.41	1.77	2.12	2.83	3.54		
.015	3.33	4.00								1.14	1.51	1.89	2.27	3.03	3.79	4.55	
.016	3.56	4.27									1.62	2.02	2.42	3.23	4.04	4.85	
.018	4.00	4.80									1.82	2.27	2.73	3.64	4.55	5.45	
.020	4.44	5.33									2.02	2.53	3.03	4.04	5.05	6.07	8.08
.022	4.89	5.87										2.78	3.33	4.44	5.56	6.67	8.89
.025	5.56	6.67											3.79	5.05	6.31	7.58	10.1
.028	6.22	7.46											4.24	5.73	7.07	8.48	11.3
.032	7.11	8.53												6.54	8.08	9.70	12.9

chart G fatigue life—1,000,000 cycles   high carbon spring steel

t	2Rn	D₂	B	C	D	E	F	G	J	K	P	R	S	U	V	W	Y
.002	.500	.600	.0203	.0305	.0407	.0508	.0610										
.003	.750	.900	.0305	.0457	.0610	.0762	.0914	.122	.152								
.004	1.00	1.20		.0610	.0813	.102	.122	.163	.203	.244	.326						
.005	1.25	1.50			.102	.127	.152	.203	.254	.305	.407						
.006	1.50	1.80			.122	.152	.183	.244	.305	.369	.488	.609					
.007	1.75	2.10				.178	.213	.285	.356	.427	.569	.711	.854				
.008	2.00	2.40					.244	.325	.406	.488	.650	.812	.975				
.010	2.50	3.00					.305	.407	.508	.610	.813	1.02	1.22	1.63			
.012	3.00	3.60						.488	.610	.732	.976	1.22	1.46	1.95			
.014	3.50	4.20							.711	.854	1.14	1.42	1.71	2.28	2.84		
.015	3.75	4.50								.915	1.22	1.52	1.83	2.43	3.04	3.68	
.016	4.00	4.80									1.30	1.63	1.95	2.60	3.25	3.90	
.018	4.50	5.40									1.46	1.83	2.19	2.93	3.65	4.39	
.020	5.00	6.00									1.63	2.03	2.44	3.25	4.06	4.88	6.50
.022	5.50	6.60										2.24	2.68	3.58	4.47	5.37	7.15
.025	6.25	7.50											3.05	4.07	5.08	6.10	8.13
.028	7.00	8.40											3.41	4.55	5.68	6.83	9.11
.032	8.00	9.60												5.20	6.50	7.80	10.4

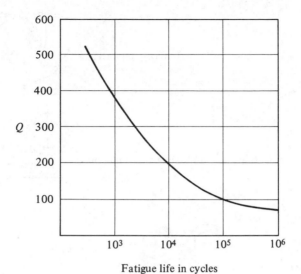

Fatigue life in cycles

Figure 4-20 / Values of Q as a function of fatigue life—SAE 1095 and stainless 301/302.

4 / *Space requirements.* For any load in the appropriate table, there is a recommended drum diameter, a natural coil diameter, material thickness, and width. If a drum is used, the diameter of the coiled spring is equal to the drum diameter plus the number of thicknesses coiled on it. Without a drum, the coil diameter is approximately equal to $2Rn$ from the table plus the thicknesses to be added.

5 / It is recommended that $1\frac{1}{2}$ turns remain on the coil in the fully extended position. Working deflections of 4 ft or more are not uncommon.

4-9 CONSTANT-FORCE SPRINGS—MOUNTING DESIGN

Three types of mountings, Fig. 4-21, are recommended by the manufacturer of Neg'ator® springs. In spindle mounting, Fig. 4-21(a), the spring will hold to a bushing by its own natural gripping action as long as sufficient material remains on the bushing at maximum extension. In very long extensions where a cumulative misalignment is possible, it is recommended that the inner coil be fastened to the drum or that a flanged drum is em-

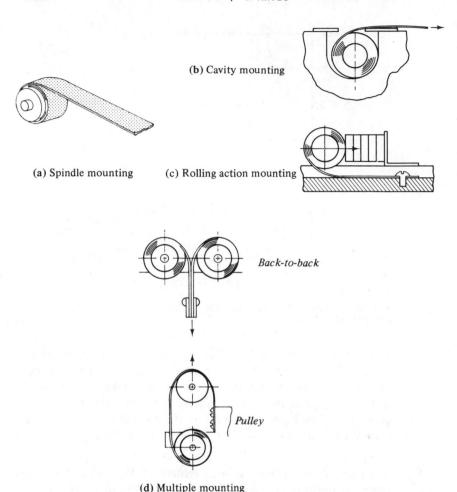

(b) Cavity mounting

(a) Spindle mounting   (c) Rolling action mounting

Back-to-back

Pulley

(d) Multiple mounting

Figure 4-21 / Mountings for constant-force springs. (Courtesy of Hunter Spring Company, A Division of Ametek, Inc., from The Hunter Neg'ator and Other Spring Devices, Bulletin 310-67)

ployed. Needless to remark, proper bearing alignment of the drum is important.

Cavity mounting, Fig. 4-21(b), while economical, has the disadvantage of high friction losses. A variation of the cavity mount is the rolling action mount illustrated in Fig. 4-21(c); the coil is free to translate, and its action pushes against a movable member.

Several variations of multiple mounting, a method used to increase the available force, are back-to-back and pulley mounting, shown in Fig. 4-21(d). Back-to-back mounting of equal springs doubles the force and cancels the end-moments that accompany a "single-spring" design. Pulley mounting doubles the force available but tends to reduce the life of the spring—by doubling the force, however, the available extension is reduced by a factor of 2.

EXAMPLE 4-7

An oven door, Fig. 4-22, which is part of an automated 24-hr per day drying operation is to open and close every 45 min. The door weighs 60 lb and has a travel of 15 in. while closing and opening. (a) Select the required number of drum-supported, high-carbon steel, constant-force springs for a required life of 4 yrs; and (b) determine the required length of the spring.

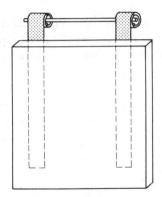

Figure 4-22 / Oven door life (Example 4-7).

solution

(a) Total operating cycles; opening and closing is equivalent to two cycles.

$$\text{No. of cycles} = 4\,\text{yrs} \times 365\,\frac{\text{days}}{\text{yr}} \times 24\,\frac{\text{hrs}}{\text{day}} \times 60\,\frac{\text{min}}{\text{hr}} \times 2\,\frac{\text{cycles}}{45\,\text{min}}$$

$$= 93,400\,\text{cycles}$$

Use Chart F in Table 4-1: Since no single spring is capable of supporting the load of 60 lb, multiple springs will have to be used. Also, back-to-back mounting is desirable, so an even number of springs should be selected. While cost is not a factor, the fewer the parts, the better.

A 4-in. wide spring with a thickness of 0.025 in. will support 10.1 lb; thus

$$\frac{60}{10.1} \approx 6 \text{ springs}$$

.(b)   The length L of each spring involves the desired travel, the $1\frac{1}{2}$ turns that are to remain on the drum, and an allowance of $\frac{1}{2}$ turn for fastening. Thus

$$L = 15 + (1.5 \times \pi \times 5.87) + (0.5 \times \pi \times 6.67)$$
$$= 53.1 \text{ in.: Use 60 in.}$$

4-10   DYNAMIC LOADING

With one exception, the two systems illustrated in Fig. 4-23 are similar— the weights have equal magnitudes, and the springs are identical. The difference? Weight W, Fig. 4-23(a), is released suddenly; the spring re-

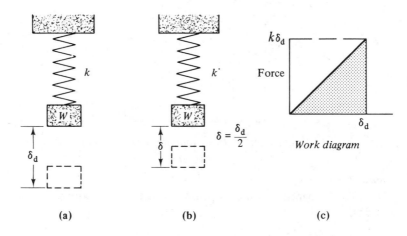

(a)   (b)   (c)

Figure 4-23 | Static and dynamic loads.

sponds to the impact with a maximum *dynamic* deflection of δ_d; and the weight then bobs up and down with a constant frequency. In contrast, Fig. 4-23(b), the hand of man is involved, and the weight is gently lowered until it is just supported by the spring; the weight is motionless, and the spring has a static deflection of δ—just one-half of δ_d. Reasons for the difference in deflection are easy to explain. In the first case, when the weight is released, it accelerates to a maximum velocity (which it reaches at the midpoint of its travel) and then decelerates until it momentarily comes to rest—the upward force that the spring exerts at this instant is greater than the downward gravity force W. This difference in force propels the weight upward. Mathematically, in moving downward, the weight loses potential energy equal to $W\delta_d$ while the spring gains an equal amount of potential energy $k\delta_d^2/2$, shown by the shaded area of the force–deflection diagram, Fig. 4-23(c):

$$W\delta_d = \tfrac{1}{2}k\delta_d^2$$

where
$$\delta_d = 2W/k \qquad\qquad \textit{(4-12)}$$

In the case of the slow release, descent is at a constant rate, and the laws of statics apply:

$$\sum F = 0$$
$$W = k\delta$$
$$\delta = W/k \qquad\qquad \textit{(4-13)}$$

For dynamic loading, the natural frequency f in cycles per second (cps) is given by

$$f = \frac{1}{2\pi}\sqrt{\frac{kg}{W}} \qquad\qquad \textit{(4-14)}$$

where g is the gravitational acceleration (32.2 fps² or 386.4 ips²). Since the ratio W/k is equal to the static deflection δ, the frequency equation can be expressed as

$$f = \frac{1}{2\pi}\sqrt{\frac{g}{\delta}} \qquad\qquad \textit{(4-15)}$$

While unrelated at this point, it is interesting to note that the frequency given by Eq. (4-15) is the same as that for a simple pendulum that has a length equal to the static deflection.

$$f = \frac{1}{2\pi}\sqrt{\frac{g}{\delta}} = \frac{1}{2\pi}\sqrt{\frac{g}{l}} \qquad\qquad \textit{(4-16)}$$

Throughout this discussion, the spring was assumed to be weightless—a reasonable assumption if the weight of the spring is small compared to the load ($\frac{1}{5}$ or less). By adding $\frac{1}{3}$ of the weight of the spring W_S to the weight of the load, a more precise value of frequency is obtained:

$$f = \frac{1}{2\pi} \sqrt{\frac{kg}{W + W_S/3}} \qquad (4\text{-}17)$$

EXAMPLE 4-8

A spring mass system similar to that shown in Fig. 4-23 has the following characteristics: $k = 100$ lb/in., $W = 10$ lb, and $W_S = 4$ lb. Determine the natural frequency of the system, correcting for the weight of the spring, and the percent error in frequency if the weight of the spring is ignored.

solution

$$f = \frac{1}{2\pi} \sqrt{\frac{kg}{W + W_S}} = \frac{1}{2\pi} \sqrt{\frac{100 \times 386.4}{10 + \frac{4}{3}}} = 9.31 \text{ cps} \quad [\text{Eq. (4-17)}]$$

$$f = \frac{1}{2\pi} \sqrt{\frac{kg}{W}} = \frac{1}{2\pi} \sqrt{\frac{100 \times 386.4}{10}} = 9.90 \text{ cps} \quad [\text{Eq. (4-14)}]$$

$$\% \text{ error} = \frac{9.90 - 9.31}{9.31} \times 100 = 6.34\%$$

4-11   VIBRATION ISOLATION

When a machine or engine is firmly attached to a supporting structure, vibrations and accompanying forces that occur are transmitted to the support. Conversely, if the support vibrates, these vibrations and forces are transmitted to the machine. Spring support, depending upon design, can either magnify or minimize these forces. The percent of the cyclic force transmitted by the machine to the structure or by the structure to the machine is given by

$$\% \text{ transmissibility} = \frac{1}{(f_d/f)^2 - 1} \times 100 \qquad (4\text{-}18)$$

where f_d is the frequency of the disturbing force, and f is the natural frequency of the spring-mass system. Equation (4-18) can also be expressed in terms of the static deflection of the system:

$$\% \text{ transmissibility} = \frac{1}{\dfrac{(2\pi f_d)^2 \delta}{g} - 1} \qquad (4\text{-}19)$$

The graph of Eq. (4-18), shown in Fig. 4-24, vividly depicts the effects of *resonance*—a situation where the natural frequency of the spring-

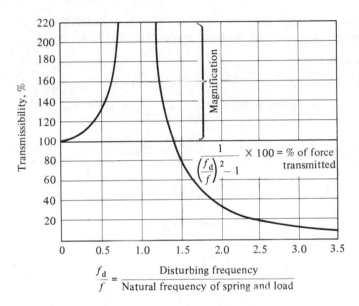

Figure 4-24 / Resonance. [*Reprinted by permission from* Design Handbook—Springs, Custom Metal Parts (*Bristol, Conn.: Associated Spring Corporation, 1967*), 59]

mass system equals or appreciably equals the disturbing frequency—the "don't march in cadence on the bridge" syndrome. Thus, for vibration isolation, the natural frequency should be as far removed as possible from the disturbing frequency.

EXAMPLE 4-9

An 1800 rpm motor, which has a 2 oz-in. unbalance, and its mounting plate are shown in Fig. 4-25. The assembly weighs 120 lb and is supported by four similar springs. Determine: (a) the spring constants of

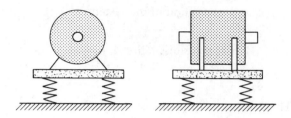

Figure 4-25 / Electric motor mounting (Example 4-9).

each of the springs if the transmissibility is to be limited to 20%; and
(b) the magnitude of the force transmitted to the support.

solution
(a)

$$\% \text{ transmissibility} = \frac{1}{(f_d/f)^2 - 1} \times 100 \quad [\text{Eq. (4-18)}]$$

where
$$f_d = \frac{1800}{60} = 30 \text{ cps}$$

$$0.2 = \frac{1}{(30/f)^2 - 1}$$

Rearranging terms and solving for f gives

$$\left(\frac{30}{f^2}\right) - 1 = \frac{1}{0.2} = 5$$

$$\frac{30}{f^2} = 6$$

$$f = \sqrt{150}$$

$$f = \frac{1}{2\pi}\sqrt{\frac{k_e g}{W}} = \sqrt{150} \quad [\text{Eq. (4-14)}]$$

where k_e = equivalent spring constant of the system:

$$k_e = \frac{(2\pi)^2 \times 150 \times 120}{386.4} = 1837 \text{ lb/in.}$$

Since the four equal springs are in *parallel*, the constant of each spring
is

$$k = \frac{k_e}{4} = \frac{1837}{4} = 459 \text{ lb/in.}$$

(b) The disturbing force is the centrifugal force caused by unbalance —2 oz at an eccentricity of 1 in. or, depending on your point of view, 1 oz at an eccentricity of 2 in.

By the centrifugal force equation:

$$F = \frac{Wr\omega^2}{g}$$

where Wr is the weight unbalance in lb-ft, ω is the frequency in rad/sec (1 rpm $= 2\pi/60$ rad/sec), and g is the acceleration of gravity in fps². Thus,

$$F = \frac{2}{16 \times 12} \times \frac{1}{32.2} \times \left(1800 \times \frac{2\pi}{60}\right)^2$$

$$= 11.5 \text{ lb}$$

The percent force transmitted to the support is, therefore,

$$F_{\text{transmitted}} = 0.2 \times 11.5 = 2.30 \text{ lb}$$

4-12 PHYSICAL AND MECHANICAL PROPERTIES OF SPRING MATERIALS

Cost, availability, and physical, electrical, and chemical properties are just a few of the major concerns in the selection of an appropriate spring material. When a large number of springs are involved, material choice is very important. If a small number are needed, availability rather than cost is the deciding factor. Obviously, the least expensive route is the "stock spring"—a spring that is a catalog item—and there are literally thousands of these available.

Table 4-2 lists the properties of some of the more common spring materials, and Table 4-3 lists the weight of spring steel wire. Figure 4-26 graphically represents the relative cost of spring wires using the cost of music wire as a base. A similar graph, Fig. 4-27, gives the ultimate strength of various materials as a function of wire diameter, and Fig. 4-28 illustrates the variation in ultimate strength of tempered spring steel as a function of Rockwell hardness.

Lastly, Table 4-4 gives a comparison of the frequently encountered and often perplexing gage systems.

TABLE 4-2

**PROPERTIES OF COMMON
SPRING MATERIALS¹**

Common name, specification	E 10⁶ psi	G 10⁶ psi	ρ density lb./in.³	Electrical conductivity % IACS	Sizes available in. Min.	Max.	Fatigue applications	Relative strength	Max. service temp. °F
High-carbon steel wires									
Music ASTM A228	30	11.5	0.284	7	0.004	0.250	Excellent	High	250
Hard-drawn ASTM A227	30	11.5	0.284	7	0.028	0.625	Poor	Medium	250
Oil-tempered ASTM A229	30	11.5	0.284	7	0.020	0.625	Poor	Medium	300
Valve-spring ASTM A230	30	11.5	0.284	7	0.050	0.250	Excellent	High	300
Alloy-steel wires									
Chrome-vanadium AISI 6150 . .	30	11.5	0.284	7	0.032	0.438	Excellent	High	425
Chrome-silicon AISI 9254	30	11.5	0.284	5	0.035	0.375	Fair	High	475
Silicon-manganese AISI 9260 .	30	11.5	0.284	4.5	0.025	0.375	Fair	High	450
Stainless-steel wires									
Martensitic AISI 410, 420	28	11	0.280	2.5	0.003	0.500	Poor	Low	500
Austenitic AISI 301, 302	28	10	0.286	2	0.005	0.375	Good	Medium	600
Precipitation-hardening 17–7 PH	29.5	11	0.286	2	0.030	0.500	Good	High	700
Nickel-chrome A286	29	10.4	0.290	2	0.016	0.200	—	Low	950

Common name, specification	E 10^6 psi	G 10^6 psi	ρ density lb./in.^{3}	Electrical conductivity % IACS	Sizes available in. Min.	Sizes available in. Max.	Fatigue applications	Relative strength	Max. service temp. °F
Copper-base alloy wires									
Phosphor-bronze ASTM B159	15	6.3	0.320	18	0.004	0.500	Good	Medium	200
Silicon-bronze ASTM B99	15	6.4	0.308	7	0.004	0.500	Fair	Low	200
Beryllium-copper ASTM B197	19	6.5	0.297	21	0.003	0.500	Excellent	High	400
Nickel-base alloys—wire and strip									
Inconel 600	31	11	0.307	1.5	0.004	0.500	Fair	Low	700
Inconel X750	31	11	0.298	1	0.004	0.563	Fair	Low	1100
Ni span C 902	27	9.6	0.294	1.6	0.004	0.500	Fair	Medium	200
High-carbon steel strip									
AISI 1050	30	11.5	0.284	7	0.010	0.125	Poor	Low	200
AISI 1065	30	11.5	0.284	7	0.003	0.125	Fair	Medium	200
AISI 1075	30	11.5	0.284	7	0.003	0.125	Good	High	250
AISI 1095	30	11.5	0.284	7	0.003	0.125	Excellent	High	250
Stainless-steel strip									
Austenitic AISI 301, 302	28	10	0.286	2	0.003	0.063	Good	Medium	600
Precipitation-hardening 17–7 PH	29.5	11	0.286	2	0.003	0.125	Good	High	700
Copper-base alloy strip									
Phosphor-bronze ASTM B103	15	6.3	0.320	18	0.003	0.188	Good	Medium	200
Beryllium-copper ASTM B194	19	6.5	0.297	21	0.003	0.375	Excellent	High	400

[1]*Courtesy of The Associated Spring Corporation*

TABLE 4-3

WEIGHT OF SPRING STEEL WIRE

WIRE DIA. (in.)	WT/100 FT (lb)	LENGTH/LB (ft)
0.004	0.004	23350.2
0.005	0.007	14944.1
0.006	0.010	10377.9
0.007	0.013	7624.6
0.008	0.017	5837.5
0.009	0.022	4612.4
0.010	0.027	3736.0
0.011	0.032	3087.6
0.012	0.039	2594.5
0.013	0.045	2210.7
0.014	0.052	1906.1
0.015	0.060	1660.5
0.016	0.069	1459.4
0.017	0.077	1292.7
0.018	0.087	1153.1
0.020	0.107	934.0
0.022	0.130	771.9
0.024	0.154	648.6
0.026	0.181	552.7
0.028	0.210	476.5
0.031	0.257	388.8
0.034	0.309	323.2
0.037	0.366	272.9
0.041	0.450	222.3
0.045	0.542	184.5
0.050	0.669	149.4
0.054	0.781	128.1
0.058	0.900	111.1
0.063	1.062	94.1
0.070	1.312	76.2
0.075	1.506	66.4
0.080	1.713	58.4
0.087	2.026	49.4
0.095	2.416	41.4
0.105	2.951	33.9
0.115	3.540	28.2
0.125	4.182	23.9
0.135	4.878	20.5
0.148	5.863	17.1
0.162	7.025	14.2

TABLE 4-3 (Cont.)

WIRE DIA. (in.)	WT/100 FT (lb)	LENGTH/LB (ft)
0.177	8.386	11.9
0.192	9.867	10.1
0.207	11.469	8.7
0.225	13.550	7.4
0.250	16.729	6.0
0.278	20.686	4.8
0.306	25.063	4.0
0.331	29.326	3.4
0.362	35.076	2.9
0.394	41.551	2.4
0.431	49.721	2.0
0.462	57.131	1.8
0.500	66.916	1.5

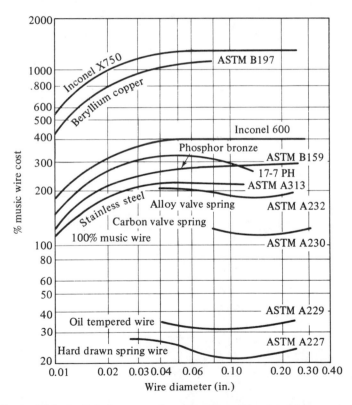

Figure 4-26 / Relative cost of spring wires. [Reprinted by permission from Design Handbook—Springs, Custom Metal Parts (Bristol, Conn.: Associated Spring Corporation, 1967), 43]

235

TABLE 4-4
WIRE AND SHEET METAL GAGES[1]

No.	United States Standard* — Uncoated steel sheets and light plates — Weight lb per sq ft	United States Standard* — Approx. thickness inches	American Steel and Wire Co. and Washburn and Moen — Steel wire	American or Brown and Sharpe Wire — Non-ferrous sheets and wire	New Birmingham Standard Sheet and Hoop — Iron and steel sheets and hoops	British Imperial or English Legal Standard — Wire	Birmingham or Stubs Iron Wire — Strips, bands, hoops and wire	No.
7/0's	20.00	.4902	.4900	.5800	.6666	.500		7/0's
6/0's	18.75	.4596	.4615		.625	.464		6/0's
5/0's	17.50	.4289	.4305	.5165	.5883	.432	.500	5/0's
4/0's	16.25	.3983	.3938	.4600	.5416	.400	.454	4/0's
3/0's	15.00	.3676	.3625	.4096	.500	.372	.425	3/0's
2/0's	13.75	.3370	.3310	.3648	.4452	.348	.380	2/0's
0	12.50	.3064	.3065	.3249	.3964	.324	.340	0
1	11.25	.2757	.2830	.2893	.3532	.300	.300	1
2	10.625	.2604	.2625	.2576	.3147	.276	.284	2
3	10.00	.2451	.2437	.2294	.2804	.252	.259	3
4	9.375	.2298	.2253	.2043	.250	.232	.238	4
5	8.75	.2145	.2070	.1819	.2225	.212	.220	5
6	8.125	.1991	.1920	.1620	.1981	.192	.203	6
7	7.50	.1838	.1770	.1443	.1764	.176	.180	7
8	6.875	.1685	.1620	.1285	.1570	.160	.165	8
9	6.25	.1532	.1483	.1144	.1398	.144	.148	9
10	5.625	.1379	.1350	.1019	.1250	.128	.134	10
11	5.00	.1225	.1205	.0907	.1113	.116	.120	11
12	4.375	.1072	.1055	.0808	.0991	.104	.109	12
13	3.75	.0919	.0915	.0720	.0882	.092	.095	13
14	3.125	.0766	.0800	.0641	.0785	.080	.083	14
15	2.8125	.0689	.0720	.0571	.0699	.072	.072	15

Thickness, inches

16	2.50	.0613	.0625	.0508	.0625	.064	.065	16
17	2.25	.0551	.0540	.0453	.0556	.056	.058	17
18	2.00	.0490	.0475	.0403	.0495	.048	.049	18
19	1.75	.0429	.0410	.0359	.0440	.040	.042	19
20	1.50	.0368	.0348	.0320	.0392	.036	.035	20
21	1.375	.0337	.0318	.0285	.0349	.032	.032	21
22	1.25	.0306	.0286	.0253	.0313	.028	.028	22
23	1.125	.0276	.0258	.0226	.0278	.024	.025	23
24	1.00	.0245	.0230	.0201	.0248	.022	.022	24
25	.875	.0214	.0204	.0179	.0220	.020	.020	25
26	.75	.0184	.0181	.0159	.0196	.018	.018	26
27	.6875	.0169	.0173	.0142	.0175	.0164	.016	27
28	.625	.0153	.0162	.0126	.0156	.0148	.014	28
29	.5625	.0138	.0150	.0113	.0139	.0136	.013	29
30	.50	.0123	.0140	.0100	.0123	.0124	.012	30
31	.4375	.0107	.0132	.0089	.0110	.0116	.010	31
32	.4062	.0100	.0128	.0080	.0098	.0108	.009	32
33	.375	.0092	.0118	.0071	.0087	.0100	.008	33
34	.3438	.0084	.0104	.0063	.0077	.0092	.007	34
35	.3125	.0077	.0095	.0056	.0069	.0084	.005	35
36	.2812	.0069	.0090	.0050	.0061	.0076	.004	36
37	.2656	.0065	.0085	.0045	.0054	.0068		37
38	.25	.0061	.0080	.0040	.0048	.0060		38
39	.2344	.0057	.0075	.0035	.0043	.0052		39
40	.2188	.0054	.0070	.0031	.0039	.0048		40

*U.S. Standard gage is officially a weight gage (in ounces per sq ft) based on wrought iron at 480 lb per cu ft. The values tabulated above give the thickness of steel (at 489.6 lb per cu ft) that will approximate the respective weights. The other gages are officially thickness gages.

Plates — over 6" to 48" wide, $\frac{1}{4}$" and thicker; over 43" wide, $\frac{3}{16}$" and thicker.   Sheets — 24" to 48" wide, under $\frac{1}{4}$" thick; over 48" wide, under $\frac{3}{16}$" thick.
Strip — 23 $\frac{15}{16}$" and narrower, under $\frac{1}{4}$" thick.

[1]*Courtesy of the American Institute of Steel Construction*

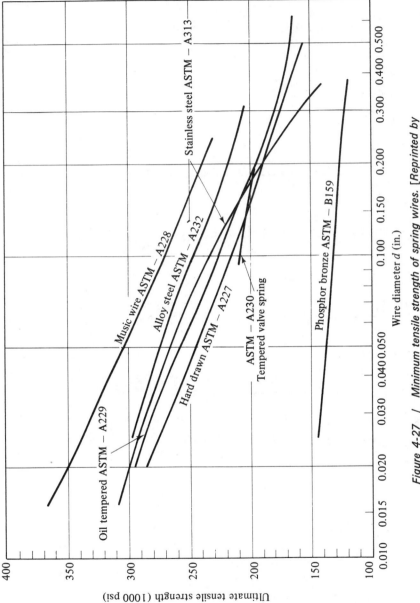

Figure 4-27 / Minimum tensile strength of spring wires. [Reprinted by permission from Design Handbook—Springs, Custom Metal Parts (Bristol, Conn.: Associated Spring Corporation, 1967), 42]

238

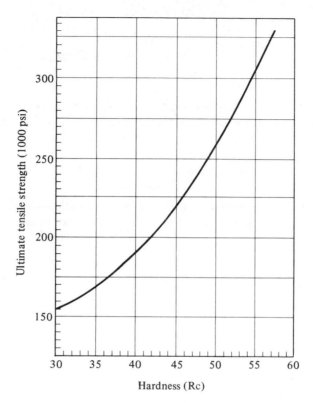

Figure 4-28 / Tensile strength versus hardness of quenched and tempered spring steel. [Reprinted by permission from Design Handbook—Springs, Custom Metal Parts *(Bristol, Conn.: Associated Spring Corporation, 1967), 42]*

QUESTIONS AND PROBLEMS

P4-1 A helical compression spring made of 0.200-dia music wire has a mean coil diameter of 1 in. Determine the load-carrying capacity if the shear stress is not to exceed 50,000 psi by: (a) using Eq. (4-2) assuming that the term $d/4R$ is negligible; and (b) using Eq. (4-2) in its entirety.

P4-2 Determine the shear stress, deflection, and spring constant of a phosphor-bronze ($G = 6.3 \times 10^6$) helical extension spring that has the following properties:

$$\text{Applied load} = 400 \text{ lb}$$
$$\text{Mean coil radius} = 2 \text{ in.}$$
$$\text{Wire diameter} = 0.30 \text{ in.}$$
$$\text{Active coils} = 10$$

P4-3   Three 10-turn helical compression springs made of $\frac{1}{4}$-in.-diameter
steel wire are parallel connected. The springs, which are used to
exert equal axial forces F on a clutch plate, have a mean radius
of 1.25 in., a free length of 8 in., and a compressed length of 6 in.
Determine the maximum force exerted against the clutch plate
and the maximum stress in the wire.

P4-4   A helical compression spring whose constant is 600 lb/in. is cut
into three equal parts, and the three pieces are then combined in
parallel. What is the spring constant of the combination?

P4-5   Two coil springs are nested as shown. The following data apply:

	OUTER SPRING	*INNER SPRING*
Free length	6 in.	5.75 in.
Mean spring diameter	1.5 in.	1 in.
Wire diameter	0.10	0.20
Active turns	10	12
Material	Phosphor-bronze $G = 6.3 \times 10^6$	Music wire $G = 12 \times 10^6$

Plot a curve that will show the variation of force with deflection
for a maximum compression of $\delta =$ in.

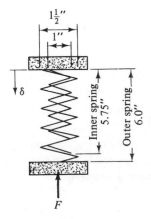

Figure P4-5

P4-6   A beam of negligible weight is supported by two helical springs as shown. Find the deflection t of a force $F = 400$ lb acting vertically downward from point A.

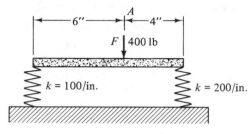

Figure P4-6

P4-7   Find the equivalent spring constant for each of the combinations shown. Assume all springs have equal constants k.

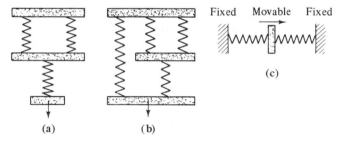

Figure P4-7

P4-8   Plot a graph of force F versus displacement for the variable pitch spring described in Example 4-4, and determine the energy stored within the spring when all coils are completely closed.

P4-9   Approximate the spring constant k of the hourglass spring shown.

Figure P4-9

The spring is made from 0.06-in.-dia steel wire and has 10 active coils.

P4-10   Compute the numerical values of Wahl's correction factor K_w for spring indices of 4, 6, 8, 10, and 12.

P4-11   Discuss some of the design ramifications of each of the following:
(a)   a spring whose index is 14;
(b)   two nested springs that are coiled in the same direction;
(c)   a spring whose outside diameter is 1.25 in. that is to operate in a 1.32-in.-dia hole; and
(d)   a compression spring having a free length of 6 in. and a diameter of 1 in.

P4-12   A 200-coil torsional projector screen spring is made from No. 14 (Washburn and Moen) music wire. The mean diameter of the helix is 1.5 in. Assume a factor of safety of 5 based on the ultimate tensile strength of 280,000 psi, and find:
(a)   the stress at the inner edge of the helix (allow for stress concentration due to curvature); and
(b)   the torque that the roller can exert after unwinding 4 revolutions from the most highly stressed condition.

P4-13   For static service, it is not unusual to use a design stress near or equal to the yield point of the spring material. Discuss some of the ramifications of "overdesign"—the use of high factors of safety for static service.

P4-14   Plot a graph that will give the stress concentration factors of Eq. (4-9) for R/d ratios from 4 to 10 in incremental values. Estimate, from your graph, the concentration factor for $R/d = 6.3$. Next, compute the precise value, and determine the percent error.

P4-15   A load P is applied to three leaf springs connected as shown. If the constants for the three springs are $k_1 = k$, $k_2 = 2k$, and $k_3 = 4k$, find the equivalent constant of the system.

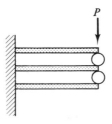

Figure P4-15

P4-16   Two spring steel leaf springs, interconnected by a cam as shown, are initially free of any bending stress. Compute the energy stored in the systems if the cam is rotated $\frac{1}{4}$-turn counterclockwise.

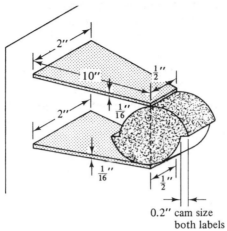

Figure P4-16

P4-17   Refer to Fig. 4-15, and show for deflections that Case (c) is equivalent to Case (b) when:
(a) $b = b_o$; and
(b) $b = o$.

P4-18   Two leaf springs are interconnected by a helical spring as shown. Derive an equation for the equivalent spring constant.

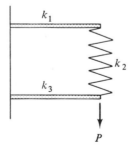

Figure P4-18

P4-19   Estimate the fatigue life in terms of the number of cycles of expected operation for each of the following:
(a)   A spring counterbalanced home-type garage door that carries a 15-yr warranty. The garage houses two cars—one

used for pleasure and the other for business. Four people drive in the family.

(b)   A window shade roller containing a torsional spring as the counterbalancing element. Make your own assumptions regarding life expectancy.

(c)   An automobile bumper spring suspension system capable of sustaining a 5-mph impact.

P4-20   The constant-force spindle-mounted spring A, shown, is used to maintain a pressure force of 1.5 lb between linkage B and the eccentric C. Select an appropriate high-carbon steel spring from the tables in this chapter, and determine the required length. The eccentric rotates at a speed of 4 rpm and is scheduled to operate eight hours per day and five days per week. Design for a life expectancy of 2 yrs.

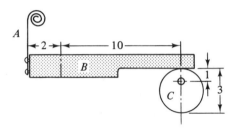

Figure P4-20

P4-21   Select an appropriate constant-force spring that will act as an automotive seat belt retractor. Base the computations on your own design criteria regarding force, length, and life expectancy.

P4-22   Weight $W = 5$ lb is dropped from a height $h = 2$ in. onto a compression spring $k = 10$ lb/in., as shown. Determine the maximum deflection of the spring assuming that the spring does not completely close.

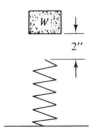

Figure P4-22

P4-23 A spring-mass system has the following characteristics: $k = 200$ lb/in., $W = 15$ lb, and a spring weight of $W_s = 8$ lb. Determine the natural frequency of the system.

P4-24 Gravitational acceleration on the moon is 5.47 fps². By what factor would the natural frequency of a spring-mass system differ on the moon relative to its frequency on the earth?

P4-25 A weight W statically suspended from a helical spring results in a deflection δ. By what factor would δ change on Mars, where the gravitational acceleration is 12.9 fps²?

P4-26 Springs are used to check the recoil of large guns. After firing and recoiling, shock absorbers allow the barrel to return to its original firing position without oscillation. Find the proper spring constant for a gun weighing 2500 lb if the recoil velocity at the instant of firing is 100 fps and if the recoil distance is 5 ft.

P4-27 A compressor operating at 600 rpm and weighing 300 lb is supported by four springs of stiffness k lb/in. each. Find k if the transmitted shaking force is restricted to 10%.

P4-28 An aircraft radio weighing 28 lb is supported by four springs rated at 100 lb/in. each. What percent of the engine vibration is transmitted to the radio at an engine speed of:
(a) 2000 rpm; and
(b) 1500 rpm?

P4-29 A cantilever rectangular leaf spring (Fig. 4-15) measures 50 mm wide by 125 mm long, with a thickness of 2 mm. The material is a steel ($E = 200 \times 10^9$ N/m²) with an ultimate tensile strength of 10^9 N/m². Using a factor of safety of 8, determine the maximum end load that can safely be sustained by this spring and the deflection produced by this load.

SUGGESTED DESIGN PROJECTS

DP4-1 You have been asked to improve on the conventional mousetrap by replacing the traditional double-torsion spring with a single compression spring. Suggest a suitable design with recommendation for spring-rate based on measurements of the characteristics of the "conventional" trap.

DP4-2 Design and build a prototype of a toy car powered only by the energy of a mousetrap spring. Your "bill of materials" must consist only of a mousetrap, $\frac{1}{8}$-in. thick hardboard, glue, string, and coat-hanger wire. The design should include calculations of efficiency.

DP4-3 Design a spring-operated batting trainer that will thrust a baseball vertically upward into the strike zone. The trainer should be capable of holding 10 balls in reserve and should be cocked and triggered by foot action.

DP4-4 Design an automobile bumper system employing four helical springs that will be capable of absorbing a 7-mph impact. Assume a gross weight—car plus passengers—of 4500 lb.

CHAPTER 5

FASTENING, JOINING, AND SEALING

"I was very much occupied with other matters, and while we all congratulated ourselves on what we had accomplished, and knew we had an interesting and novel apparatus, we generally regarded it more or less as a curiosity with no very large practicable possibilities."

THOMAS ALVA EDISON
*1922 Reflection on His
Invention of the Motion
Picture Camera*

5-1   INTRODUCTION

Fastening expense—once considered trivial—has become a critical factor in the overall cost of production. The modern automobile contains in the neighborhood of 10,000 *different* fastening devices—nuts, bolts, screws, pins, solders, weldments, and adhesives. These devices require highly specialized tools and machines to produce, and they require equally specialized equipment to put in place. Many problems face the industry; two—one old and one new—are most serious. The old problem involves standardization, or really the lack of rigid standards. Gunsmiths and clockmakers as early as the 17th century recognized the need for *interchangeability*—the Civil War nurtured this recognition. The problem not only still exists, it has been magnified. World trade, a necessity for existence, is prompting the new problem—the monumental task of changing from U.S. standards to the metric system, a change that is destined to take place, and it will start with the fastening industry.

5-2   NOMENCLATURE OF BOLTS, NUTS, AND SCREWS

Every industry eventually develops a language that uniquely satisfies its need for communication; thus, various "set" terms are used to describe the features and dimensions of threaded fasteners. Figure 5-1 illustrates the accepted definitions of the terms and phrases that follow.

Major Diameter.   Largest diameter of a screw thread.

Minor or Root Diameter.   Smallest diameter of a screw thread.

Pitch.   Distance, measured parallel to the screw axis, between a point on the screw thread and a corresponding point on the next thread. Pitch is equal to the reciprocal of the number of threads per inch.

Pitch Diameter.   Effective diameter of thread; the diameter of an imaginary line drawn through the thread profile such that the widths of the thread and grooves are equal—a length approximately midway between the major and minor diameters.

Lead.   Distance that the screw advances in a twist of one revolution. In a *single* thread, the lead and pitch are equal. In a *double* thread, the lead is twice the pitch. In a triple thread, the lead is three times the pitch, and so on.

Thread Length.   Length of the threaded portion of the shank—in standard bolts there is usually a direct relationship between thread length and either shank length or thread diameter. For example, the standard thread length of a 6-in.-or-shorter hexagon-head machine bolt is 2 diameters plus $\frac{1}{4}$ in. If the same bolt is longer than 6 in., the thread length is usually 2 diameters plus $\frac{1}{2}$ in.

Shank Diameter.   Full or undersized as shown; full shanks are characteristic of bolts having *cut* threads, while undersized shanks are the product of the *roll* threading operation.

Minimum Transverse Shear Area.   Minimum area of threaded section—root diameter area.

Tensile Shear Area.   A theoretical area used for computing tensile strength—the area of an imaginary circle whose diameter is the mean of the pitch and root diameters. An empirical formula for

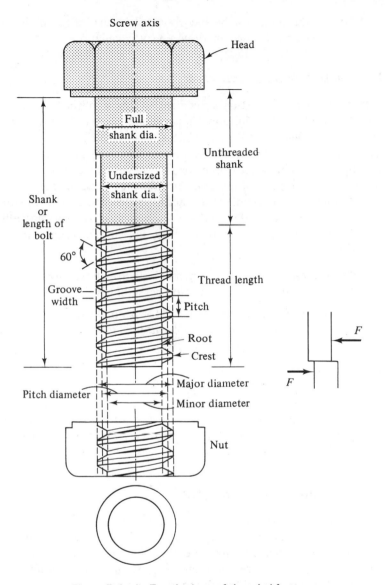

Figure 5-1 / Terminology of threaded fasteners.

computing this area A_t is given by

$$A_t = \frac{\pi}{4}\left(D - \frac{0.974}{n}\right)^2 \qquad (5\text{-}1)$$

where D is the major diameter and n is the number of threads per inch. A 5–15% added tensile safety factor, depending upon the size of the bolt, will be incorporated in the design if the root diameter area is used in preliminary calculations.

5-3   THREAD FORMS AND FITS

In 1948, thread standardization agreements were established between the United States, Great Britain, and Canada; the Unified Forms—Unified Coarse Thread (UNC) and Unified Fine Thread (UNF)—were the result. Other thread forms exist, of course, but at least the problem of interchangeability has somewhat diminished. Table 5-1 briefly describes the geometry and application of the more prominent thread forms.

Varying amounts of manufacturing tolerance and operating allowance for a given style of thread are distinguished by *class*; the suffixes "A" and "B" are used to designate internal and external threads, respectively.

Classes 1A and 1B.   A rough commercial-quality thread used where a loose fit is tolerable and spin-on-assembly is desirable.

Classes 2A and 2B. A recognized standard for normal use; most commercial bolts, screws, and nuts fall within this category.

Classes 3A and 3B.   Used where "close-fit" is an important requirement.

Class 4.   An obsolete classification.

Class 5.   A force fit for semipermanent assembly.

Table 5-2 lists the dimensional features of Classes 2A and 3A of the Unified Thread Form; in a similar manner, Table 5-3 describes the DIN (Deutche Industrie Norm) Metric Thread.

5-4   BOLT AND SCREW DETAILS

There are virtually thousands of types and styles of bolts and screws; orderly classification, however, can be made according to head style, shoulder form, unthreaded shank style, thread type, and point. Figure 5-2 describes some of the more common head, shoulder, and point styles, while Fig. 5-3 shows some of the frequently encountered bent forms, studs, and set screws. Forms of nuts, washers, and pins are shown in Fig. 5-4.

5-5   BOLT STRESSES

In metal-to-metal connections the tensile force produced in the bolt, through tightening, remains constant—the force is not diminished by an opposing external load unless, of course, the opposing load exceeds the

TABLE 5-1
THREAD FORMS AND CHARACTERISTICS

Classification	Physical form	General use
Unified Coarse Thread (UNC)		Widely used for most threaded products — bolts, nuts and screws — made in the United States, Canada and the United Kingdom.
Unified Fine Thread (UNF)		Same form as UNC except a greater number of threads per inch are incorporated in a specific diameter. Used for automotive and aircraft work where limited wall thickness requires a fine thread.
American Standard Buttress		Thread forms capable of withstanding high axial stress.
American Standard Acme		Used primarily for converting rotational motion to translational motion; double and triple leads are frequently encountered in this type of thread.
American Standard Stub Acme		A variation of the Standard Acme used where space is a critical factor.
British Standard Whitworth		Crest and root are rounded; a general thread form used in the United Kingdom.
European Metric		A thread form similar to the unified series; the thread has greater depth, however. This system has been adopted by most European countries using the metric systems, by Japan and the U.S.S.R.

TABLE 5-2

UNIFIED THREAD SERIES—EXTERNAL DIMENSIONS OF CLASS 2A AND 3A THREADS[1]

Size and threads per inch	Series designation	Class	Allowance	Major diameter			Pitch diameter			Nominal minor diameter
				Limits		Tolerance	Limits		Tolerance	
				Maximum	Minimum		Maximum	Minimum		
2-56	UNC	2A	0.0006	0.0854	0.0813	0.0041	0.0738	0.0717	0.0021	0.0635
		3A	0.0000	0.0860	0.0819	0.0041	0.0744	0.0728	0.0016	0.0641
2-64	UNF	2A	0.0006	0.0854	0.0816	0.0038	0.0753	0.0733	0.0020	0.0662
		3A	0.0000	0.0860	0.0822	0.0038	0.0759	0.0744	0.0015	0.0668
3-48	UNC	2A	0.0007	0.0983	0.0938	0.0045	0.0848	0.0825	0.0023	0.0727
		3A	0.0000	0.0990	0.0945	0.0045	0.0855	0.0838	0.0017	0.0734
3-56	UNF	2A	0.0007	0.0983	0.0942	0.0041	0.0867	0.0845	0.0022	0.0764
		3A	0.0000	0.0990	0.0949	0.0041	0.0874	0.0858	0.0016	0.0771
4-40	UNC	2A	0.0008	0.1112	0.1061	0.0051	0.0950	0.0925	0.0025	0.0805
		3A	0.0000	0.1120	0.1069	0.0051	0.0958	0.0939	0.0019	0.0813
4-48	UNF	2A	0.0007	0.1113	0.1068	0.0045	0.0978	0.0954	0.0024	0.0857
		3A	0.0000	0.1120	0.1075	0.0045	0.0985	0.0967	0.0018	0.0864
5-40	UNC	2A	0.0008	0.1242	0.1191	0.0051	0.1080	0.1054	0.0026	0.0935
		3A	0.0000	0.1250	0.1199	0.0051	0.1088	0.1069	0.0019	0.0943
5-44	UNF	2A	0.0007	0.1243	0.1195	0.0048	0.1095	0.1070	0.0025	0.0964
		3A	0.0000	0.1250	0.1202	0.0048	0.1102	0.1083	0.0019	0.0971
6-32	UNC	2A	0.0008	0.1372	0.1312	0.0060	0.1169	0.1141	0.0028	0.0980
		3A	0.0000	0.1380	0.1320	0.0060	0.1177	0.1156	0.0021	0.0997
6-40	UNF	2A	0.0008	0.1372	0.1321	0.0051	0.1210	0.1184	0.0026	0.1065
		3A	0.0000	0.1380	0.1329	0.0051	0.1218	0.1198	0.0020	0.1073
8-32	UNC	2A	0.0009	0.1631	0.1571	0.0060	0.1428	0.1399	0.0029	0.1248
		3A	0.0000	0.1640	0.1580	0.0060	0.1437	0.1415	0.0022	0.1257
8-36	UNF	2A	0.0008	0.1632	0.1577	0.0055	0.1452	0.1424	0.0028	0.1291
		3A	0.0000	0.1640	0.1585	0.0055	0.1460	0.1439	0.0021	0.1299

TABLE 5-2 (Cont.)

Size and threads per inch	Series designation	Class	Allowance	Major diameter			Pitch diameter			Nominal minor diameter
				Limits		Tolerance	Limits		Tolerance	
				Maximum	Minimum		Maximum	Minimum		
10–24	UNC	2A	0.0010	0.1890	0.1818	0.0072	0.1619	0.1586	0.0033	0.1379
		3A	0.0000	0.1900	0.1828	0.0072	0.1629	0.1604	0.0025	0.1389
10–32	UNF	2A	0.0009	0.1891	0.1831	0.0060	0.1688	0.1658	0.0030	0.1508
		3A	0.0000	0.1900	0.1840	0.0060	0.1697	0.1674	0.0023	0.1517
12–24	UNC	2A	0.0010	0.2150	0.2078	0.0072	0.1879	0.1845	0.0034	0.1639
		3A	0.0000	0.2160	0.2088	0.0072	0.1889	0.1863	0.0026	0.1649
12–28	UNF	2A	0.0010	0.2150	0.2085	0.0065	0.1918	0.1886	0.0032	0.1712
		3A	0.0000	0.2160	0.2095	0.0065	0.1928	0.1904	0.0024	0.1722
1/4–20	UNC	2A	0.0011	0.2489	0.2408	0.0081	0.2164	0.2127	0.0037	0.1876
		3A	0.0000	0.2500	0.2419	0.0081	0.2175	0.2147	0.0028	0.1887
1/4–28	UNF	2A	0.0010	0.2490	0.2425	0.0065	0.2258	0.2225	0.0033	0.2052
		3A	0.0000	0.2500	0.2435	0.0065	0.2268	0.2243	0.0025	0.2062
5/16–18	UNC	2A	0.0012	0.3113	0.3025	0.0087	0.2752	0.2712	0.0040	0.2431
		3A	0.0000	0.3125	0.2028	0.0087	0.2764	0.2734	0.0030	0.2443
5/16–24	UNF	2A	0.0011	0.3114	0.3042	0.0072	0.2843	0.2806	0.0037	0.2603
		3A	0.0000	0.3125	0.3053	0.0072	0.2854	0.2827	0.0027	0.2614
3/8–16	UNC	2A	0.0013	0.3737	0.3643	0.0094	0.3331	0.3287	0.0044	0.2970
		3A	0.0000	0.3750	0.3656	0.0094	0.3344	0.3311	0.0033	0.2983
3/6–24	UNF	2A	0.0011	0.3739	0.3667	0.0072	0.3468	0.3430	0.0038	0.3228
		3A	0.0000	0.3750	0.3678	0.0072	0.3479	0.3450	0.0029	0.3239
7/16–14	UNC	2A	0.0014	0.4361	0.4253	0.0103	0.3897	0.3850	0.0047	0.3485
		3A	0.0000	0.4375	0.4272	0.0103	0.3911	0.3876	0.0035	0.3499
7/16–20	UNF	2A	0.0013	0.4362	0.4281	0.0081	0.4037	0.3995	0.0042	0.3749
		3A	0.0000	0.4375	0.4294	0.0081	0.4050	0.4019	0.0031	0.3762

[1]Courtesy ITT Harper Inc., Morton Grove, Illinois.

TABLE 5-2 (Cont.)

Size and threads per inch	Series designation	Class	Allowance	Major diameter Limits Maximum	Major diameter Limits Minimum	Major diameter Tolerance	Pitch diameter Limits Maximum	Pitch diameter Limits Minimum	Pitch diameter Tolerance	Nominal minor diameter
1/2–13	UNC	2A	0.0015	0.4985	0.4876	0.0109	0.4485	0.4435	0.0050	0.4041
		3A	0.0000	0.5000	0.4891	0.0109	0.4500	0.4463	0.0037	0.4056
1/2–20	UNF	2A	0.0013	0.4987	0.4906	0.0081	0.4662	0.4619	0.0043	0.4374
		3A	0.0000	0.5000	0.4919	0.0081	0.4675	0.4643	0.0032	0.4387
9/16–12	UNC	2A	0.0016	0.5609	0.5495	0.0114	0.5068	0.5016	0.0052	0.4587
		3A	0.0000	0.5625	0.5511	0.1114	0.5084	0.5045	0.0039	0.4603
9/16–18	UNF	2A	0.0014	0.5611	0.5524	0.0087	0.5250	0.5205	0.0045	0.4929
		3A	0.0000	0.5625	0.5538	0.0087	0.5264	0.5230	0.0034	0.4943
5/8–11	UNC	2A	0.0016	0.6234	0.6113	0.0121	0.5644	0.5589	0.0055	0.5119
		3A	0.0000	0.6250	0.6129	0.0121	0.5660	0.5619	0.0041	0.5135
5/8–18	UNF	2A	0.0014	0.6236	0.6149	0.0087	0.5875	0.5828	0.0047	0.5554
		3A	0.0000	0.6250	0.6163	0.0087	0.5889	0.5854	0.0035	0.5568
3/4–10	UNC	2A	0.0018	0.7482	0.7353	0.0129	0.6832	0.6773	0.0059	0.6255
		3A	0.0000	0.7500	0.7371	0.0129	0.6850	0.6806	0.0044	0.6255
3/4–16	UNF	2A	0.0015	0.7485	0.7391	0.0094	0.7079	0.7029	0.0050	0.6718
		3A	0.0000	0.7500	0.7406	0.0094	0.7094	0.7056	0.0038	0.6733
7/8–9	UNC	2A	0.0019	0.8731	0.8592	0.0139	0.8009	0.7946	0.0063	0.7368
		3A	0.0000	0.8750	0.8611	0.0139	0.8028	0.7981	0.0047	0.7387
7/3–14	UNF	2A	0.0016	0.8734	0.8631	0.0103	0.8270	0.8216	0.0054	0.7858
		3A	0.0000	0.8750	0.8647	0.0103	0.8286	0.8245	0.0041	0.7874
1–8	UNC	2A	0.0020	0.9080	0.9830	0.0150	0.9168	0.9100	0.0068	0.8446
		3A	0.0000	1.0000	0.9850	0.0150	0.9188	0.9137	0.0051	0.6466
1–12	UNF	2A	0.0018	0.9882	0.9868	0.0103	0.9441	0.9382	0.0059	0.8060
		3A	0.0000	1.0000	0.9886	0.0103	0.9459	0.9415	0.0044	0.8978

254

TABLE 5-2 (Cont.)

Size and threads per inch	Series designation	Class	Allowance	Major diameter Limits Maximum	Major diameter Limits Minimum	Major diameter Tolerance	Pitch diameter Limits Maximum	Pitch diameter Limits Minimum	Pitch diameter Tolerance	Nominal minor diameter
1-1/3-7	UNC	2A	0.0022	1.1228	1.1064	0.0164	1.0300	1.0228	0.0072	0.9475
		3A	0.0000	1.1250	1.1085	0.0164	1.0322	1.0268	0.0054	0.9497
1-1/8-8	UN	2A	0.0021	1.1229	1.1079	0.0150	1.0417	1.0348	0.0069	0.9895
		3A	0.0000	1.1250	1.1100	0.0150	1.0438	1.0386	0.0052	0.9716
1-1/8-12	UNF	2A	0.0018	1.1232	1.1118	0.0114	1.0691	1.0631	0.0060	1.0210
		3A	0.0000	1.1250	1.1136	0.0114	1.0709	1.0664	0.0045	1.0228
1-1/4-7	UNC	2A	0.0022	1.2478	1.2314	0.0164	1.1550	1.1476	0.0074	1.0725
		3A	0.0000	1.2500	1.2336	0.0164	1.1572	1.1517	0.0055	1.0747
1-1/4-8	UN	2A	0.0021	1.2479	1.2329	0.0150	1.1667	1.1597	0.0070	1.0945
		3A	0.0000	1.2500	1.2350	0.0150	1.1688	1.1635	0.0053	1.0966
1-1/4-12	UNF	2A	0.0018	1.2482	1.2368	1.0114	1.1941	1.1879	0.0062	1.1460
		3A	0.0000	1.2500	1.2386	1.0114	1.1959	1.1913	0.0046	1.1478
1-3/8-8	UNC	2A	0.0024	1.3726	1.3544	0.0182	1.2643	1.2583	0.0080	1.1681
		3A	0.0000	1.3750	1.3568	0.0182	1.2667	1.2607	0.0060	1.1705
1-3/8-8	UN	2A	0.0022	1.3728	1.3578	0.0150	1.2916	1.2844	0.0072	1.2194
		3A	0.0000	1.3750	1.3600	0.0150	1.2938	1.2884	0.0054	1.2216
1-3/8-12	UNF	2A	0.0019	1.3731	1.3617	0.0114	1.3190	1.3127	0.0083	1.2709
		3A	0.0000	1.3750	1.3636	0.0114	1.3209	1.3162	0.0047	1.2728
1-1/2-6	UNC	2A	0.0024	1.4976	1.4794	0.0182	1.3893	1.3812	0.0081	1.2931
		3A	0.0000	1.5000	1.4818	0.0182	1.3917	1.3856	0.0061	1.2955
1-1/2-8	UN	2A	0.0022	1.4978	1.4828	0.0150	1.4166	1.4093	0.0073	1.3444
		3A	0.0000	1.5000	1.4850	0.0150	1.4188	1.4133	0.0055	1.3466
1-1/2-12	UNF	2A	0.0019	1.4913	1.4867	0.0114	1.4440	1.4376	0.0064	1.3959
		3A	0.0000	1.5000	1.4836	0.0114	1.4459	1.4411	0.0048	1.3978

TABLE 5-2 (Cont.)

Size and threads per inch	Series designation	Class	Allowance	Major diameter			Pitch diameter			Nominal minor diameter
				Limits		Tolerance	Limits		Tolerance	
				Maximum	Minimum		Maximum	Minimum		
1-5/8-8	UN	2A	0.0022	1.6228	1.6078	0.0150	1.5416	1.5342	0.0074	1.4694
		3A	0.0000	1.6250	1.6100	0.0150	1.5438	1.5382	0.0056	1.4716
1-3/4-5	UNC	2A	0.0027	1.7473	1.7268	0.0205	1.6174	1.6085	0.0089	1.5019
		3A	0.0000	1.7500	1.7295	0.0205	1.6201	1.6134	0.0067	1.5016
1-3/4-8	UN	2A	0.0023	1.7477	1.7329	0.0150	1.6865	1.6590	0.0075	1.5943
		3A	0.0000	1.7500	1.7350	0.0150	1.6688	1.6632	0.0053	1.5966
1-7/8-8	UN	2A	0.0023	1.8727	1.8577	0.0150	1.7915	1.7838	0.0077	1.7193
		3A	0.0000	1.8750	1.8600	0.0150	1.7938	1.7881	0.0057	1.7216
2-4-1/2	UNC	2A	0.0020	1.9971	1.9751	0.0220	1.8528	1.8433	0.0095	1.7345
		3A	0.0000	2.0000	1.9780	0.0220	1.8557	1.8486	0.0071	1.7274
2-8	UN	2A	0.0023	1.9977	1.9827	0.0150	1.9165	1.9037	0.0078	1.8443
		3A	0.0000	2.0000	1.9850	0.0150	1.9188	1.9130	0.0058	1.8466

TABLE 5-3
DIN-13 METRIC THREAD (DEUTCHE INDUSTRIE NORM) EXTERNAL DIMENSIONS[1]

Size (mm)	Size (inch)	Pitch (mm)	Pitch (inch)	Class of Fit	Major Diameter Maximum (mm)	Major Diameter Maximum (inch)	Major Diameter Minimum (mm)	Major Diameter Minimum (inch)	Pitch Diameter Maximum (mm)	Pitch Diameter Maximum (inch)	Pitch Diameter Minimum (mm)	Pitch Diameter Minimum (inch)	Minor Diameter Nominal (mm)	Minor Diameter Nominal (inch)
M2	.0787	.4	.1574	Close	2.000	.0787	1.900	.0748	1.740	.0685	1.690	.0665	1.480	.0583
				Medium	2.000	.0787	1.900	.0748	1.740	.0685	1.690	.0653	1.480	.0583
				Free	2.000	.0787	–	–	–	–	–	–	–	–
M2.3	.0905	.4	.1574	Close	2.300	.0905	2.200	.0866	2.040	.0803	1.990	.0783	1.780	.0701
				Medium	2.300	.0905	2.200	.0866	2.040	.0803	1.960	.0771	1.780	.0701
				Free	2.000	.0905	–	–	–	–	–	–	–	–
M2.6	.1024	.45	.1771	Close	2.600	.1024	2.488	.0979	2.308	.0909	2.258	.0889	2.016	.0794
				Medium	2.600	.1024	2.488	.0979	2.308	.0909	2.228	.0877	2.016	.0794
				Free	2.600	.1024	–	–	–	–	–	–	–	–
M3	.1181	.5	.1968	Close	3.000	.1181	2.880	.1134	2.675	.1053	2.625	.1033	2.350	.0925
				Medium	3.000	.1181	2.880	.1134	2.675	.1053	2.595	.1022	2.350	.0925
				Free	3.000	.1181	–	–	–	–	–	–	–	–
M3.5	.1378	.6	.2362	Close	3.500	.1378	3.360	.1323	3.110	.1224	3.047	.1199	2.720	.1071
				Medium	3.500	.1378	3.360	.1323	3.110	.1224	3.010	.1185	2.720	.1071
				Free	3.500	.1378	–	–	–	–	–	–	–	–
M4	.1575	.7	.2755	Close	4.000	.1575	3.850	.1516	3.545	.1396	3.482	.1371	3.090	.1216
				Medium	4.000	.1575	3.850	.1516	3.545	.1396	3.445	.1356	3.090	.1216
				Free	4.000	.1575	–	–	–	–	–	–	–	–
M5	.1969	.8	.3149	Close	5.000	.1969	4.820	.1898	4.480	.1764	4.426	.1743	3.960	.1559
				Medium	5.000	.1969	4.820	.1898	4.480	.1764	4.380	.1724	3.960	.1559
				Free	5.000	.1969	4.820	.1898	4.480	.1764	4.320	.1701	3.960	.1559

[1]Courtesy ITT Harper Inc., Morton Grove, Illinois.

TABLE 5-3 (Cont.)

Size		Pitch (Distance Between Threads)		Class of Fit	Major Diameter				Pitch Diameter				Minor Diameter Nominal	
					Maximum		Minimum		Maximum		Minimum			
mm	inch	mm	inch		mm	inch	mm	inch	mm	inch	mm	inch	mm	inch
M6	.2362	1	.0393	Close	6.000	.2362	5.776	.2274	5.350	.2106	5.279	.2078	4.700	.1850
				Medium	6.000	.2362	5.776	.2274	5.350	.2106	5.238	.2062	4.700	.1850
				Free	6.000	.2362	5.776	.2274	5.350	.2106	5.170	.2035	4.700	.1850
M7	.2756	1	.0393	Close	7.000	.2756	6.776	.2668	6.350	.2500	6.279	.2472	5.700	.2244
				Medium	7.000	.2756	6.776	.2668	6.350	.2500	6.238	.2456	5.700	.2244
				Free	7.000	.2756	6.776	.2668	6.350	.2500	6.170	.2429	5.700	.2244
M8	.3149	1.25	.0492	Close	8.000	.3149	7.750	.3051	7.188	.2830	7.117	.2802	6.376	.2510
				Medium	8.000	.3149	7.750	.3051	7.188	.2830	7.076	.2786	6.376	.2510
				Free	8.000	.3149	7.750	.3051	7.188	.2830	7.008	.2759	6.376	.2510
M9	.3543	1.25	.0492	Close	9.000	.3543	8.750	.3445	8.188	.3224	8.117	.3196	7.376	.2094
				Medium	9.000	.3543	8.750	.3445	8.188	.3224	8.076	.3179	7.376	.2094
				Free	9.000	.3543	8.750	.3445	8.188	.3224	8.008	.3465	7.376	.2094
M10	.3939	1.5	.0590	Close	10.000	.3937	9.720	.3827	9.026	.3553	8.936	.3518	8.052	.3170
				Medium	10.000	.3937	9.720	.3827	9.026	.3553	8.886	.3498	8.052	.3170
				Free	10.000	.3937	9.720	.3827	9.026	.3553	8.802	.3465	8.052	.3170
M11	.4330	1.5	.0590	Close	11.000	.4330	10.720	.4220	10.026	.3947	9.936	.3912	9.052	.3564
				Medium	11.000	.4330	10.720	.4220	10.026	.3947	9.886	.3892	9.052	.3564
				Free	11.000	.4330	10.720	.4220	10.026	.3947	9.802	.3576	9.052	.3564
M12	.4724	1.75	.0688	Close	12.000	.4724	11.660	.4567	10.863	.4277	10.763	.4237	9.726	.3829
				Medium	12.000	.4724	11.600	.4567	10.863	.4277	10.703	.4214	9.726	.3829
				Free	12.000	.4724	11.600	.4567	10.863	.4277	10.613	.4178	9.726	.3829

TABLE 5-3 (Cont.)

Size		Pitch (Distance Between Threads)		Class of Fit	Major Diameter				Pitch Diameter				Minor Diameter Nominal	
					Minimum		Maximum		Minimum		Maximum			
mm	inch	mm	inch		mm	inch	mm	inch	mm	inch	mm	inch	mm	inch
M14	.5511	2	.0781	Close	14.000	.5511	13.525	.5325	12.701	.5000	12.601	.4961	11.402	.4489
				Medium	14.000	.5511	13.525	.5325	12.701	.5000	12.541	.4937	11.402	.4489
				Free	14.000	.5511	13.525	.5325	12.701	.5000	12.451	.4902	11.402	.4489
M16	.6299	2	.0781	Close	16.000	.6299	15.525	.6112	14.701	.5787	14.601	.5748	13.402	.5276
				Medium	16.000	.6299	15.525	.6112	14.701	.5787	14.541	.5724	13.402	.5276
				Free	16.000	.6299	15.525	.6112	14.701	.5787	14.451	.5689	13.402	.5276
M18	.7086	2.5	.0984	Close	18.000	.7086	17.440	.6866	16.376	.6447	16.276	.6408	14.752	.5808
				Medium	18.000	.7086	17.440	.6866	16.376	.6447	16.216	.6384	14.752	.5808
				Free	18.000	.7086	17.440	.6866	16.376	.6447	16.126	.6349	14.752	.5808
M20	.7874	2.5	.0984	Close	20.000	.7874	19.440	.7654	18.376	.7234	18.276	.7195	16.752	.6595
				Medium	20.000	.7874	19.440	.7654	18.376	.7234	18.216	.7172	16.752	.6596
				Free	20.000	.7874	19.440	.7654	18.376	.7234	18.126	.7136	16.752	.6595
M22	.8661	2.5	.0984	Close	22.000	.8661	21.440	.8441	20.376	.8022	20.276	.7983	18.752	.7383
				Medium	22.000	.8661	21.440	.8441	20.376	.8022	20.216	.7959	18.752	.7383
				Free	22.000	.8661	21.440	.8441	20.376	.8022	20.126	.7924	18.752	.7383
M24	.9448	3	.1181	Close	24.000	.9448	23.400	.9213	22.051	.8681	21.926	.8632	20.102	.8557
				Medium	24.000	.9448	23.400	.9213	22.051	.8681	21.851	.8603	20.102	.8557
				Free	24.000	.9448	23.400	.9213	22.051	.8681	21.736	.8543	20.102	.8557
M27	1.0629	3	.1181	Close	27.000	1.0629	26.400	1.0394	25.051	.9862	24.962	.9813	23.102	.9095
				Medium	27.000	1.0629	26.400	1.0394	25.051	.9862	24.851	.9784	23.102	.9095
				Free	27.000	1.0629	26.400	1.0394	25.051	.9862	24.736	.9739	23.102	.9095

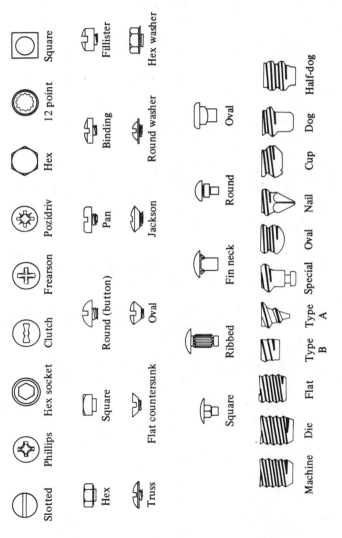

Figure 5-2 / Types of screw heads and points.

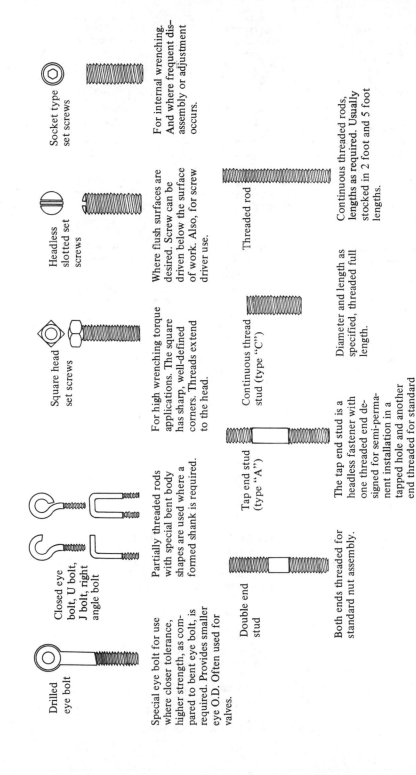

Drilled eye bolt

Special eye bolt for use where closer tolerance, higher strength, as compared to bent eye bolt, is required. Provides smaller eye O.D. Often used for valves.

Closed eye bolt, U bolt, J bolt, right angle bolt

Partially threaded rods with special bent body shapes are used where a formed shank is required.

Square head set screws

For high wrenching torque applications. The square has sharp, well-defined corners. Threads extend to the head.

Headless slotted set screws

Where flush surfaces are desired. Screw can be driven below the surface of work. Also, for screw driver use.

Socket type set screws

For internal wrenching. And where frequent disassembly or adjustment occurs.

Double end stud

Both ends threaded for standard nut assembly.

Tap end stud (type "A")

The tap end stud is a headless fastener with one threaded end designed for semi-permanent installation in a tapped hole and another end threaded for standard nut assembly.

Continuous thread stud (type "C")

Diameter and length as specified, threaded full length.

Threaded rod

Continuous threaded rods, lengths as required. Usually stocked in 2 foot and 5 foot lengths.

Figure 5-3 / Special types of fasteners.

261

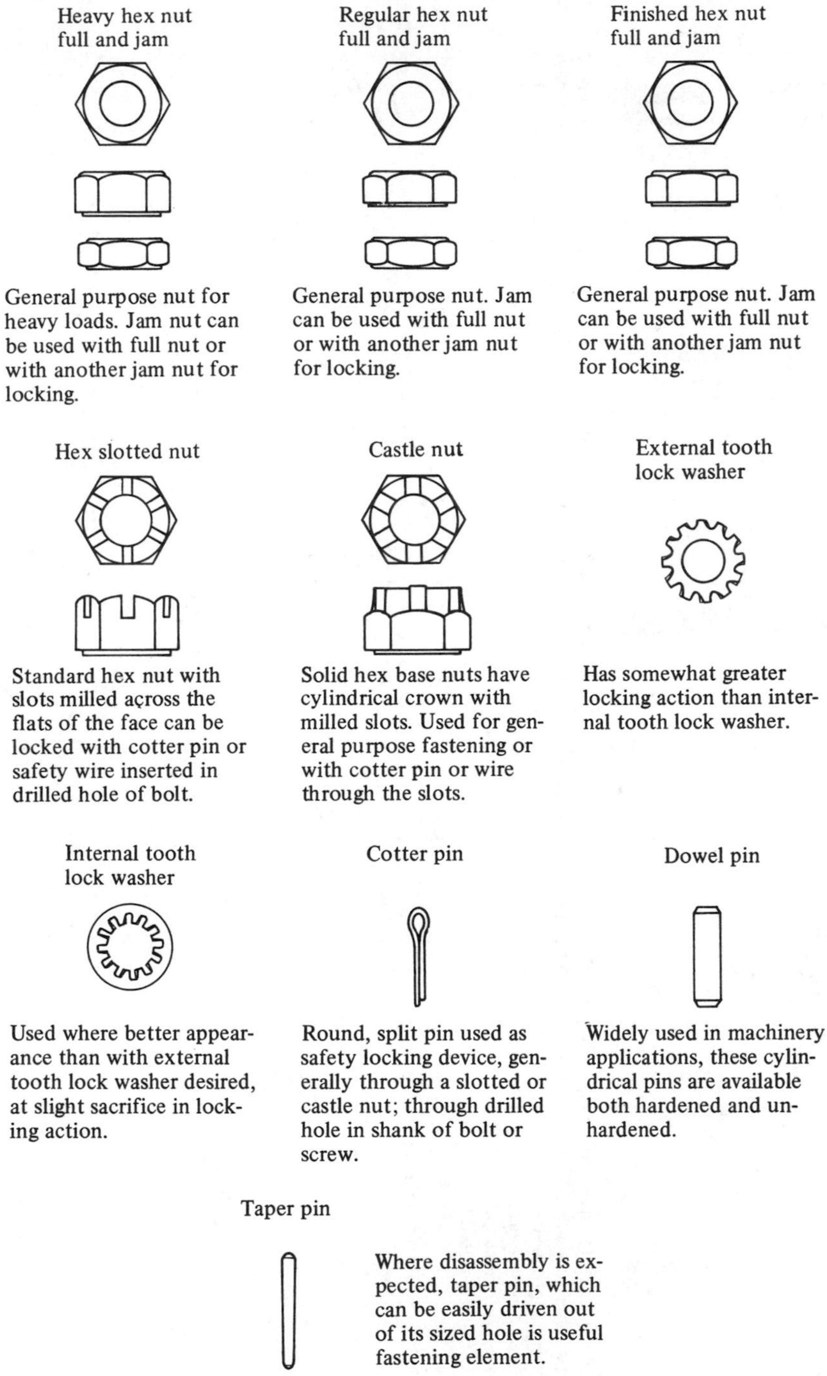

Heavy hex nut
full and jam

General purpose nut for heavy loads. Jam nut can be used with full nut or with another jam nut for locking.

Regular hex nut
full and jam

General purpose nut. Jam can be used with full nut or with another jam nut for locking.

Finished hex nut
full and jam

General purpose nut. Jam can be used with full nut or with another jam nut for locking.

Hex slotted nut

Standard hex nut with slots milled across the flats of the face can be locked with cotter pin or safety wire inserted in drilled hole of bolt.

Castle nut

Solid hex base nuts have cylindrical crown with milled slots. Used for general purpose fastening or with cotter pin or wire through the slots.

External tooth
lock washer

Has somewhat greater locking action than internal tooth lock washer.

Internal tooth
lock washer

Used where better appearance than with external tooth lock washer desired, at slight sacrifice in locking action.

Cotter pin

Round, split pin used as safety locking device, generally through a slotted or castle nut; through drilled hole in shank of bolt or screw.

Dowel pin

Widely used in machinery applications, these cylindrical pins are available both hardened and unhardened.

Taper pin

Where disassembly is expected, taper pin, which can be easily driven out of its sized hole is useful fastening element.

Figure 5-4 / Fastener components.

initial tightening load. Metal-to-metal connections are called *rigid* joints—
the bolt, stretched through tightening, retains the same amount of stretch
until a balance exists between the external load and the internal bolt force.
Such is not the case in a *flexible* joint—a gasketed connection—Fig. 5-5(a).
For analysis, assume the bolt to be tightened to an initial load F_o. The
free-body diagram, Fig. 5-5(b), shows a balance between the compressive

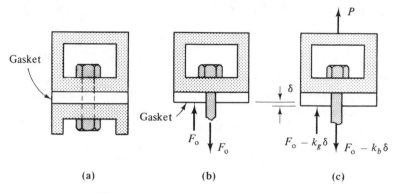

Figure 5-5 / Flexible joint with gasket.

force F_o on the gasket and the tensile load F_o on the bolt. Next, an external
force P is applied to the member, as shown in Fig. 5-5(c); the force in the
gasket diminishes by $k_g\delta$, and the bolt force increases by $k_b\delta$, where k_g and
k_b are the spring constants of the gasket and bolt respectively, and δ is the
equal change in length of both the bolt and gasket.

$$k_g = \frac{E_g A_g}{t}$$

$$k_b = \frac{E_b A_b}{L_b}$$

The equations of static equilibrium apply:

$$\Sigma F = 0$$
$$P = (F_o - k_g\delta) - (F_o + k_b\delta)$$
$$P = (k_b + k_g)\delta$$
$$\delta = \frac{P}{k_b + k_g}$$

The final bolt force F_f becomes

$$F_f = k_b\delta + F_o = \frac{k_b P}{k_b + k_g} + F_o \qquad (5\text{-}2)$$

Expressed in terms of the moduli of elasticity, areas and lengths, Eq. (5-2) becomes

$$F_f = \left(\frac{\dfrac{E_b A_b}{L_b}}{\dfrac{E_b A_b}{L_b} + \dfrac{E_g A_g}{t}} \right) P + F_o \qquad (5\text{-}3)$$

If the coefficient of P is expressed as a constant C, the former equation becomes

$$F_f = CP + F_o \qquad (5\text{-}4)$$

where C may have values lying between 0 and 1. For soft, spongy gaskets, $E_g A_g/t$ is insignificant compared to $E_b A_b/L_b$ and $C = 1$. For hard, thin gaskets or no gasket at all, $E_g A_g/t$ approaches infinity and $C = 0$; the bolt force F_f is then equal to the initial force F_o.

There are limitations to Eq. (5-4), if the term CP exceeds the initial bolt force, the joint separates, and the equation is no longer valid.

EXAMPLE 5-1

Find the final stress in the shank and in the thread root of a single $\frac{1}{2}$-13 UNC Class 3 steel bolt used to join the connection shown in Fig. 5-5. The following data apply:

$$
\begin{aligned}
\text{Bearing surface:} &\quad 2 \text{ sq. in.}\\
\text{Initial tightening stress}\\
\text{in shank of bolt:} &\quad 40{,}000 \text{ psi}\\
\text{Gasket thickness:} &\quad \tfrac{1}{8} \text{ in.}\\
\text{Gasket modulus of elasticity:} &\quad 0.25 \times 10^6 \text{ psi}\\
\text{External load:} &\quad 1000 \text{ lb}
\end{aligned}
$$

solution

Bolt shank area A:

$$A = \frac{\pi D^2}{4} = \frac{\pi \times \overline{0.50}^2}{4} = 0.196 \text{ in.}^2$$

Tensile area A_t at thread root by Eq. (5-1):

$$A_t = \frac{\pi}{4} \left(D - \frac{0.974}{n} \right)^2$$

$$= \frac{\pi}{4} \left(0.50 - \frac{0.974}{13} \right)^2 = 0.142 \text{ in.}^2$$

Initial force in bolt:

$$F_o = \sigma A = 40,000 \times 0.196 = 7840 \text{ lb}$$

Spring constants of bolt and gasket:

$$\frac{E_b A_b}{L_b} = \frac{30 \times 10^6 \times 0.196}{2} = 2.94 \times 10^6 \text{ lb/in.}$$

$$\frac{E_g A_g}{t} = \frac{0.25 \times 10^6 \times (2 - 0.196)}{0.125} = 3.61 \times 10^6 \text{ lb/in.}$$

Constant C by Eq. (5-3):

$$C = \frac{2.94 \times 10^6}{(2.94 \times 10^6) + (3.61 \times 10^6)} = 0.449$$

Final bolt force by Eq. (5-4):

$$\begin{aligned} F_f &= CP + F_o \\ &= (0.449 \times 1000) + 7840 \\ &= 8289 \text{ lb} \end{aligned}$$

Final stress in shank:

$$\sigma_b = F_f/A = 8289/0.196 = 42,300 \text{ psi}$$

Final stress in thread root:

$$\sigma_t = F_f/A_t = 8289/0.142 = 58,400 \text{ psi}$$

5-6 JOINT DESIGN—GENERAL CONSIDERATIONS

Experience and general know-how are two factors that cannot be used as coefficients in design formulas. For example, which is better, fewer small bolts or one large bolt in a given connection? Part of the answer lies in the relative costs of small versus large—four small bolts may be cheaper than one large bolt—small holes are less expensive to fabricate. Secondly, when a greater number of small bolts are used, compressive forces are more evenly distributed, particularly when gaskets are involved in the design. There is a limit, of course—a thousand clock screws would not make a reasonable replacement for two $\frac{1}{4}$-in. machine bolts.

Another general rule in good connection design is the avoidance, whenever possible, of combined loading—tension, bending, and shear.

Consider the three joints illustrated in Fig. 5-6. In Case I, the bolts are subjected to both tensile (tightening) forces and shear forces. Shear is eliminated in Case II by the "step" in the joining member; in a similar manner, well-fitted dowel pins, Case III, eliminate the shear load on the bolts.

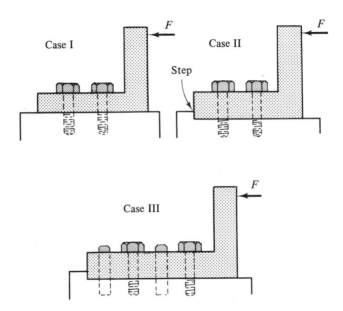

Figure 5-6 / Shear and tension in combined loading.

5-7   ECCENTRIC LOADING

Consider the gusset plate bolted to a machine frame in Fig. 5-7(a). Shear forces differ from bolt to bolt because of *direct* and *twisting* loads. The equivalent static loading on the bolts, Fig. 5-7(b), consists of a moment $M = 5P$ lb-in. and a direct load P, both acting through the centroid O of the bolt area. Viewing the free-body diagram of the bolts, Fig. 5-7(c), the direct shear forces are simply $P/3$, while the sum of the moments of P_1, P_2, and P_3 about O must balance the moment M. To proceed with the analysis, the centroid O is found by "area-moments" taken about the left bolt.

$$\bar{Y} = \frac{A_1 y_1 + A_2 y_2 + A_3 y_3}{\sum A}$$

$$= \frac{(A_1 \times O) + (A_2 \times 2) + (A_3 \times 9)}{A_1 + A_2 + A_3}$$

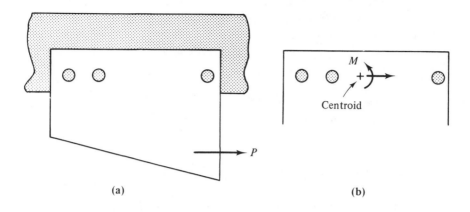

(a)   (b)

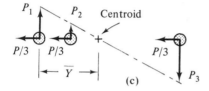

(c)

Figure 5-7 / Eccentrically loaded bolted connection.

If the bolts have equal cross-sectional areas, the distance to the centroid becomes $\bar{Y} = 9/3 = 3$ in. Summation of force moments about O gives

$$M = 5P = (3 \times P_1) + (1 \times P_2) + (6 \times P_3)$$

It is apparent from the drawing that the bolt to the far right is the most severely loaded; by geometry, P_1 and P_2 can be expressed in terms of P_3:

$$\frac{P_1}{3} = \frac{P_3}{6}; \qquad P_1 = \frac{P_3}{2}$$

and

$$\frac{P_2}{1} = \frac{P_3}{6}; \qquad P_2 = \frac{P_3}{6}$$

Substitution gives

$$5P = \frac{3P_3}{2} + \frac{P_3}{6} + 6P_3 = 7.67P_3$$

$$P_3 = \frac{5}{7.67} = 0.652P$$

The resultant shear force P_s on bolt number 3 is

$$P_s = \sqrt{(P/3)^2 + (0.652P)^2}$$
$$= 0.74P$$

With the tightening load given, Mohr's circle (Chap. 1) would be used to determine maximum tensile and shear stresses.

5-8   RIVETED CONNECTIONS

The traditional hammered hot, steel rivet, as a fastening device, is rapidly becoming a historical curiosity. Evolution, in the form of automation and design ingenuity, however, has produced new forms of rivets that are initially cheaper than screws, less costly to set, and extremely versatile. Rivets do have their limitations—size for size, they are weaker than screws and bolts, and rivet disassembly is difficult.

Some of the more common types of rivets are shown in Fig. 5-8.

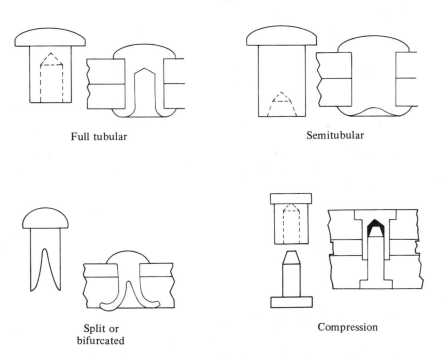

Full tubular   Semitubular

Split or
bifurcated   Compression

Figure 5-8 / Types of standard rivets.

Styles of blind rivets—rivets that are used when only one side of a joint is accessible—are described in Fig. 5-9.

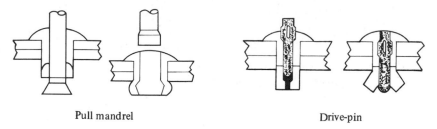

Pull mandrel   Drive-pin

Figure 5-9 | Types of blind rivets, used where one side of the work is inaccessible.

5-9 MULTIPLE CONNECTORS

When the line of action of a load P passes through the centroid of a group of multiple connectors (rivets or bolts), as in Fig. 5-10(a), it can be assumed that each connector carries an equal share of the load. Shear and bearing stresses are evenly distributed so that, for shear,

$$\tau = \frac{P}{nA_s} \qquad (5\text{-}5)$$

and for bearing,

$$\sigma_b = \frac{P}{nA_b} \qquad (5\text{-}6)$$

where A_s and A_b are shear and bearing areas respectively, and n is the number of connectors.

Tensile stresses in the joined members vary with bolt or rivet geometry, Fig. 5-10(d); bearing stresses on the members are numerically the same as on the connectors.

EXAMPLE 5-2

Determine the safe load P and the *efficiency* of the bolted lap-joint shown in Fig. 5-11. The bolts are standard $\frac{1}{2}$-13 UNC. The following permissible stresses apply: $\tau = 15,000$ psi, $\sigma_b = 35,000$ psi, and $\sigma_t = 25,000$ psi. (Neglect bending.)

solution

Since five bolts equally help support the load, each supports a force of $P/5$.

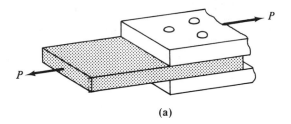

(a)

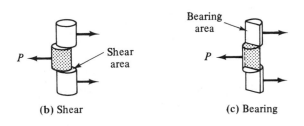

(b) Shear   (c) Bearing

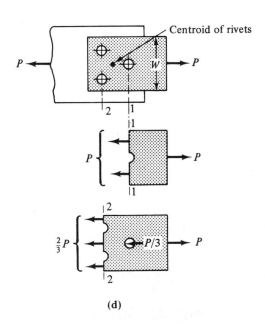

(d)

Figure 5-10 / Multiple connectors.

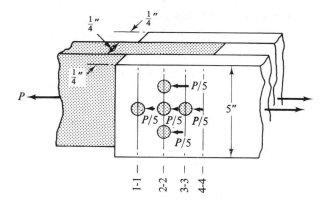

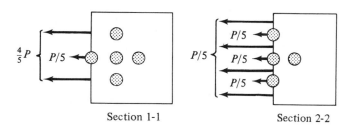

Section 1-1   Section 2-2

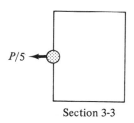

Section 3-3

Figure 5-11 / Bolted connection (Example 5-2).

Permissible bolt load (shear):

$$\frac{P_s}{5} = \tau A_s$$

$$P_s = 15{,}000 \times \pi/4 \times \overline{0.5}^2 \times 5$$

$$= 14{,}700 \text{ lb}$$

Permissible bolt load (bearing):

$$\frac{P_b}{5} = \text{stress} \times \text{projected area}$$

$$P_b = 35{,}000 \times \tfrac{1}{2} \times \tfrac{1}{4} \times 5$$

$$= 21{,}900 \text{ lb}$$

Permissible tensile load in connected member:

At section 1–1, $P_t = \sigma_t A_t$

$$= 25{,}000 \times (5 - 0.5) \times 0.25$$

$$= 28{,}100 \text{ lb}$$

At section 2–2, $\tfrac{4}{5} P_t = \sigma_t A_t$

$$P_t = \tfrac{5}{4} \times 25{,}000 \times [5 - (3 \times 0.5)] \times 0.25$$

$$= 27{,}300 \text{ lb}$$

At section 3–3, $\dfrac{P_t}{5} = \sigma_t A_t$

$$P_t = 5 \times 25{,}000 \times (5 - 0.5) \times 0.25$$

$$= 141{,}000 \text{ lb}$$

At section 4–4, $P_t = \sigma_t A_t$

$$= 25{,}000 \times 5 \times 0.25$$

$$= 31{,}300 \text{ lb}$$

Permissible values of load P in ranking order are:

BOLT	
shear:	14,700 lb
bearing:	21,900 lb
MEMBER	
section 2–2:	27,300 lb
section 1–1:	28,100 lb
section 4–4:	31,300 lb
section 3–3:	141,000 lb

Efficiency is a measure of the strength of the joint compared to the strength of the joined member, hence

$$\text{Eff} = \frac{14{,}700}{31{,}300} \times 100 = 47\%$$

5-10 WELDED CONNECTIONS

Gas, arc, and resistance welding technology has advanced to the point that welding is probably the most important single method of permanently joining metallic components. There are no size limits to welding; the joint is as strong as, or stronger than, the parent metal, and the process can be used with equal dependability in the shop or in the field.

Some of the more common types of welded joints are shown in Fig. 5-12. The terminology for a fillet weld is given in Fig. 5-13.

The throat of a fillet weld is the section of minimum area. For equal legs l, the shear area A_s in terms of weld length L is

$$A_s = lL \sin 45° = 0.707 \, lL$$

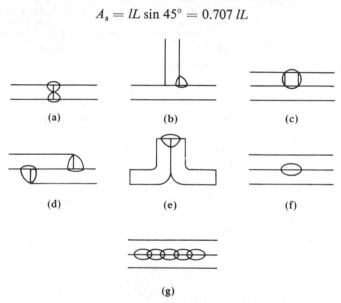

Figure 5-12 / Types of welded joints: (a) butt weld, (b) fillet weld, (c) plug weld, (d) lap weld, (e) edge weld, (f) spot weld, (g) seam weld.

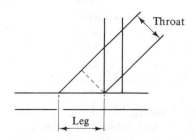

Figure 5-13 / Terminology for a fillet weld.

Usually when welding steels, either E60xx arc welding rods, such as 6010, 6011, 6027, etc., or E70xx rods such as 7010, 7018, etc., are used for butt and fillet welds. The first two numbers of the rod designation give the minimum tensile stress of the deposited weld metal. Thus 6010 or 6014 rods have a guaranteed minimum tensile strength of 60,000 psi, and 7010 or 7018 rods have 70,000 psi. Steels with yield strengths below 40,000 psi, such as A7, A36, or A373 structural steels, all with yield strengths below 40,000 psi, are usually welded with 60xx rods; steels with yield strengths above 40,000 psi, such as A441, are welded with 70xx rods. Allowable shear stress on the throat of a fillet deposited by these rods is given in the following:

ROD TYPE	ALLOWABLE STRESS ON THROAT	ALLOWABLE STRESS ON LEG
60xx	18,000 psi	12,700 psi
70xx	21,000 psi	14,800 psi

These allowables correspond to 800 psi and 925 psi per inch of weld length per $\frac{1}{16}$ in. of leg for 60xx and 70xx rods. For example, a fillet weld with a $\frac{3}{16}$-in. leg deposited with 60xx rods has an allowable stress of 3×800 or 2400 lb per inch of weld length; a $\frac{1}{4}$-in. leg deposited from 70xx rods has an allowable stress of 3700 lb per inch of weld length. These allowable stresses are shear stresses, since a fillet weld will always fail in shear. The allowable shear loads per inch of weld for various fillet sizes are:

FILLET SIZE (in.)	DESIGN LOAD (lb per in.)	
	60xx RODS	70x RODS
1/8	1600	1850
3/16	2400	2775
1/4	3200	3700
5/16	4000	4625
3/8	4800	5550
1/2	6400	7400
5/8	8000	9250
3/4	9600	11,100

For transverse loading, Fig. 5-14, the throat of the weld is still assumed to be stressed in shear, even though the welded member, itself, is in tension. The same table of strengths can be used by simply assuming a

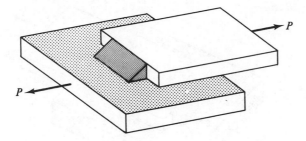

Figure 5-14 | The tension load is taken in shear on the throat of the weld.

30% stronger joint. Thus, a double-welded transverse joint with $\frac{1}{4}$-in. fillets (60xx rods) would have a design strength of

$$3200 \times 2 \times 1.3 = 8320 \text{ lb per in. of weld}$$

Actual testing has shown that 100% *plate strength welds* can be achieved by welding fillets on both sides of the plate and for full plate lengths in accordance with the following:

PLATE THICKNESS (in.)	FILLET SIZE (in.)
3/16	5/32
1/4	3/16
5/16	1/4
3/8	5/16
7/16	3/8
1/2	7/16
5/8	1/2
3/4	5/8
1	3/4

EXAMPLE 5-3

Two plates are joined by three sections of $\frac{1}{4}$-in. fillet weld and use 6027 welding rods as shown in Fig. 5-15(a). What length L is required if the joint is to have 100% plate strength? Assume an allowable tensile strength of the plate to be 20,000 psi.

solution

The "working" load P is first determined.

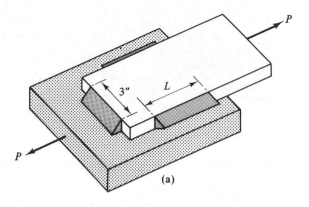

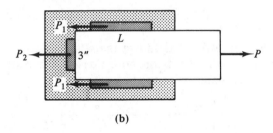

Figure 5-15 / Welded connection (Example 5-3).

$$P = \sigma A = 20,000 \times 5 \times \tfrac{3}{8} = 37,500 \text{ lb}$$

Equations of static equilibrium are next applied to the free-body diagram, Fig. 5-15(b),

$$\sum F = 0$$
$$2P_1 + P_2 = P$$

Substitution of numerical values gives

$$(2 \times 3200L) + (3200 \times 3) = 37,500$$
$$L = 4.35$$

Use $4\tfrac{1}{2}$ inches.

For any plate thickness, there is a minimum size leg of the fillet weld that should be deposited. These minimum sizes are:

PLATE THICKNESS	MINIMUM FILLET LEG
0–1/2″	3/16
9/16–3/4″	1/4
13/16–1$\frac{1}{2}$″	5/16
over 1$\frac{1}{2}$–2$\frac{1}{4}$″	3/8
over 2$\frac{1}{2}$–6″	1/2

The reason for these minima is the following. Steels are highly conductive to heat. If a weld that is too small is deposited on a heavy plate, the great mass of cold plate quenches the weld metal, making it hard and somewhat brittle. A minimum leg size is accompanied by a minimum amount of welding heat input serving to delay this sudden cooling effect.

At the other extreme, too large a fillet weld should be avoided. Large fillet welds require excessive heat input in order to be deposited, resulting in distortion of the weldment. It is preferable to deposit a smaller fillet using a longer weld length to obtain the required shear strength. For example, a leg length of $\frac{1}{2}$ in. on a $\frac{3}{4}$-in. plate thickness is preferable to a fillet leg equal to the plate thickness.

5-11 SEALS

Leakage control is critically involved in practically all mechanical systems, from the prevention of the kitchen faucet drip to the protection of vehicles and their occupants against the rigors of outer space. Sealing problems are not new, they are just becoming more difficult. Mechanical systems are required to function under a variety of sometimes opposing conditions: high temperatures, low temperatures, high pressures, vacuum, static loading, dynamic loading, chemical attack, electrochemical attack, a moment of use, a lifetime of use.

As with all fields, a language of the industry has evolved. The most frequently encountered terms and phrases associated with seals are defined as follows:

Absorption. Physical mechanism by which one substance attracts and takes up another substance within itself.

Adsorption. Physical mechanism by which one substance attracts another substance to its surface and through molecular forces causes the incident substance to adhere thereon.

Antirad.   A material that inhibits radiation damage.

Back-up Ring.   Washer-like device installed on the high-pressure side of an O-ring.

Compression Set.   Failure of a material to return to its original shape after release of a compressive load.

Durometer.   Instrument for measuring hardness of rubber and rubber-like materials.

Elastomer.   Any of various polymers having elastic properties of natural rubber.

Gasket.   Device used to retain fluids under pressure or to seal out foreign matter—normally refers to a static seal.

Gland.   Cavity into which a seal is installed—includes the groove and mating surface which together confine the seal.

Hermetic Seal.   An airtight seal.

Leakage Rate.   Rate at which a gas or liquid passes, by diffusion, and escapes a given barrier.

Media.   Fluids or gases being sealed.

Oil Swell.   Increase in volume of a material due to absorption of oil.

O-Ring.   Belt of circular cross section which effects a seal through squeeze and pressure.

Packing.   Flexible material used to seal pressurized fluids, normally under dynamic conditions.

Packing Groove.   Cavity, in one member of a concentric joint, that accommodates a packing.

Polymer.   Material formed by chemically joining several monomers (molecules bound together as a unit).

Rubber, Natural.   Raw or crude rubber obtained from vegetable sources.

Rubber, Synthetic.   Manmade (synthesized) elastomers.

Seal.   Device used to prevent leakage of media.

Static Seal.   Seal that functions between mating parts that do not move relative to one another.

Swell.   Volume increase in seal caused by contact with media.

Virtual Leak.   Apparent traceable leak in a vacuum system.

Wiper Ring.   Device used to remove excess fluid or contaminants from reciprocating member before reaching seals or packing.

5-12 *STATIC SEALING*

Four main static sealing methods are predominant in modern design—flat gaskets, O-ring-in-groove seals, Gask-O-Seals®,[1] and air or heat or catalyst-curing formed-in-place elastomers. The oldest, simplest, and least costly is the flat gasket (Fig. 5-16), which has been made from a variety of materials, such as paper, asbestos, cork, or rubber. Flat gaskets, however, are subject to disintegration due to absorption, sudden rupture, seepage, and cold flow.

The O-ring-in-groove seal (Fig. 5-16), while more costly than the flat gasket, affords a seal with little or no leakage. It is best suited to circular grooves, since the O-ring itself is an elastic circular member.

In the Gask-O-Seal® (Fig. 5-16), an elastomer is permanently held in a metal retainer. Under pressure of assembly, Fig. 5-17, the elastic

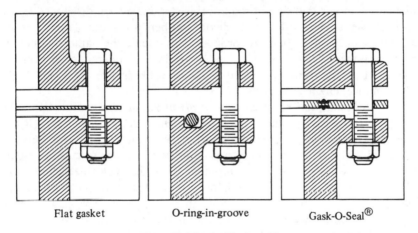

Flat gasket   O-ring-in-groove   Gask-O-Seal®

Figure 5-16 / Static seals.

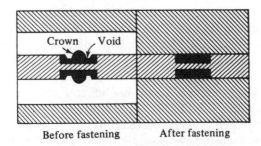

Before fastening   After fastening

Figure 5-17 / Assembly of a Gask-O-Seal.®

---

[1]Gask-O-Seal®; Parker Seal Co., Culver City, California.

portion is deformed from a round to a square or oblong shape. Bolt loads are relatively light, ranging from 30 to 150 pounds per linear inch of seal.

Formed-in-place seals have become increasingly popular, particularly when unique physical properties are called for in a given design. The most prominent sealing materials are the silicone rubbers—an elastomer which can both withstand temperature extremes and can function in hostile environments.

Three basic types of silicone rubbers[2] in use can be characterized by the method used to cure them:

Heat-vulcanized silicone rubbers are cross-linked by the application of heat after the addition of a peroxide vulcanizing agent or catalyst. Vulcanization can take place in a mold, in a hot air vulcanizer, or in an autoclave. The reaction is fast: with small parts, vulcanization takes about 2 or 3 minutes. Minimum vulcanization temperature is about 240°F. Although many silicone rubbers require a post cure to optimize properties, silicone materials that do not require this step have been developed in the past few years. Heat-vulcanized silicone rubbers are available in a variety of stocks or gums. They are processed in much the same manner as organic rubbers. They may be molded, extruded, calendered, or dispersion-coated. Usually, heat-vulcanized silicone rubbers are used for high-performance, engineered parts.

Silicone rubbers which cure in air at room temperature after the addition of a catalyst are called *room-temperature-vulcanizing* (RTV) *rubbers*. These materials are liquids and pastes available in a wide range of consistencies. Because they need not be heated in order to cure, two-part RTVs require less processing equipment than heat-vulcanized materials. Their physical properties, although lower than those of the heat-vulcanized materials, have been vastly improved in recent years. Two-part silicone rubbers are widely used for making molds for tools and prototype parts, and as electrical potting and encapsulating materials.

One-part RTV silicones do not require the addition of a catalyst. Moisture in the air triggers the curing reaction as soon as the material is forced from caulking guns or tubes. One-part RTVs are also available in a wide range of consistencies. The more fluid materials are used as coatings, while the more viscous ones are widely used as caulks and sealants. These sealants are highly weather-resistant and function well even at extreme temperatures. Their rubbery nature has made them especially useful in bonding and sealing windshields, attaching rear-view mirrors, and in sealing drip rail joints on vinyl-covered tops.

Silicone rubbers have generally poor physical properties at room

---

[2]Detailed information through the courtesy of the Dow Corning Corporation, Midland, Michigan.

temperature but may be outstanding in low- and high-temperature applications. Strength, resistance to tearing, and elongation of the silicones are all low when compared with the same properties of other rubbers, and they "cold flow" (deform or compress) excessively. Therefore, while outstandingly successful in many applications, they are sometimes ineffective substitutes in more routine applications of gasketing.

The characteristics of the heat-vulcanized silicone rubbers are summarized in Table 5-4. The Durometer hardness numbers indicate that silicone rubbers are generally soft. Everyone is familiar with the hardness of an automobile tire: it has a Durometer hardness of 70. The silicone rubbers in the table are in the range of 50 to 60, indicating a soft rubber. Similarly, silicone rubber tensile strengths of 1000 psi or less compare unfavorably with almost all other rubbers. Compression set (cold flow) as high as 30 or 40% indicates that these rubbers can rapidly become permanently deformed by pressure. However, oil resistance and service temperature range are far beyond the performance obtainable in other rubbers.

5-13 DYNAMIC SEALS

As with many machine components, the engineer is seldom required to initiate a seal design "from scratch." Hundreds of manufacturers carry as stock items an almost endless variety of special-purpose seals. Off-the-shelf components, such as gears, screws, seals, belts, bearings, and transmissions, when applicable, save time, energy, and money. The designer's major role involves first, a knowledgeable statement of requirements, and second, a search for the appropriate item. Major factors in seal selection involve:

1 / Compatibility between the seal and the surrounding media.

2 / Temperature or temperature range anticipated.

3 / Pressure or pressure fluctuations to be encountered.

4 / Motion—rotational or translational.

5 / Relative velocities between components.

6 / Life expectancy.

5-14 O-RINGS

Of the several types of dynamic seals (it is also used as a static seal), perhaps the O-ring is the most versatile and useful. It is also one of the few engineering components capable of accommodating abuse and faulty design.

TABLE 5-4

PROPERTIES OF TYPICAL SILASTIC® SILICONE RUBBERS[1]

Classification	Major Uses	Typical Properties							Oil Resistance (percent swell 70 hrs. at 300 F)		
		Service Temperature Range (degrees F)	Durometer Hardness, (Shore A) (points)	Tensile Strength (psi)	Elongation (percent)	Tear Strength (Die B) (pp)	Compression Set (percent after 70 hrs. at 300 F)	Brittle Point (degrees F)	ASTM No. 1	ASTM No. 3	Silastic Silicone Rubber
General purpose	Clamp covers, closures, boots, diaphragms, bellows, gasketing	−70 to 500	50	950	300	80	30	−100	+8	+55	50
High performance	Shock mounts, boots, dust covers, form-ing blankets, belting, gaskets, seals, O-rings, sheet goods, diaphragms	−0 to 450	50	1350	600	190	29	−115	+7	+42	55U

[1]SILASTIC®—Brand name for silicone rubber made by Dow Corning Corporation, Midland, Michigan.

TABLE 5-4 (Cont.)

Typical Properties

Classification	Major Uses	Service Temperature Range (degrees F)	Durometer Hardness (Shore A) (points)	Tensile Strength (psi)	Elongation (percent)	Tear Strength (Die B) (pp)	Compression Set (percent after 70 hrs. at 300 F)	Brittle Point (degrees F)	Oil Resistance (percent swell, 70 hrs. at 300 F) ASTM No. 1	Oil Resistance ASTM No. 3	Silastic Silicone Rubber
Extreme high temperature	Gaskets for oven and autoclave doors; processing hot glass and metal	−70 to 600	80	700	175	85	35	−100	+6	+35	82
Extreme low temperature	Aircraft seals and gaskets, missile seals, bellows	−130 to 500	50	950	400	100	21	−178	+10	+65	651
Low compression set	O-rings, grommets, high temperature gaskets, rubber-covered rolls	−70 to 500	60	975	230	70	12	−100	+6	+33	746U
Fuel, oil and solvent resistant	O-rings, oil seals, tubing for hydraulic systems, boots	−70 to 350	60 / 50	1100 / 1250	200 / 500	70 / 265	40 / 20	−90 / −87	+1 / Nil	+5 / +4	LS-63U / LS-2332U

While most O-rings are of nitrile rubber with a circular cross section, other rubbers and other cross sections are sometimes employed. Some variants of the O-ring are shown in Fig. 5-18. The rectangular ring gives slightly higher friction than the O-ring but is well suited to low-speed applications. The advantage of the quad-ring and H-ring over the O-ring is the resistance of these shapes to being twisted by the reciprocating rod against which these rings seal.

Harder O-rings are used for high-pressure sealing. A Durometer hardness of 70 is suited to low-pressure applications, but for high pressures such as 3000 psi, a Durometer hardness of 90 would be more resistant to distortion and extrusion.

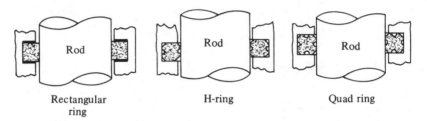

Figure 5-18 | Dynamic seals that serve the same function as an O-ring.

5-15 DESIGN OF O-RING GLANDS

The groove into which an O-ring fits is called a *gland*. The gland may be cut into either the male or the female member if the seal is a dynamic seal. For either static or dynamic sealing, the following procedure is used to design an O-ring seal.

For illustration, assume that an O-ring is to be seated in the gland of a piston in a hydraulic cylinder. The inside diameter of the cylinder must be known, as well as the clearance between piston and cylinder. The cross-sectional diameter of the O-ring is chosen. Two rules are followed:

1 /   The depth of the gland should be such as to squeeze the O-ring between 20 and 30% of its diameter.

2 /   The width of the O-ring gland should be the diameter of the O-ring plus twice the above squeeze.

Suppose the I.D. of the cylinder is 4.000 in., the clearance between piston and cylinder is 0.008 in. on diameter (0.004 in. on radius), and the O-ring is 0.25 in. in diameter. The ring O.D. of the O-ring will be 4 inches. Suppose we select a squeeze of 20%. Then 20% of 0.25 = 0.050 in. The O-ring must be compressed to 0.250–.050 or 0.200 in.

Then the depth of the gland must be 0.200 in. less the radial clearance:

$$0.200 - 0.004 = 0.196 \text{ in.}$$

The width of the gland must be

$$0.250 + (2 \times 0.05) = 0.350 \text{ in.}$$

See Fig. 5-19 for further details of the gland. The sharp edges must be removed ("break corners"), and a slight taper is advantageous.

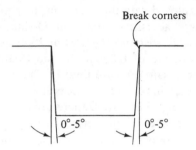

Figure 5-19 / O-ring gland.

If the gland were cut into the cylinder wall, the procedure would be identical.

The surface finish of the part on which the O-ring rubs must be a standard ground finish (a surface finish of about 32 micro-inches). If the finish is rough, the O-ring will be abraded, but if the finish is too smooth, oil will not attach to it, and the O-ring will then lack lubrication. O-rings must be oil-lubricated.

O-ring dimensions change slightly in service through two effects which oppose each other: compression set and swell. Compression set is the flattening of the compressed O-ring over a prolonged period of time. Compression set increases with temperature but can be roughly estimated as 10% of the cross-sectional diameter. Volume change or swell is the increase in size of a rubber occurring as a result of contact with and absorption of hydraulic fluid. Twenty percent is the maximum allowable swell for O-ring rubbers.

If the compression set is 12% and the swell is 15% for the above O-ring example, the final squeeze would be

$$\begin{aligned}
&\text{initial squeeze} + \text{swell} - \text{compression set} \\
&= \quad 20\% \quad + 15\% - \quad 10\% \quad = 25\%
\end{aligned}$$

Twenty-five percent is within the maximum allowable squeeze of 30%.

Two O-ring accessories are sometimes employed for the support of the O-ring against high pressures. Both are flat or contoured rings of rubber, leather, or felt. If the accessory ring is located in the O-ring groove on the high-pressure side of the O-ring, it is referred to as a *back-up ring*. The combination of O-ring and back-up ring can withstand much higher pressures than an O-ring alone, since a portion of the total pressure is sustained by the back-up ring. If the same ring is located in the O-ring groove on the low-pressure side of the O-ring, the accessory ring is referred to as an *anti-extrusion ring*, since it will prevent the O-ring from extruding into the clearance space. Sometimes both a back-up and an anti-extrusion ring are employed with the O-ring in the gland (Fig. 5-20).

Without these accessory rings, an O-ring with a Durometer hardness of 70 can easily sustain pressures of 2000 psi over long periods with a diametral clearance of 0.008 in. If the clearance is reduced to 0.004 in. on diameter, a safe pressure limit is 3000 psi. With a diametral clearance of 0.004 in., it will require a static pressure of about 20,000 psi to blow or extrude the O-ring. The O-ring therefore is an exceedingly accommodating and reliable device. It will even seal after pieces of the O-ring are broken off. It must, however, never be installed or operated without lubrication.

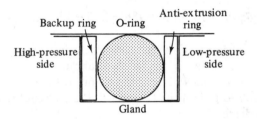

Figure 5-20 / O-ring gland with supporting rings.

5-16 EXCLUSION SEALS

The exclusion seal is usually a wiper or scraper. Figure 5-21 shows a wiper against the rod of a hydraulic cylinder. As its name implies, the purpose of this type of seal is to exclude, that is, to prevent the entry of foreign matter into machine parts. The wiper of this figure protects the rod seals of the cylinder from damage by abrasive particles that may be drawn into the cylinder by the rod. Such rod wipers are frequently made of a suitable synthetic rubber, such as polyurethane, which is tough, strong, and abrasion-resistant, and has an unusually low friction coefficient for a rubber.

Another type of exclusion seal is the V-ring of Fig. 5-22. The V-ring is an annular rubber with a tapered sealing lip extending conically

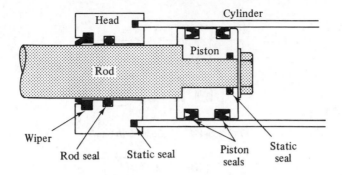

Figure 5-21 / Seals in a hydraulic cylinder.

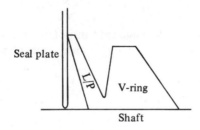

Figure 5-22 / V-ring seal attached to a shaft.

outward. Sealing pressure of the lip against the counter-surface is kept low to reduce frictional heat and wear.

A sealing wiper is a double-ended seal that serves as a wiper at one end and a seal at the other. Such a component is shown in Fig. 5-23.

The types of exclusion seals appropriate to sealed ball and roller bearings are discussed in Chap. 9.

Figure 5-23 / Double-ended wiper/seal combination.

5-17 LABYRINTH SEALS

The labyrinth seal of Fig. 5-24 has a positive clearance and thus a small leakage rate. The leakage is kept to a very small amount by the labyrinthine path of small cross section through which the fluid must move.

The number of rings required in the labyrinth can be found from the following empirical formula if the leakage rate is first decided.

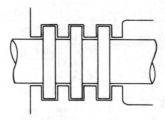

Figure 5-24 / Labyrinth seal.

$$N = \frac{(40P - 2600)\left(\dfrac{W}{A}\right)}{540\left(\dfrac{W}{A}\right) - P}$$

where N = number of rings
         W = permissible leakage rate in pounds fluid per second
         A = cross section of leakage path, square inches
            = $C(\pi D)$
         C = clearance, inches
         D = diameter, inches
         P = absolute pressure

5-18 MECHANICAL END-FACE SEALS

A mechanical end-face seal is a seal that prevents escape of fluid at a rotating shaft by the use of two faces in contact, one mounted on the rotating shaft and the other on the housing. Spring pressure keeps the two surfaces in contact, as in Fig. 5-25. The contacting faces must be carefully

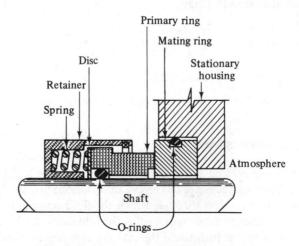

Figure 5-25 / Mechanical end-face seal.

lapped to be ultraflat if any leakage is to be prevented. Such a seal causes no shaft wear, has no leakage, and can seal against pressure.

5-19 OTHER SEALS AND APPLICATIONS

Figure 5-21 is a view of a hydraulic cylinder showing the dynamic seals between piston and cylinder, a compression packing sealing the rod of the cylinder, and a rod wiper for protection of the rod. A variety of seals and wipers are available for such locations and functions, though the most common types of dynamic seals are lip seals, U-seals, cup seals, V-ring packings, and O-rings. The first three types are self-energizing, meaning that pressure expands them to seal more firmly. Several of these seals are illustrated in Fig. 5-26. The self-energizing seal of course can seal in only

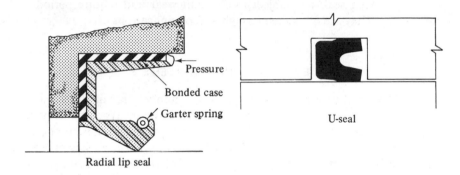

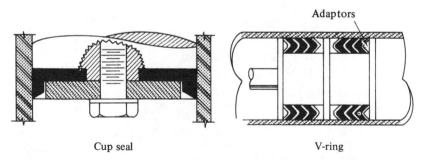

Figure 5-26 / Various types of seals.

one direction; a double-acting cylinder requires that its piston be sealed for both directions of movement.

Self-energizing (pressure-energized) seals are assembled with a slight interference fit for positive sealing effect. Such seals must be made of soft materials such as felt, rubber, and leather. They are easily damaged and in time will wear out. But they are inexpensive and readily replaced. The design of the gland should make provision for ease of replacement.

The self-energizing effect produces higher friction forces at higher pressures. The U-seal can be used only to 1500 psi; at higher pressures, friction force is sufficient to abrade the lip.

The packing type of seal of Fig. 5-26 uses a number of chevron rings of asbestos, fiber, leather, or other soft material, packed into the gland under light pressure, the opening of the "vee" pointing in the direction of pressure. Though this type of seal is somewhat self-energizing, the friction force does not increase very greatly at higher pressures. Such packing seals are well-suited to low- or high-pressure applications. Their disadvantage is a tendency to score the shaft, because foreign particles are readily embedded in this type of packing. A hard shaft must therefore be used against such packing.

Packing seals are subject to constant wear and require periodic adjustment. They are not suited to high-speed applications.

QUESTIONS AND PROBLEMS

P5-1   An additional tensile safety factor is implied in Eq. (5-1)—normally it would be assumed that the minimum (root) area of a threaded bolt would be the proper area to use in computations involving tensile strength. Experimentation, however, has proved this assumption overly conservative. Determine the magnitude of this safety factor for the following threads:
(a) $\frac{1}{4}$-20 UNC Class 2A
(b) $\frac{1}{4}$-28 UNF Class 2A
(c) $1\frac{1}{4}$-7 UNC Class 3A
(d) $1\frac{1}{4}$-12 UNC Class 3A
(e)   M-7 DIN Metric medium-fit
(f)   M-27 DIN Metric close-fit

P5-2   Calculate the approximate weight per 100 of the steel socket head cap screw shown.

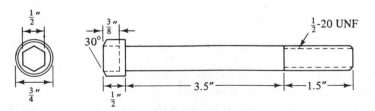

Figure P5-2

P5-3   Two $\frac{3}{8}$-24 UNC Class 2A steel bolts, stressed to 15,000 psi in their unthreaded shank, are used to join the rigid machine members *A* and *B* as shown. Determine (a) the average pressure between the mated surfaces; and (b) the separation force *P*.

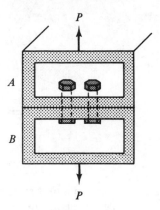

Figure P5-3

P5-4   Determine the approximate added tensile stress in the bolts of Prob. 5-3 if the nuts are each tightened an additional quarter turn.

P5-5   Assume that a $\frac{3}{16}$-in. gasket ($G = 40 \times 10^4$ psi) is used between members *A* and *B* in the pressure connection shown. The four steel M10 DIN Metric medium-fit cap screws are tightened to an initial stress of 15,000 psi each. Plot a graph that will show bolt stress as a function of load *P*.

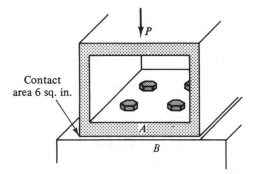

Figure P5-5

P5-6   The circular gasket shown consists of a hard elastomer ($E = 0.5 \times 10^6$ psi) sandwiched between two rings of a magnesium alloy ($E = 6.3 \times 10^6$ psi). Determine the effective spring constant of the composite section.

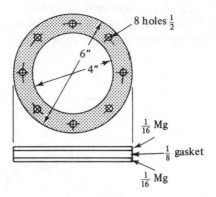

Figure P5-6

P5-7 A gasketed inspection hole in a steel pressure vessel is capped as shown; the lid is held in place with twelve $\frac{3}{8}$-in. steel bolts having an initial tightening load of 2500 lb each. Determine the bolt load if, after capping, the vessel is pressurized to 100 psi. $E = 10^5$ for the gasket.

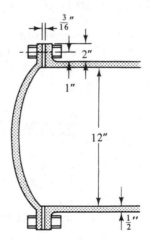

Figure P5-7

P5-8 In tightening, a tensile load of 2000 lb and a torsional load of 1000 in.-lb are applied to a M15 DIN Metric medium-fit bolt. Determine the principal stresses developed in the thread root.

P5-9 Compare the load-carrying capacity of the two bolt configurations shown. Five-eighth-in.-shank diameter bolts are used in each case, and the working stress in shear is 20,000 psi.

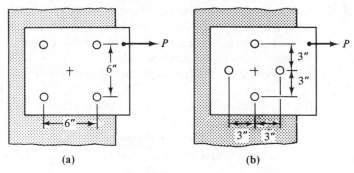

Figure P5-9

P5-10   A steel bar and an aluminum bar, bolted to rigid end members
as shown, are fastened at their free ends by a 1-in.-diameter steel
pin. Determine the shearing stress in the pin and the appropriate
brass and steel bolt (UNC) sizes if the temperature of the joined
bars drops 50°F.

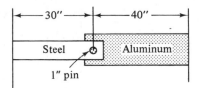

Steel area: 1 sq in.

Aluminum area: 2 sq in.

Figure P5-10

P5-11   Two steel sections are joined with 60xx as shown. Determine the
minimum weld lengths if the U-member carries a centroidal load
P that imparts a stress of 20,000 psi in the section.

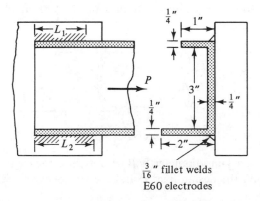

$\frac{3}{16}$ " fillet welds
E60 electrodes

Figure P5-11

P5-12 A 6 × 8-in. rectangular steel tube is joined to a vertical member with 6-in. top and bottom welds as shown. The effect of the moment is to produce a tensile stress on the top weld and a compressive stress on the bottom weld.

(a) Determine the leg width of the welds using 6027 welding rods.

(b) Determine the leg width of the welds using 7024 rods.

(c) If 70xx rods are 10% more expensive than 60xx rods, which rod provides the lower weld cost?

(d) Is the amount of weld metal deposited in a fillet weld proportional to leg length?

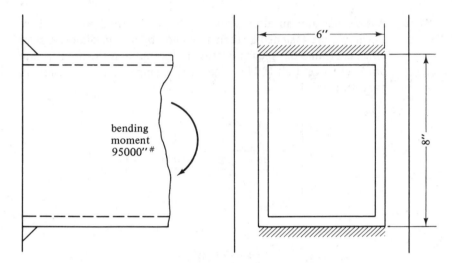

bending
moment
95000″#

6″

8″

Figure P5-12

P5-13 Telescoping tubes are to be connected to form a joint as shown. The designer must select one of three methods: blind riveting, welding, or bolting. Discuss the factors involved in reaching a decision.

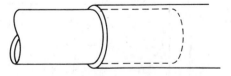

Figure P5-13

P5-14 Seven ½-in.-diameter bolts are used to secure the plates as shown. Find the design load P and the efficiency of the connection if the following stresses apply for both the bolts and the members:

Shear: 15,000 psi

Bearing: 40,000 psi

Tension: 20,000 psi

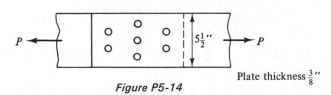

Figure P5-14

Plate thickness $\frac{3}{8}''$

P5-15   Plot a graph that will show the permissible load *P* that the connection illustrated can carry as a function of eccentricity 0 in ≤ *e* ≤ 10 in. The shear load on the most severely loaded bolt must not exceed 1000 lb.

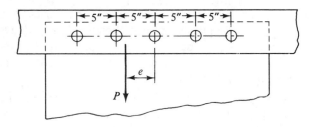

Figure P5-15

P5-16   Determine the elongation when a $\frac{1}{2}$-13 UNC Class 2A steel bolt is tightened, as shown, to produce a direct tensile stress of 30,000 psi. Assume that the effective length of the threaded section includes the first, second, third, and one-half of the fourth engaged thread in the nut.

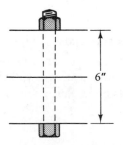

Figure P5-16

P5-17 Explain why the tensile load is always equal to the tightening load when two rigid members are joined, regardless of the separating force, as long as the rigid members do not part.

P5-18 For the following worked examples in this chapter, convert all data to SI units and rework the examples:
(a) Example 5-1
(b) Example 5-2
(c) Example 5-3

P5-19 For the following problems, convert all units to SI units and solve the problems:
(a) P5-9
(b) P5-11
(c) P5-12
(d) P5-14

SUGGESTED DESIGN PROJECTS

DP5-1 A typical piston seal installation for an *oscillating-piston/fixed-cylinder* hydraulic assembly is shown in Fig. 5-21. Devise a laboratory test set-up that will provide data for the determination of frictional energy losses as a function of piston speeds, cylinder pressure, and media.

DP5-2 As a designer, part of your job involves the searching and pricing of off-the-shelf components for a particular machine. List at least five manufacturers of each of the following:
(a) O-Rings
(b) Radical lip seals
(c) Clearance seals
(d) Asbestos packing
(e) Diaphragm seals
(f) Self-sealing screws
(g) Locknuts
(h) Cap screws
(i) Blind rivets
(j) Weld screws and studs

DP5-3 Using whatever library research material you can find, prepare a table that will show heat, water, flame, electrical, acid, and oil resistances (rate as poor, fair, good, or excellent) of the following sealing materials:
(a) Natural rubber
(b) Nitrile
(c) Butadiene

(d)   Ethylene propylene
(e)   Isoprene
(f)   Polyurethane
(g)   Silicone rubber
(h)   Teflon®[3]

DP5-4 As a designer, you are assigned to develop from scratch a *blind bolt* to be used to production-join heavy-gage aircraft-grade aluminum. The bolt will be used in the *wet wing* or fuel tank area of commercial aircraft. The head must be flush with the outer surface, the bolt must be self-sealing, installation must be accomplished by one person from one side of the structure, and the disassembly must be accomplished without damage to the structure.

DP5-5 Design a small bench-type hydraulic press complete with hydraulic cylinder, with a capacity of approximately 10 tons.

The press requires a clear opening between top and bottom platens of 24 inches. This distance is fixed: the top platen does not move. The platens should have an overall area of 30″ × 30″. The platens must be designed for rigidity, therefore do not use a thinner plate than $1\frac{1}{2}$ in. The tension rods tying the two platens together must also be rigid; make them of 2-in. tubing with an adequate wall thickness.

The hydraulic cylinder is mounted vertically on the top platen. Hydraulic pressure must not greatly exceed 1500 psi. From a catalog of seamless hydraulic tubing, select a suitable tubing I.D. and wall thickness. The wall thickness is selected by calculating minimum wall thickness from the formula for a thin cylindrical shell:

$$\text{minimum wall thickness} = \frac{\text{O.D.} \times \text{psi}}{2 \times \text{allowable stress}}$$

Seamless hydraulic tubing has an ultimate tensile stress of 65,000 psi. Using a factor of safety of 3 gives an allowable stress for use in this formula of 20,000 psi (higher pressure, or the use of air instead of oil would require a larger factor of safety). Find a suitable tubing size such that no machining or grinding must be performed on the inside diameter.

Decide the clearance between piston and cylinder. Seal the piston with two O-rings. The piston O.D. must be ground: allow the grinder operator a total range on diameter of 0.001 in.

---

[3]Teflon® Registered Trademark of E. I. DuPont De Nemours & Co., Inc.

Make the piston and rod of 1040 plain carbon steel $1\frac{1}{4}\phi$. Seal the cylinder tube to the end heads of the cylinder with O-rings or other suitable static seal. For O-rings and other types of seals, you must consult catalogs for details and available sizes.

Complete the design, giving detail drawings for all glands.

CHAPTER 6

BELT AND CHAIN DRIVES

APRIL 22. The next morning I began to consider of means to put this resolve in execution; but I was at a great loss about my tools. I had three large axes, and abundance of hatchets, but with much chopping and cutting knotty hard wood they were all full of notches, and dull; and though I had a grindstone, I could not turn it and grind my tools too. This caused me as much thought as a statesman would have bestowed upon a grand point of politics, or judge upon the life and death of a man. At length, I contrived a wheel with a string, to turn it with my foot, that I might have both my hands at liberty.

DANIEL DEFOE
*Life and Adventures
of Robinson Crusoe*

6-1 *INTRODUCTION*

Various means of transmitting power historically followed the development of various means of producing power—from the singular strength of a foot or an arm to the pressures of wind or water to the heat of nuclear reactors. Obviously, a great deal is owed to that lonely stranger who decided that round was easier to push than square. Not too many years ago, entire machine lines were run from a single power source by a series of belts, pulleys, and jack shafts. Belt and chain drives are still an important method of transmitting power—the automotive fan belt and the bicycle chain are classic examples.

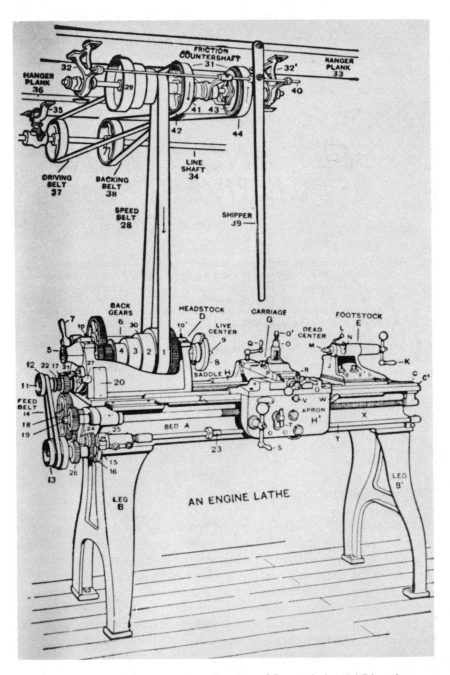

Figure 6-1 | An engine lathe. (Courtesy of Boston Industrial Education
Institute Book Company)

6-2   FLAT BELTS—BASIC RELATIONSHIP

While one of the oldest forms of power transmission, flat belts still have several unique advantages over other methods:

1 /   Flexibility of design;

2 /   Wide range of power capabilities;

3 /   Can be made "endless," a desirable feature in conveyor applications; and

4 /   Cushioning effect against load fluctuations.

Power transmission by a flat belt–pulley combination is accomplished by frictional forces created between the belt and pulley as shown in Fig. 6-2(a). The expression for the ratio of *tight-side tension* T_1 to *slack-side tension* T_2 at the point of slipping, Fig. 6-2(b), is given by

$$\frac{T_1}{T_2} = e^{\mu\beta} \tag{6-1}$$

where $e = 2.718$, the base of the Naperian logarithm

$\mu =$ coefficient of friction between the belt and pulley

$\beta =$ belt contact angle in radians

$T_1, T_2 =$ tight- and slack-side tensile forces respectively in the belt expressed in like units of force, usually pounds

Equation (6-1) applies to situations where the belt and drive are moving at a relatively slow speed. At high velocities there is a tendency for the belt to leave the pulley due to centrifugal effects; the contact forces between the belt and pulley are reduced. To incorporate centrifugal effects, Eq. (6-1) is modified to read

$$\frac{T_1 - T_c}{T_2 - T_c} = e^{\mu\beta} \tag{6-2}$$

where $$T_c = \frac{wV^2}{3600g} \tag{6-3}$$

and $w =$ belt weight in pounds per lineal foot

$V =$ belt speed in feet per minute

$g =$ gravitational acceleration $= 32.2$ fps^2

Note that Eq. (6-2) reverts to Eq. (6-1) when T_c has a negligible magnitude; Eq. (6-2) becomes invalid when $T_c = T_2$.

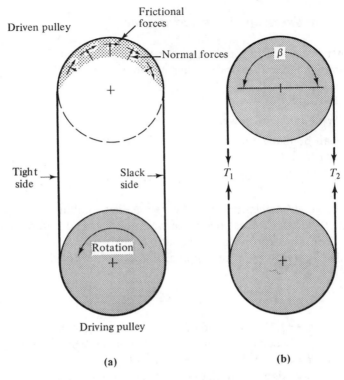

Figure 6-2 / Power transmission by a flat belt.

Horsepower, which is defined as the rate of doing 550 ft-lb of work per second, or $550 \times 60 = 33,000$ ft-lb of work per minute, is related to angular velocity by the equation

$$hp = \frac{M_e \omega}{550} \qquad (6\text{-}4)$$

where M_e = effective torque in ft-lb per sec
        ω = angular velocity in radians per sec

Since belt speed V in ft-per-sec is a function of angular velocity ω of the pulley and the radius r of the pulley, it follows that

$$V = r\omega$$

$$hp = \frac{M_e V}{550r} \qquad (6\text{-}5)$$

where V, M_e, and r are in units of fps, ft-lb, and ft respectively. Referring to Fig. 6-3, the equivalent torque produced by the belt pull is

$$M_e = [(T_1 - T_c) - (T_2 - T_c)]r = (T_1 - T_2)r \qquad (6\text{-}6)$$

Rotation

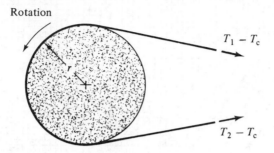

$T_1 - T_c$

$T_2 - T_c$

Figure 6-3 / Belt pull.

Equation (6-5) can, therefore, be rewritten as

$$hp = \frac{(T_1 - T_2)V}{550} \qquad (6\text{-}7)$$

Since belt speed is usually expressed in ft per min, Eq. (6-7) becomes

$$hp = \frac{(T_1 - T_2)V}{550 \times 60} = \frac{(T_1 - T_2)V}{33,000} \qquad (6\text{-}8)$$

Finally, in terms of belt friction, contact angle, tight-side tension, and belt speed, horsepower can be written as

$$hp = \frac{(T_1 - T_c)V}{33,000} \times \left(1 - \frac{1}{e^{\mu\beta}}\right) \qquad (6\text{-}9)$$

For convenience, magnitudes of the expressions $1/e^{\mu\beta}$ and $[1 - (1/e^{\mu\beta})]$ are given in the graphs, Fig. 6-4 and Fig. 6-5, for practical limits of μ and β.

EXAMPLE 6-1

A flat belt is used to transmit 25 hp to a scrap plastics crusher. The belt, which weighs 2.5 lb per lineal ft, has a contact arc of 200° on the driving pulley. Belt speed is 1000 fpm, and the coefficient of friction between the belt and pulley is 0.35. Determine, at the instant of impending belt slippage: (a) centrifugal tension T_c; (b) tight-side tension T_1; and (c) slack-side tension T_2.

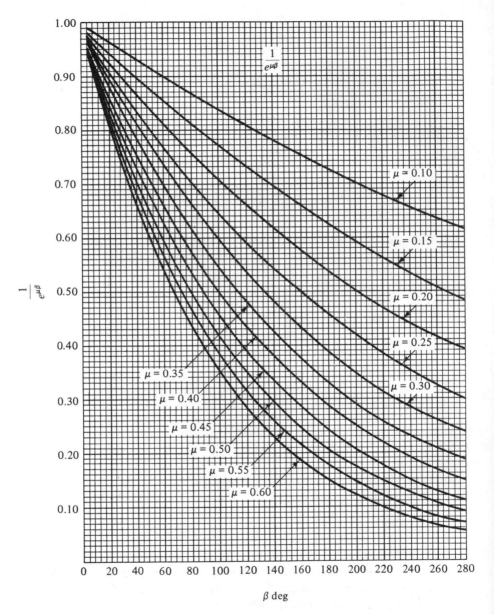

Figure 6-4 / Values of $1/e^{\mu\beta}$.

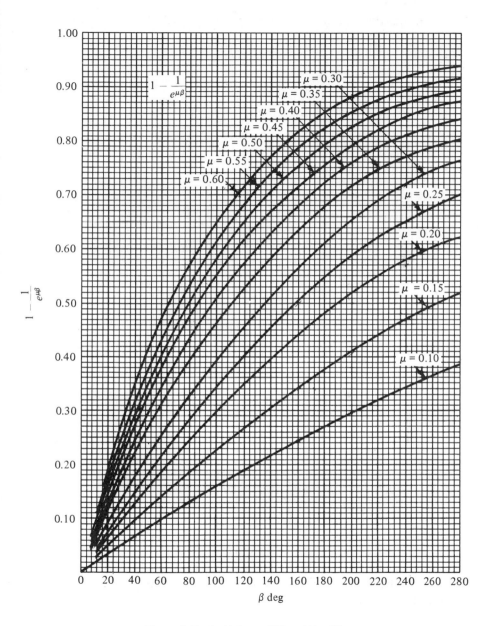

Figure 6-5 / Values of $[1 - (1/e^{\mu\beta})]$.

solution

(a) As an aid to understanding actual calculations of $e^{\mu\beta}$ and $[1 - (1/e^{\mu\beta})]$

$$\mu = 0.35 \text{ (given)}$$

$$\beta = 200° \times \frac{2\pi \text{ rad}}{360} = 3.49 \text{ rad}$$

$$e^{\mu\beta} = e^{(0.35 \times 3.49)} = e^{1.22} = 3.39$$

$$\left(1 - \frac{1}{e^{\mu\beta}}\right) = \left(1 - \frac{1}{3.39}\right) = 0.705$$

From Eq. (6-3):

$$T_c = \frac{wV^2}{3600g} = \frac{2.5 \times \overline{1000}^2}{3600 \times 32.2} = 21.57 \text{ lb}$$

(b) From Eq. (6-9):

$$\text{hp} = \frac{(T_1 - T_c)V}{33,000} \times \left(1 - \frac{1}{e^{\mu\beta}}\right)$$

$$25 = \frac{(T_1 - 21.57)1000}{33,000} \times 0.705$$

$$T_1 = 1190 \text{ lb}$$

(c) From Eq. (6-8):

$$\text{hp} = \frac{(T_1 - T_2)V}{33,000}; \qquad 25 = \frac{(1190 - T_2)1000}{33,000}$$

$$T_2 = 365 \text{ lb}$$

6-3   CONTACT ANGLES—ANGLE OF WRAP

Of major consideration in belt design, whether the design concerns flat or V-belts, is contact angle or *angle of wrap*. Consider, for example, the horse-power capacities of the three drives shown in Fig. 6-6. Assume, for the sake of illustration, that T_1, T_c, V, and μ are the same in all instances and that only the contact angle varies—40°, 180°, and 220°. So that compari-sons can be made, let $\mu = 0.35$. For Case I,

$$\mu\beta = 0.35 \times 40 \times \frac{2\pi}{360} = 0.244$$

and

$$1 - \frac{1}{e^{0.244}} = 0.307$$

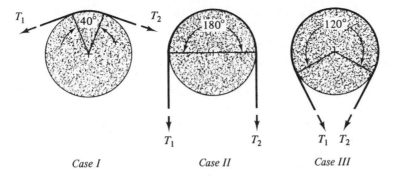

Case I   Case II   Case III

Figure 6-6 / Different angles of wrap.

For Case II, $\qquad \mu\beta = 0.35 \times 180 \times \dfrac{2\pi}{360} = 1.10$

and $\qquad\qquad\qquad\qquad 1 - \dfrac{1}{e^{1.10}} = 0.667$

For Case III, $\qquad \mu\beta = 0.35 \times 220 \times \dfrac{2\pi}{360} = 1.34$

and $\qquad\qquad\qquad\qquad 1 - \dfrac{1}{e^{1.34}} = 0.739$

Since the quantity $[(T_1 - T_c)V]/33,000$ is the same in all three cases, it follows that the pulleys can each transmit:

$$hp_I = \frac{T_1 - T_c}{33,000} \times 0.244$$

$$hp_{II} = \frac{T_1 - T_c}{33,000} \times 0.667$$

and $\qquad\qquad hp_{III} = \dfrac{T_1 - T_c}{33,000} \times 0.739$

Thus, by increasing the *angle of wrap* from 40° to 180°, hp_{II} expressed in terms of hp_I is

$$hp_{II} = \frac{0.667}{0.244} \times hp_I = 2.73\ hp_I$$

By changing from 40° to 220°, even a more dramatic increase is evident:

$$hp_{III} = \frac{0.739}{0.244} \times hp_I = 3.03\ hp_I$$

It must be realized that the formulas presented are based on the assumption that the belt is on the verge of slipping—a theoretical, rather than a practical assumption. It must also be remembered that belt and pulley characteristics are affected by a variety of things: temperature, humidity, type of application, and age are just a few of the many variables. Also of practical design concern are force and dimensional factors that affect belt life. An empirical equation, which is based on countless tests on a particular belt material, shows belt life in flexing to be defined by Eq. (6-10)

$$\text{Belt life} = C\frac{d^{5.4} \times L}{V^{0.5} \times t^{6.3} \times T_1^{4.2}} \qquad (6\text{-}10)$$

where
C = a constant
d = small pulley diameter
L = belt length
V = belt speed
t = belt thickness
T_1 = tight-side tension

The belt life equation serves to illustrate some rather important factors affecting belt life. For example, by doubling the tight-side tension, belt life is reduced by a factor of over 18; by reducing belt thickness by 20%, belt life is increased 300%.

6-4   CALCULATION OF CONTACT ANGLE AND BELT LENGTH

For the *open belt* drive shown in Fig. 6-7, the contact angles β_1 and β_2 and the belt length L may be computed by:

$$\sin \theta = \frac{D - d}{2C} \qquad (6\text{-}11)$$

$$\beta_1 = 180° + 2\theta = 180° + 2 \sin^{-1} \frac{D - d}{2C} \qquad (6\text{-}12)$$

$$\beta_2 = 180° - 2\theta = 180° - 2 \sin^{-1} \frac{D - d}{2C} \qquad (6\text{-}13)$$

$$L_{\text{exact}} = 2C \cos \theta + \frac{\pi(D + d)}{2} + \frac{\pi\theta°(D - d)}{180°} \qquad (6\text{-}14)$$

$$L_{\text{approx}} = 2C + 1.57(D + d) + \frac{(D - d)^2}{4C} \qquad (6\text{-}15)$$

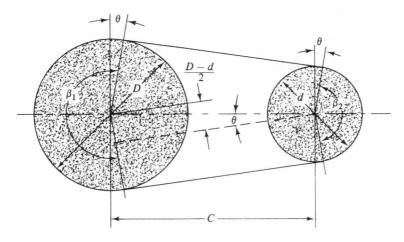

Figure 6-7 / Open belt drive.

For the *crossed belts*, Fig. 6-8, β_1, β_2, and L are given by:

$$\theta \sin = \frac{D + d}{2C} \qquad (6\text{-}16)$$

$$\beta_1 = \beta_2 = 180° + 2\theta = 180° + 2 \sin^{-1} \frac{D + d}{2C} \qquad (6\text{-}17)$$

$$L_{\text{exact}} = 2C \cos \theta + \frac{(90° + \theta°)(D + d)}{180°} \qquad (6\text{-}18)$$

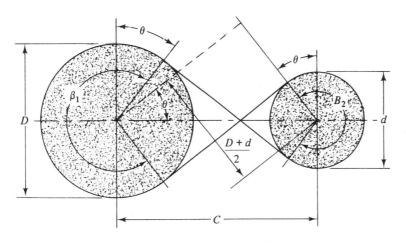

Figure 6-8 / Crossed belt drive.

EXAMPLE 6-2

For the open belt of Fig. 6-7, let $D = 40$ in., $d = 8$ in., and $C = 30$ in. Find: (a) the contact angles β_1 and β_2; (b) both the approximate and exact lengths; and (c) the percent error between L_{approx} and L_{exact}.

solution

(a)   From Eq. (6-12):

$$\beta_1 = 180° + 2\theta = 180° + (2 \times 32.23) = 244.46°$$

Eq. (6-13):

$$\beta_2 = 180° - 2\theta = 115.54°$$

(b)   From Eq. (6-14):

$$L_{exact} = 2C \cos \theta + \frac{\pi(D+d)}{2} + \frac{\pi\theta°(D-d)}{180°}$$

$$= 2 \times 30 \times \cos 32.23 + \frac{\pi(40+8)}{2} + \frac{\pi \times 32.23°(40-8)}{180°}$$

$$= 144.2 \text{ in.}$$

From Eq. (6-15):

$$L_{approx} = 2C + 1.57(D+d) + \frac{(D-d)^2}{4C}$$

$$= (2 \times 30) + 1.57(40+8) + \frac{(40-8)^2}{4 \times 30}$$

$$= 143.9$$

(c) $\% \text{ error} = \dfrac{144.2 - 143.9}{144.2} \times 100 = 0.21\%$

6-5   FLOATING IDLERS AND BEARING PRESSURES

Floating idlers are used to supply the slack-side tension to a belt necessary to prevent slippage. When idlers are not used, at least one set of bearings must be adjustable to provide a means of applying sufficient belt tension and to allow for adjustment, or *take-up*, as the belt stretches. Idlers also serve a second purpose: they increase the angle of wrap and thereby reduce the belt tensions required for a given design horsepower. In practice, the idler is usually located near the smallest-diameter pulley. It is necessary, of course, to determine the required weight of the idler if the system is gravity-controlled—other means of supplying force would have to be employed, springs for example, if the pulleys were horizontal.

     Figure 6-9 illustrates a step-by-step graphical analysis of a typical

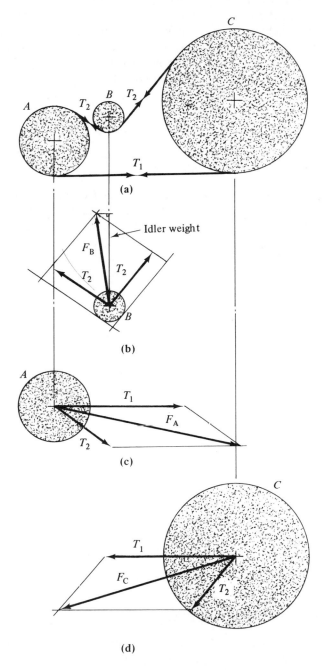

(a)

Idler weight

F_B

T_2

T_2

B

(b)

A

T_1

F_A

T_2

(c)

C

T_1

F_C

T_2

(d)

Figure 6-9 / *Graphical analysis of an idler.*

system; it is necessary, at the start, to draw a precise scaled-down layout, Fig. 6-9(a), of the pulley arrangement and to select a respectively large force scale to achieve reasonable accuracy. With the slack-side and tight-side tensions known, a parallelogram of forces is drawn, Fig. 6-9(b), and the resultant bearing force F_B is graphically determined. The vertical vector component of F_B is the required idler weight. Similar graphical techniques, Fig. 6-9(c) and Fig. 6-9(d), are used to determine bearing forces acting on each pulley.

6-6   FLAT BELT SERVICE FACTORS AND DESIGN HORSEPOWER

It often seems that reality and theory are in conflict; design formulas which are based on sound mathematics and physics fall short in predicting the perfect problem-solver—nature. Until the engineer/scientist can precisely mathematize nature, he must alter the theoretical to conform to reality. Table 6-1 lists typical corrections—service factors—in terms of the type of use and the type of power source. For *special* conditions, these service factors are increased as follows:

$$\text{Continuous operation:}\quad 0.2$$
$$\text{Frequent start, stop, or reversal:}\quad 0.1$$
$$\text{Speed-up drives:}\quad 0.2$$
$$\text{Wet or excessive oil conditions:}\quad 0.1$$

EXAMPLE 6-3

A four-cylinder, continuously operating, reciprocating pump rated at 25 hp is driven by a six-cylinder gasoline engine. The pump operates under wet conditions. Determine the design horsepower.

solution
From Table 6-1 and the aforementioned list of *special* conditions,

$$\text{Design hp} = \text{rated hp} \times \text{sum of service factors}$$
$$= 25(1.8 + 0.2 + 0.1)$$
$$= 53 \text{ hp}$$

6-7   DRIVE CONFIGURATIONS

Figure 6-10 illustrates some of the more common belt drive configurations. Of particular interest are *stepped cone pulleys*, which are used in pairs to obtain speed variations in a driven machine. A relatively narrow belt is

TABLE 6-1

FLAT BELT SERVICE FACTORS[1]

Driven machine		Driver						
		A-C motors						Engines Gas or diesel
		Squirrel cage			Wound rotor (slip ring)	Synchronous		
		Normal torque		High torque		Normal torque (150% to 249%)	High torque (250% to 400%)	
General type	Specific type	Line start	Compensator start					
Agitators	for liquids	1.0	1.0	1.2	—	—	—	—
	for semi-liquids	1.2	1.0	1.4	1.2	—	—	—
Bakery machinery	—	1.2	—	—	—	—	—	—
Brick and clay machinery	de-airing machine, granulator auger, cutting table, rolls	—	1.2	1.4	1.4	—	—	—
	mixer, dry press	—	1.2	1.6	1.4	—	—	—
	pug mill	1.5	1.3	1.8	1.5	—	—	—
Compressors	centrifugal, rotary	1.2	1.2	—	1.4	1.4	—	—
	reciprocating, 1 or 2 cyl.	1.4	1.4	—	1.5	1.5	—	—
	reciprocating, 3 or more cyl.	1.2	1.2	—	1.4	1.4	—	—
Conveyors	apron, bucket, pan, elevator	—	1.4	1.6	—	—	—	—
	belt (ore, coal, sand, etc.)	—	1.2	1.4	—	—	—	—
	flight	—	1.6	1.8	—	—	—	—
	oven, belt (light package)	—	1.0	1.1	—	—	—	—
Crushing machinery	jaw, cone crushers, crushing rolls gyratory, ball, pebble, tube mills	—	1.4	1.6	1.4	1.4	1.6	1.4

[1]Courtesy of Uniroyal Industrial Products

TABLE 6-1 (Cont.)

Driven machine		Driver						
		A-C motors				Synchronous		Engines Gas or diesel
		Squirrel cage			Wound rotor (slip ring)	Normal torque (150% to 249%)	High torque (250% to 400%)	
		Normal torque		High torque				
General type	Specific type	Line start	Compensator start					
Fans, blowers	centrifugal, induced draft, exhausters	1.2	1.2	–	1.4	–	–	1.4
	propeller, mine fans	1.6	1.6	1.6	1.6	–	1.8	1.6
	positive blowers	1.6	1.6	–	2.0	2.0	1.8	1.6
Flour, feed, cereal mill machinery	bolters, sifters, separators	1.0	1.0	–	–	–	–	–
	grinders, purifiers, reels mainline shaft, hammermills	1.4	1.4	1.6	1.4	1.4	–	1.8
Generators, exciters	–	1.2	–	–	–	–	–	1.2
Line shafts	–	1.4	1.4	–	1.4	1.4	2.0	1.6
Machine tools	grinders, milling machines boring mills, planers, shears	1.2	–	–	1.4	–	–	–
	lathes, screw machines, cam cutters shapers, drill press, drop hammers	1.0	–	–	1.2	–	–	–
Mills	pebble, rod, ball, roller	–	1.4	1.6	1.4	–	–	–
	flaking mills, tumbling barrels	–	1.6	1.6	1.4	–	–	–
Oil-field machinery	–	1.2	1.2	1.4	–	–	–	1.4

TABLE 6-1 (Cont.)

Driven machine		Driver						
		A-C motors						Engines Gas or diesel
		Squirrel cage			Wound rotor (slip ring)	Synchronous		
		Normal torque		High torque		Normal torque (150% to 249%)	High torque (250% to 400%)	
General type	Specific type	Line start	Compensator start					
Paper machinery	jordan engines	1.5	1.3	1.8	1.5	1.6	1.8	—
	beaters, paper machines	1.4	1.4	—	1.5	—	—	—
	calenders, agitators, dryers	1.2	1.2	1.4	1.2	—	—	—
Printing machinery	—	1.2	1.2	—	1.2	—	—	—
Pumps	centrigugal, gear, rotary	1.2	1.2	1.4	1.4	—	—	1.2
	reciprocating, 1 or 2 cyl	1.4	1.4	—	1.6	1.6	1.8	2.0
	3 or more	1.2	1.2	—	1.4	1.4	1.6	1.8
Rubber plant machinery	—	1.4	1.4	1.4	1.4	—	1.8	—
Sawmill machinery	log canter, log jack, cutoff saws trimmers, slashers, swing saws	1.4	1.4	—	1.4	—	—	—
	brand mill, circular, hogs, resaw	2.0	1.6	—	1.8	—	1.6	—
	planers	1.2	1.2	—	1.2	—	—	—
	edgers	1.6	1.6	—	1.6	—	1.6	—

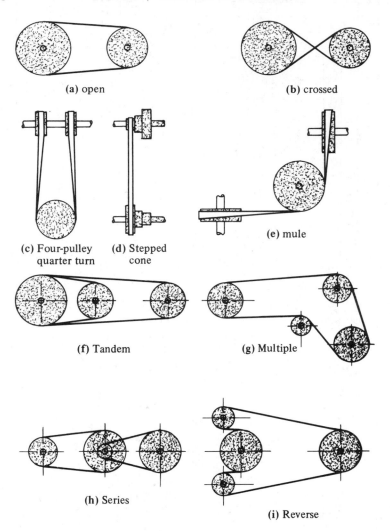

Figure 6-10 / Belt drive configurations.

used, and the belt is moved, usually while in motion, across the pulley faces by means of a *shifter fork*.

6-8   BELT DESIGNATIONS

Four basic materials, employed either singly or in combination, are used in belt manufacture: leather, rubber, plastic, and fabric. Each material has its advantages or disadvantages, depending upon the specific application. Belts can also be classified as being *cordless* or *corded*. In the cordless variety, the entire belt acts to resist tension—belts of this type are easily spliced. In a corded belt, such materials as cotton, nylon, fiberglass, or steel, impregnated in a base material, supply the required strength to the belt. Figure 6-11 shows a typical cross section of a corded belt. Belt sizes are designated, Fig. 6-12, by top width and thickness or by top width and number of plies, depending upon the belt construction.

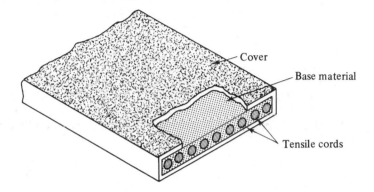

Figure 6-11 / Cross section of a corded belt.

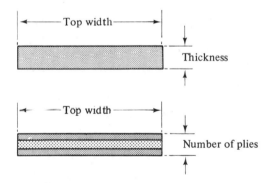

Figure 6-12 / Designation of flat-belt size.

6-9 PULLEY CONSIDERATIONS

A flat belt moving on a conical pulley, Fig. 6-13(a), has a tendency to move higher and higher on the cone because of differential tensions at edges A and B. This tendency can be utilized to keep the belt in position by *crowning* the pulley, as illustrated in Fig. 6-13(b). The circular crown causes less fatigue in the belt but is more difficult to machine than the cone crown. Usual practice is to provide a crown depth of $\frac{1}{8}$-in. per foot of pulley width.

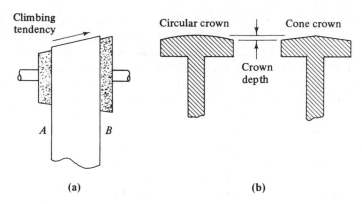

Figure 6-13 / Crowning of pulleys.

In a pair of pulleys, it is sufficient to crown only one pulley, and idlers are rarely crowned.

6-10 V-BELT DRIVES

V-belts, identified as such because of their shape, form a V-belt drive when combined with a set of two or more *sheaves* or grooved pulleys. Such drives are used in practically every industry for both light and heavy duty.

The sheave consists of a pulley with one or more wedge grooves, as shown in Fig. 6-14(a). In effect, the wedge increases the normal component of force P_n, Fig. 6-14(b), between the belt and groove sides by an amount equal to

$$P_n = \frac{P}{2 \sin \phi} \tag{6-19}$$

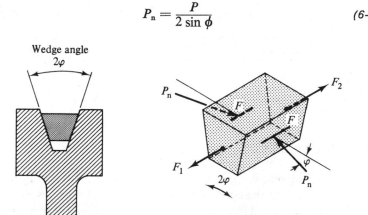

Figure 6-14 / V-belt wedge.

where ϕ is one-half of the wedge, or groove, angle. The frictional or *tractive* force F, a product of the coefficient of friction μ and the normal force P_n (which acts on both sides of the wedge), becomes

$$F = 2\mu P_n = \frac{2\mu P}{2 \sin \phi} = \frac{\mu P}{\sin \phi} = \mu_e P \qquad (6\text{-}20)$$

where $\mu_e = \mu/\sin \phi$—the equivalent coefficient of friction. Previously discussed flat-belt drive equations are valid for V-belts when μ_e is substituted for μ.

6-11 BELT STANDARDS

Figure 6-15 gives the nominal dimensions of five standard V-belt sections (designated by the letters A, B, C, D, and E) together with their approximate weight per foot and their recommended maximum *tight-side* tensions. Table 6-2 and the accompanying figure give the industries' standard groove dimensions for V-belt sheaves. The *pitch diameter* is measured to that plane in the belt that does not undergo a change in circumferential length as the belt rotates around the sheave. The pitch diameter must be used in calculations involving contact angle and belt length. Equations (6-21) and (6-22) give belt length as a function of center distance, and center distance in terms of belt length, respectively.

$$L = 2C + 1.57(D' + d') + \frac{(D' - d')^2}{4C} \qquad (6\text{-}21)$$

$$C = \frac{b + \sqrt{b^2 - 32(D' - d')^2}}{16} \qquad (6\text{-}22)$$

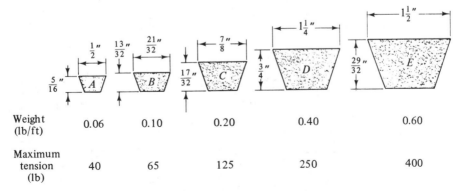

Weight (lb/ft)	0.06	0.10	0.20	0.40	0.60
Maximum tension (lb)	40	65	125	250	400

Figure 6-15 / Standard V-belt sections.

TABLE 6-2
STANDARD AND DEEP GROOVE DIMENSIONS (INCHES)[1]

Belt	Pitch diameter (inches) Minimum	Range	Groove angle (±½°)	Standard groove dimensions W	D (±0.031)	X†	Deep groove†† dimensions W	D (±0.031)	X†
A	3.0	2.6 to 5.4 Over 5.4	34° 38°	0.494 ±0.005 0.504 ±0.005	0.490	0.125	0.589 ±0.005 0.611 ±0.005	0.645	0.280
B	5.4	4.6 to 7.0 Over 7.0	34° 38°	0.637 ±0.005 0.650 ±0.005	0.580	0.175	0.747 ±0.005 0.774 ±0.005	0.760	0.355
C	9.0	7.0 to 7.99 8.0 to 12.0 Over 12.0	34° 36° 38°	0.879 ±0.007 0.887 ±0.007 0.895 ±0.007	0.780	0.200	1.066 ±0.007 1.085 ±0.007 1.105 ±0.007	1.085	0.505
D	13.0	12.0 to 12.99 13.0 to 17.0 Over 17.0	34° 36° 38°	1.259 ±0.007 1.271 ±0.007 1.283 ±0.007	1.050	0.300	1.513 ±0.007 1.541 ±0.007 1.569 ±0.007	1.465	0.715
E	21.0	18.0 to 24.0 Over 24.0	36° 38°	1.527 ±0.010 1.542 ±0.010	1.300	0.400	1.816 ±0.010 1.849 ±0.010	1.745	0.845

†Add 2 X to pd to get OD.

††Deep groove sheaves are intended for quarter-turn drives and for long center vertical shaft drives. They may also be necessary for such applications as car shakers, vibrating screens and certain types of crushers where oscillations in the center distance may occur.

[1] Courtesy of Uniroyal Industrial Products

where $b = 4L - 6.28(D' + d')$
$D' = $ pitch diameter of large sheave in inches
$d' = $ pitch diameter of small sheave in inches
$L = $ pitch length of belt
$C = $ center distance in inches

Table 6-3 lists service factors for several representative machines in terms of driver and power source. Design horsepower is the product of the service factor and the theoretical horsepower requirement. Since specific belt properties vary, manufacturers' catalogs and data guides are normally consulted when designing V-belt systems. As an illustration, however, the graphs and tables that follow are typical of the data necessary in belt and sheave selection.

Recommended V-belt cross sections for various combinations of design horsepower and small sheave rpm are obtained by referring to Fig. 6-16, while maximum horsepower ratings in terms of belt speed for multiple V-belt drives are given in Figs. 6-17 through 6-21. These ratings are valid for 180° contact angles and normal belt length. Values must be adjusted (see Tables 6-4 through 6-6) for other than normal conditions. When possible, stock sheaves should be used; Table 6-7 is a partial list of manufacturers' standards.

The following example problem illustrates the use of the tables and charts in the design of a multiple V-belt drive system.

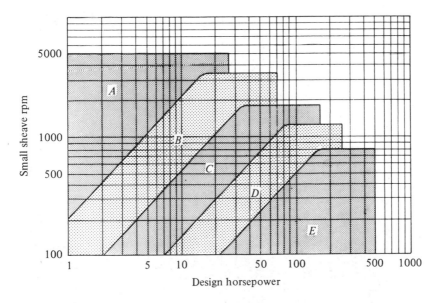

Figure 6-16 / V-belt selection graph.

TABLE 6-3
V-BELT SERVICE FACTORS[1]

Driven Machine Types	Driver Types					
	Electric Motors: AC Normal Torque Squirrel Cage and Synchronous AC Split Phase DC Shunt Wound Internal Combustion Engines			Electric Motors: AC Hi-Torque AC Hi-Slip AC Repulsion-Induction AC Single Phase Series Wound AC Slip Ring DC Compound Wound		
	Intermittent service	Normal service	Continuous service	Intermittent service	Normal service	Continuous service
Driven Machine Types noted below are representative samples only. Select a category most closely approximating your application from those listed below. If idlers are used, add the following to the service factor. Idler on slack side (inside) — None Idler on slack side (outside) — 0.1 Idler on tight side (inside) — 0.1 Idler on tight side (outside) — 0.2						
Agitators for Liquids Blowers and Exhausters Centrifugal Pumps and Compressors Fans up to 10 HP Light Duty Conveyors	1.0	1.1	1.2	1.1	1.2	1.3
Belt Conveyors for Sand, Grain, etc. Dough Mixers Fans Over 10 HP Generators Line Shafts Laundry Machinery Machine Tools Punches-Presses-Shears Printing Machinery Positive Displacement Rotary Pumps Revolving and Vibrating Screens	1.1	1.2	1.3	1.2	1.3	1.4

Driven Machine Types	Driver Types					
Driven Machine Types noted below are representative samples only. Select a category most closely approximating your application from those listed below. If idlers are used, add the following to the service factor. Idler on slack side (inside) — None Idler on slack side (outside) — 0.1 Idler on tight side (inside) — 0.1 Idler on tight side (outside) — 0.2	Electric Motors: AC Normal Torque Squirrel Cage and Synchronous AC Split Phase DC Shunt Wound Internal Combustion Engines			Electric Motors: AC Hi-Torque AC Hi-Slip AC Repulsion-Induction AC Single Phase Series Wound AC Slip Ring DC Compound Wound		
	Intermittent service	Normal service	Continuous service	Intermittent service	Normal service	Continuous service
Brick Machinery Bucket Elevators Exciters Piston Compressors Conveyors (Drag-Pan-Screw) Hammer Mills Paper Mill Beaters Piston Pumps Positive Displacement Blowers Pulverizers Saw Mill and Woodworking Machinery Textile Machinery	1.2	1.3	1.4	1.4	1.5	1.6
Crushers (Gyratory-Jaw-Roll) Mills (Ball-Rod-Tube) Hoists Rubber Calenders-Extruders-Mills	1.3	1.4	1.5	1.5	1.6	1.8
Chokable Equipment	2.0	2.0	2.0	2.0	2.0	2.0

¹Courtesy of Goodyear Tire and Rubber Company

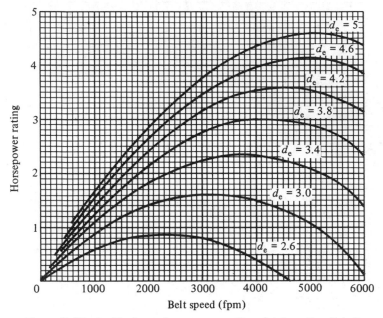

Figure 6-17 / Maximum horsepower ratings for A section V-belts.

TABLE 6-4

**SMALL DIAMETER CORRECTION FACTORS
FOR MULTIPLE V-BELTS**

SPEED RATIO RANGE	SMALL-DIAMETER FACTOR
1.000–1.019	1.00
1.020–1.032	1.01
1.033–1.055	1.02
1.056–1.081	1.03
1.082–1.109	1.04
1.110–1.142	1.05
1.143–1.178	1.06
1.179–1.222	1.07
1.223–1.274	1.08
1.275–1.340	1.09
1.341–1.429	1.10
1.430–1.562	1.11
1.563–1.814	1.12
1.815–2.948	1.13
2.949 and over	1.14

TABLE 6-5

BELT LENGTH CORRECTION FACTORS[1]

A BELT		B BELT		C BELT		D BELT		E BELT	
BELT PITCH LENGTH	CORREC-TION FACTOR	BELT PITCH LENGTH	CORREC-TION FACTOR	BELT PITCH LENGTH	CORREC-TION FACTOR	BELT PITCH LENGTH	CORREC-TION FACTOR	BELT PITCH LENGTH	CORREC-TION FACTOR
27.3	0.81	36.8	0.81	53.9	0.80	123.3	0.86	184.5	0.91
32.3	0.84	39.8	0.83	62.9	0.82	131.3	0.87	199.5	0.92
36.3	0.87	43.8	0.85	70.9	0.85	147.3	0.90	214.5	0.94
39.3	0.88	47.8	0.87	77.9	0.87	161.3	0.92	241.0	0.96
43.3	0.90	52.8	0.89	83.9	0.89	176.3	0.93	271.0	0.99
47.3	0.92	56.8	0.90	87.9	0.90	183.3	0.94	301.0	1.01
52.3	0.94	61.8	0.92	92.9	0.91	198.3	0.96	331.0	1.03
56.3	0.96	69.8	0.95	98.9	0.92	213.3	0.96	361.0	1.05
61.3	0.98	76.8	0.97	107.9	0.94	240.8	1.00	391.0	1.07
69.3	1.00	82.8	0.98	114.9	0.95	270.8	1.03	421.0	1.09
76.3	1.02	86.8	0.99	122.9	0.97	300.8	1.05	481.0	1.12
81.3	1.04	91.8	1.00	130.9	0.98	330.8	1.07	541.0	1.14
86.3	1.05	98.8	1.02	146.9	1.00	360.8	1.09	601.0	1.17
91.3	1.06	106.8	1.04	160.9	1.02	390.8	1.11		
97.3	1.08	113.8	1.05	175.9	1.04	420.8	1.12		
106.3	1.10	121.8	1.07	182.9	1.05	480.8	1.16		
113.3	1.11	129.8	1.08	197.9	1.07	540.8	1.18		
121.3	1.13	145.8	1.11	212.9	1.08	600.8	1.20		
129.3	1.14	159.8	1.13	240.9	1.11				
		174.8	1.15	270.9	1.14				
		181.8	1.16	300.9	1.16				
		196.8	1.18	330.9	1.19				
		211.8	1.19	360.9	1.21				
		240.3	1.22	390.9	1.23				
		270.3	1.25	420.9	1.24				
		300.3	1.27						

[1] Courtesy Goodyear Tire and Rubber Co., Lincoln, Nb.

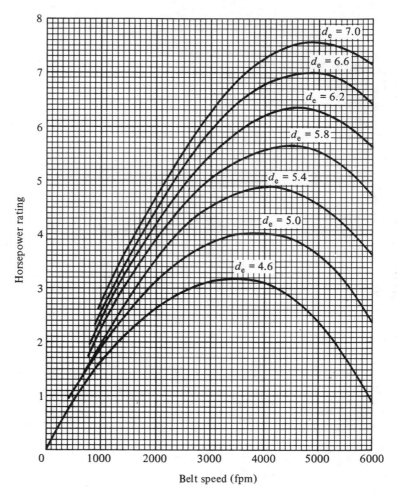

$d_e = 7.0$
$d_e = 6.6$
$d_e = 6.2$
$d_e = 5.8$
$d_e = 5.4$
$d_e = 5.0$
$d_e = 4.6$

Horsepower rating

0   1000   2000   3000   4000   5000   6000

Belt speed (fpm)

Figure 6-18 / Maximum horsepower ratings for B section V-belts.

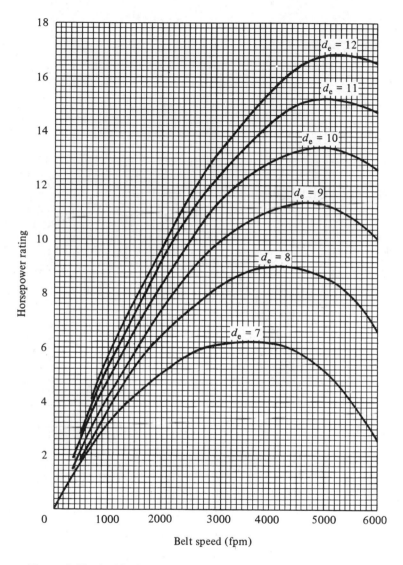

Figure 6-19 | Maximum horsepower ratings for C section V-belts.

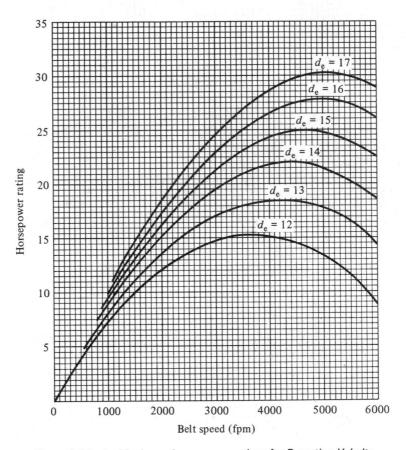

Figure 6-20 / Maximum horsepower ratings for D section V-belts.

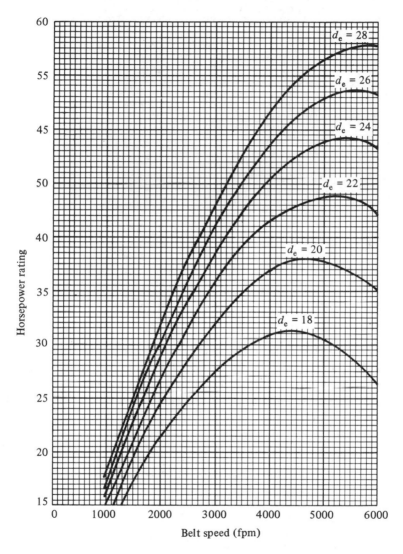

Figure 6-21 | Maximum horsepower ratings for E section V-belts.

TABLE 6-6

CONTACT ANGLE CORRECTION FACTORS

ARC OF CONTACT ON SMALL SHEAVE *(deg)*	CORRECTION FACTOR
180°	1.00
174	0.99
169	0.97
163	0.96
157	0.94
151	0.93
145	0.91
139	0.89
133	0.87
127	0.85
120	0.82
113	0.80
106	0.77
99	0.73
91	0.70
83	0.65

EXAMPLE 6-4

A 30-hp Hi-Torque 1160 rpm AC motor is used to continuously power a pulverizer at a driven speed of 400 rpm. The center distance between sheaves is 60 in., and the driven sheave has a 40-in. pitch diameter. Determine: (a) the belt size; (b) the belt length; and (c) the number of belts required. The following information is given:

1 / Type of power source = Hi-Torque AC motor
2 / Kind of application = Pulverizer
3 / Horsepower requirement of the system = 30 hp
4 / Speed of the driver = 1160 rpm
5 / Speed of the driven machine = 400 rpm
6 / Design center distance between sheaves = 60 in.
7 / Pitch diameter of driven sheave = 40 in.

solution

(a)   Table 6-3 gives the appropriate service factor of 1.6:

$$\text{Design horsepower} = 1.6 \times 30 = 48$$

TABLE 6-7
STOCK SHEAVE PITCH DIAMETERS

A	B	C	D	E
3.0	3.4	7.0	12.0	Special
3.2	3.6	7.5	13.0	Order
3.4	3.8	8.0	13.5	
3.6	4.0	8.5	14.0	
3.8	4.2	9.0	14.5	
4.0	4.4	9.5	15.0	
4.2	4.6	10.0	15.5	
4.6	5.0	10.5	16.0	
4.8	5.2	11.0	18.0	
5.0	5.4	12.0	20.0	
5.2	5.6	13.0	22.0	
5.6	6.0	14.0	27.0	
6.0	6.4	16.0	33.0	
6.4	6.8	18.0	40.0	
7.0	7.4	20.0	48.0	
8.2	8.6	24.0	50.0	
9.0	9.4	30.0	—	
10.6	11.0	36.0	—	
12.0	12.4	44.0	—	
15.0	15.4	50.0	—	
18.0	18.4	—	—	
—	20.0	—	—	
—	25.0	—	—	
—	30.0	—	—	
—	38.0	—	—	

Figure 6-16 indicates that a D cross section is to be used.

The speed ratio $= 1160/400 = 2.9$, and the size of the small sheave $= 40/2.9 = 13.8$-in. pitch diameter. Use a 14-in. stock sheave.

(b) The belt pitch length is computed using Eq. (6-21).

$$L = 2C + 1.57(D' + d') + \frac{(D' - d')^2}{4C}$$

$$= (2 \times 60) + 1.57(40 + 14) + \frac{(40 - 14)^2}{4 \times 60}$$

$$= 207.6 \text{ in.}$$

(c)   From Table 6-4 the small diameter factor is 1.13, which gives an equivalent diameter d_e of

$$d_e = 14 \times 1.13 = 15.8 \text{ in.}$$

The belt speed in fpm is computed by:

$$\text{Belt speed} = r\omega$$

where r = small sheave radius in ft
          ω = small sheave angular velocity in radians per min

Thus, $\text{Belt speed} = \dfrac{14}{2 \times 12} \times 1160 \times 2\pi = 4250 \text{ fpm}$

From Fig. 6-20, by interpolation, a D cross section traveling at approximately 4250 fpm with a $d_e = 15.8$ can transmit 29 hp per belt. The smaller contact angle is found by Eq. (6-13):

$$\beta_2 = 180° - 2 \sin^{-1} \frac{D - d}{2C} = 180 - 2 \sin \frac{40 - 14}{2 \times 60}$$
$$= 155°$$

Tables 6-5 and 6-6 respectively are used to find the length correction factor and the arc of contact correction factor:

$$\text{Length correction factor} = 0.96$$
$$\text{Arc correction factor} = 0.94$$

The horsepower per belt, therefore, is

$$\text{hp/belt} = 29 \times 0.96 \times 0.94 = 26.2$$

Since 48/26.2 = 1.83, two belts are required.

6-12   BELT TENSIONING

Initial and sustained belt tensioning is a critical aspect in a drive system. Too much tension shortens belt life and adds unnecessary forces to bearings and drive shafts. Too little tension results in belt slip, heat generation, and energy loss—factors which also decrease belt life. Ideally, belts should be tensioned to "just" that value that will avoid slippage when the drive is operating at maximum load.

The *pivoted motor mount*, Fig. 6-22, offers a unique solution to the tensioning problem. As illustrated in the figure, the driving motor weight W together with the mounting bracket weight w_m are balanced against the belt pulls. Proper balancing is attained when the motor is

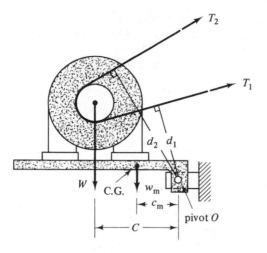

Figure 6-22 / Pivoted motor mount for belt tension.

positioned so that maximum belt pulls can be maintained. Once positioned, the pressure between the belt and sheave is fixed regardless of whether the system is operating at capacity or standing idle. The sum of the belt pulls decreases, however, as the power transmitted decreases; the belt life correspondingly increases. For the practical purpose of keeping distances to a minimum, the tight side of the belt should be closest to the pivot.

EXAMPLE 6-5

Assume the following data apply to the drive system of Fig. 6-22:

1 / At maximum load, $T_1 = 400$ lb, and $T_2 = 100$ lb.

2 / Motor weight $W = 150$ lb.

3 / Mounting bracket weight $w_m = 50$ lb, with a center of gravity at $c_m = 10$ in.

4 / Belt pull distances $d_1 = 4$ in., and $d_2 = 12$ in.

Determine: (a) the motor position C; (b) the belt pulls when the system is operating at $\frac{1}{3}$ maximum load; and (c) the belt pulls under *no load* conditions.

solution

(a) Moments are balanced about the pivot point O.

$$\Sigma M_0 = O$$
$$WC + w_m c_m = T_1 d_1 + T_2 d_2$$
$$C = \frac{(400 \times 4) + (100 \times 12) - (50 \times 10)}{150} = 15.33 \text{ in.}$$

(b) Let T_1' and T_2' represent the tight-side and slack-side tensions at $\frac{1}{3}$ maximum load. Taking moments about the pivot O gives

$$4T_1' + 12T_2' = (50 \times 10) + (150 \times 15.33)$$

or $\qquad\qquad T_1' + 3T_2' = 700 \qquad\qquad\qquad\qquad\qquad (1)$

At $\frac{1}{3}$ maximum load,

$$T_1' - T_2' = \tfrac{1}{3}(T_1 - T_2) = \tfrac{1}{3}(400 - 100) = 100 \qquad\qquad (2)$$

The simultaneous solution of equations (1) and (2) gives

$$T_1' = 250 \text{ lb} \qquad \text{and} \qquad T_2' = 150 \text{ lb}$$

(c) Under *no load* conditions, tight-side and slack-side tensions are equal, hence the use of T_1'' and T_2'':

$$T_1''(4 + 12) = (50 \times 10) + (150 \times 15.33)$$
$$T_1'' = T_2'' = 175 \text{ lb}$$

6-13 CHAIN DRIVES

Chain drives, Fig. 6-23, consist of an endless series of connected links that mesh with toothed wheels, called *sprockets*. These sprockets, in turn, are keyed to the shafts of the driving and driven machine. Unique features of a roller chain are its "hinge action" during engagement with the sprocket and the absence of rubbing action between the rollers and the sprocket teeth. Chain drives also maintain a positive speed ratio between driving and driven components, so tension on the slack side is not necessary. Because of positive link engagement, minimum required arc of contact is less for chains than for belts; thus, chain drives can operate with shorter shaft center distances.

6-14 SPROCKET NOMENCLATURE

Sprocket diameter designations are shown in Fig. 6-24. These designations and other sprocket features are described in the following paragraphs.

The *pitch diameter* is the diameter of the circle that passes through the center of the link pins as the chain engages the sprocket. Since the chain pitch is measured on a straight line between centers of adjacent

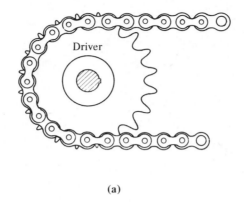

(a)

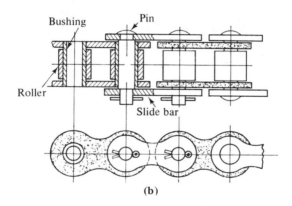

(b)

Figure 6-23   /   Power chain.

pins, the chain pitch lines form a series of chords on the pitch circle. The pitch diameter is a function of the chain pitch (chordal length) and the number of teeth in the sprocket; thus

$$D = \frac{p}{\sin\left(\frac{180°}{n}\right)} \quad \text{and} \quad R = \frac{d}{2} \qquad (6\text{-}23)$$

where D = pitch diameter in inches
R = pitch radius in inches
p = chain pitch in inches
n = number of teeth in sprocket

As an example, the required pitch diameter of a 20-tooth sprocket used

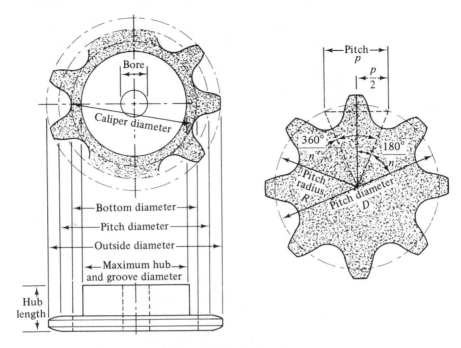

Figure 6-24 / Sprockets.

with a No. 100 standard roller chain ($1\frac{1}{4}$-in. pitch length) is

$$D = \frac{1.25}{\sin\left(\dfrac{180°}{20}\right)} = \frac{1.25}{0.1564} = 7.9906 \text{ in.}$$

and
$$R = \frac{7.9906}{2} = 3.9953 \text{ in.}$$

The **bottom diameter** is the diameter of a circle tangent to the curve at the bottom of the tooth gap and is equal to the difference between the pitch diameter and the roller diameter.

Used as a quality control measurement, the **caliper diameter** is the same as the bottom diameter for a sprocket with an even number of teeth. For a sprocket with an odd number of teeth, it is defined as the distance from the bottom of one tooth gap to that of the nearest opposite tooth gap.

To ensure clearance for standard link plates, the maximum **hub and groove diameter**, *MHD*, is given by

$$MHD = p\left[\cot\left(\frac{180°}{n}\right) - 1\right] - 0.030 \qquad (6\text{-}24)$$

Thus, for $p = 1\frac{1}{4}$ in., and $n = 20$,

$$MHD = 1.25\left[\cot\left(\frac{180°}{20}\right) - 1\right] - 0.030$$
$$= 1.25[6.3138 - 1] - 0.030$$
$$= 6.612 \text{ in.}$$

6-15 STANDARD ROLLER CHAINS

Table 6-8 lists the dimensions, strengths, and weights of American Standard *single-width* roller chains. Multiple strand chains, Fig. 6-25, increase power transmission capacity very nearly in proportion to the number of strands. At elevated speeds, it is possible to use shorter pitch chains and thereby reduce noise and weight. Multiple roller chains (double, triple, and quadruple) are available in all standard pitches and are manufacturers' stock items.

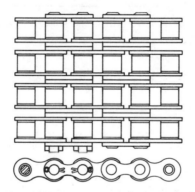

Figure 6-25 / Multiple strand chain.

6-16 DESIGN PRACTICE

Chain and sprocket selection is based on horsepower and type of drive, speeds and sizes of shafting, and surrounding conditions. Table 6-9 lists service factors as a function of the type of driver and the type of load, atmospheric conditions and hours of operation. Multiple strand factors are given in Table 6-10. Thus, a four-strand roller chain, Fig. 6-25, has a horsepower capacity of 3.3 times that of a single-strand chain.

$$\text{hp table rating} = \frac{\text{hp to be transmitted} \times \text{product of service factors}}{\text{multiple strand factor}}$$

(6-25)

TABLE 6-8

DIMENSIONS, STRENGTHS, AND WEIGHTS OF ASA STANDARD, SINGLE-WIDTH CHAINS[1]

ASA Chain No.	Pitch	Roller Width E	Roller Diam. H	Over-All Riveted A	From Pin End to C.L. B	From Pin Head to C.L. C	Side Plate Thickness EE	Side Plate Height F	Pin Diam. G	Average Ultimate Strength Lbs.	Average Weight per Foot Lbs.
25[1]	$\frac{1}{4}$	$\frac{1}{8}$	0.130	0.31	0.19	0.15	0.030	0.23	0.0905	925	0.085
35[1]	$\frac{3}{8}$	$\frac{3}{16}$	.200	.47	.34	.23	.050	.36	.141	2,100	.22
41	$\frac{1}{2}$	$\frac{1}{4}$	.306	.51	.37	.26	.050	.39	.141	2,000	.28
40	$\frac{1}{2}$	$\frac{5}{16}$	$\frac{5}{16}$	.65	.42	.32	.060	.46	.156	3,700	.41
50	$\frac{5}{8}$	$\frac{3}{8}$	.400	.79	.56	.40	.080	.59	.200	6,100	.68
60	$\frac{3}{4}$	$\frac{1}{2}$	$\frac{15}{32}$	.98	.64	.49	.094	.68	.234	8,500	.96
80	1	$\frac{5}{8}$	$\frac{5}{8}$	1.28	.74	.64	.125	$\frac{7}{8}$	.312	14,500	1.7
100	$1\frac{1}{4}$	$\frac{3}{4}$	$\frac{3}{4}$	1.54	.91	.77	.156	$1\frac{5}{32}$	.375	24,000	2.7
120	$1\frac{1}{2}$	1	$\frac{7}{8}$	1.94	1.14	.97	.187	$1\frac{3}{8}$	.437	34,000	4.0
140	$1\frac{3}{4}$	1	1	2.08	1.22	1.04	.218	$1\frac{5}{8}$	.500	46,000	5.2
160	2	$1\frac{1}{4}$	$1\frac{1}{8}$	2.48	1.46	1.24	.250	$1\frac{7}{8}$	.562	58,000	6.8
180	$2\frac{1}{4}$	$1\frac{13}{32}$	$1\frac{13}{32}$	2.81	1.74	1.40	.281	$2\frac{1}{8}$	.687	76,000	9.1
200	$2\frac{1}{2}$	$1\frac{1}{2}$	$1\frac{9}{16}$	3.02	1.86	1.51	.312	$2\frac{5}{16}$	.781	95,000	10.8

DIMENSIONS—INCHES

[1] *Non-roller*

TABLE 6-9

SERVICE FACTORS FOR ROLLER CHAINS

Type of Driven Load	Type of input power		
	Internal Combustion Engine with Hydraulic Drive	Electric Motor or Turbine	Internal Combustion Engine with Mechanical Drive
Smooth	1.0	1.0	1.2
Moderate Shock	1.2	1.3	1.4
Heavy Shock	1.4	1.5	1.7

Atmospheric conditions	Relatively clean and moderate temperature	1.0
	Moderately dirty and moderate temperature	1.2
	Exposed to weather, very dirty, abrasive, mildly corrosive and reasonably high temperatures	1.4

Daily Operating Range	8–10 Hours	1.0
	10–24 Hours	1.4

Both chain length and center distances are usually given in "pitches"— obviously, the required length of chain must consist of a whole number of links. The equations that follow give either the chain length in pitches when the center distance is known or the center distance, again in pitches, when the length is known.

$$L = 2C + \frac{N+n}{2} + \frac{(N-n)^2}{4\pi^2 C} \qquad (6\text{-}26)$$

$$C = \frac{L - \frac{N+n}{2} + \sqrt{\left(L - \frac{N+n}{2}\right)^2 - 8\frac{(N-n)^2}{4\pi^2}}}{4} \qquad (6\text{-}27)$$

where L = length of chain in pitches
C = center distance in pitches
N = number of teeth on large sprocket
n = number of teeth on small sprocket

TABLE 6-10

MULTIPLE STRAND FACTORS

NUMBER OF STRANDS	MULTIPLE-STRAND FACTOR
2	1.7
3	2.5
4	3.3

As with belts, chain speed V in fpm is given by

$$V = \frac{\pi \times D \times \text{rpm}}{12} \qquad (6\text{-}28)$$

where D = pitch diameter in inches.

Recommended horsepower ratings, Table 6-11, apply to lubricated single-pitch, single-strand, standard ASA roller chains. For multiple strands, the ratings must be multiplied by the *strand factor*; since the tables refer to a service factor of 1, the values must be reduced accordingly through the use of Eq. (6-25). Table 6-11 also indicates the recommended type of lubrication. Specifically there are four types:

I / *Manual lubrication.* Oil is periodically applied with a brush or oil can at least once every 8 hours.

II / *Drip lubrication.* Oil drops are continually directed between the link plate edges from a drip lubricator.

III / *Bath lubrication.* With bath lubrication the lower strand of chain runs through a sump of oil in the drive housing. The oil level should reach the pitch line of the chain at its lowest point during operation.

IV / *Oil stream lubrication.* The lubricant is applied by a circulating pump that supplies the chain with a continuous stream of oil.

EXAMPLE 6-6

A four-strand No. 100 roller chain, which is powered through a speed reducer by an electric motor, operates 8 hours per day under moderate shock and clean conditions. The driving sprocket has 15 teeth and rotates at 400 rpm; the large sprocket has 24 teeth. The sprockets are separated by 30 pitches. Determine: (a) the horsepower transmitted; (b) the pitch diameter of each sprocket; (c) the chain speed; (d) the chain length; (e) the working load in the chain and the safety factor based on the ultimate strength; and (f) the type of lubrication required.

TABLE 6-11

HORSEPOWER RATINGS OF STANDARD, SINGLE-STRAND ROLLER CHAINS[1]

ASA No. 25 — $\frac{1}{4}$" Pitch

No. of Teeth Small Spkt.	Revolutions per minute—small sprocket																			
	100	500	900	1200	1800	2500	3000	3500	4000	4500	5000	5500	6000	6500	7000	7500	8000	8500	9000	10000
11	0.054	0.23	0.39	0.50	0.73	0.98	1.15	1.32	1.42	1.19	1.01	0.88	0.77	0.68	0.61	0.55	0.50	0.46	0.42	0.36
12	0.059	0.25	0.43	0.55	0.80	1.07	1.26	1.45	1.62	1.36	1.16	1.00	0.88	0.78	0.70	0.63	0.57	0.52	0.48	0.41
15	0.075	0.32	0.54	0.70	1.01	1.36	1.61	1.85	2.08	1.89	1.62	1.40	1.23	1.09	0.98	0.88	0.80	0.73	0.67	0.57
18	0.092	0.39	0.66	0.86	1.23	1.66	1.95	2.25	2.53	2.49	2.12	1.84	1.62	1.43	1.28	1.16	1.05	0.96	0.88	0.75
21	0.108	0.46	0.78	1.01	1.46	1.96	2.31	2.65	2.99	3.13	2.68	2.32	2.04	1.80	1.61	1.46	1.32	1.21	1.11	0.95
24	0.125	0.53	0.90	1.17	1.69	2.26	2.67	3.07	3.45	3.83	3.27	2.83	2.48	2.20	1.97	1.78	1.61	1.47	1.35	1.16
30	0.159	0.68	1.15	1.49	2.14	2.88	3.39	3.90	4.40	4.89	4.57	3.96	3.47	3.08	2.76	2.49	2.26	2.06	1.89	1.62

Lubrication zones (diagonal bands): Type I, Type II, Type III, Type IV

ASA No. 35 — $\frac{3}{8}$" Pitch

No. of Teeth Small Spkt.	Revolutions per minute—small sprocket																			
	100	500	900	1200	1800	2500	3000	3500	4000	4500	5000	5500	6000	6500	7000	7500	8000	8500	9000	10000
11	0.18	0.78	1.32	1.72	2.47	3.32	2.93	2.32	1.90	1.59	1.36	1.18	1.04	0.92	0.82	0.74	0.67	0.61	0.56	0.48
12	0.20	0.86	1.45	1.89	2.72	3.65	3.35	2.66	2.17	1.82	1.56	1.35	1.18	1.05	0.94	0.85	0.77	0.70	0.64	0.55
15	0.26	1.09	1.85	2.40	3.45	4.64	4.66	3.70	3.03	2.54	2.17	1.88	1.65	1.46	1.31	1.18	1.07	0.98	0.90	0.77
18	0.31	1.33	2.25	2.92	4.20	5.65	6.13	4.87	3.98	3.34	2.85	2.47	2.17	1.92	1.72	1.55	1.41	1.29	1.18	1.01
21	0.37	1.57	2.66	3.45	4.97	6.68	7.73	6.13	5.02	4.21	3.59	3.11	2.73	2.42	2.17	1.96	1.77	1.62	1.49	1.27
24	0.43	1.81	3.08	3.98	5.74	7.71	9.09	7.49	6.13	5.14	4.39	3.80	3.34	2.96	2.65	2.39	2.17	1.98	1.82	1.55
30	0.54	2.31	3.91	5.07	7.30	9.81	11.6	10.5	8.57	7.18	6.14	5.32	4.67	4.14	3.70	3.34	3.03	2.77	2.54	2.17

Lubrication zones (diagonal bands): Type I, Type II, Type III, Type IV

[1]Courtesy of Cullman Wheel Company

TABLE 6-11 (Cont.)

ASA No. 40 — $\frac{1}{2}$" Pitch

Revolutions per minute—small sprocket

(Type I, Type II, Type III, Type IV band labels appear diagonally across the table, from low to high rpm.)

No. of Teeth Small Spkt.	50	200	400	600	900	1200	1800	2400	3000	3500	4000	4500	5000	5500	6000	6500	7000	7500	8000	9000
11	0.23	0.80	1.50	2.16	3.11	4.03	4.66	3.03	2.17	1.72	1.41	1.18	1.01	0.87	0.77	0.68	0.61	0.55	0.50	0
12	0.25	0.88	1.65	2.37	3.42	4.43	5.31	3.45	2.47	1.96	1.60	1.34	1.15	1.00	0.87	0.77	0.69	0.62	0.57	0
15	0.32	1.12	2.10	3.02	4.35	5.64	7.43	4.82	3.45	2.74	2.24	1.88	1.60	1.39	1.22	1.08	0.97	0.87	0	
18	0.39	1.37	2.55	3.68	5.30	6.86	9.76	6.34	4.54	3.60	2.95	2.47	2.11	1.83	1.60	1.42	1.27	1.15	0	
21	0.46	1.62	3.02	4.34	6.26	8.11	11.7	7.99	5.72	4.54	3.71	3.11	2.66	2.30	2.02	1.79	1.60	1.45	0	
24	0.54	1.87	3.48	5.02	7.23	9.36	13.5	9.76	6.99	5.54	4.54	3.80	3.25	2.81	2.47	2.19	1.96	0		
30	0.68	2.38	4.43	6.38	9.20	11.9	17.2	13.6	9.76	7.75	6.34	5.31	4.54	3.93	3.45	0				

ASA No. 41 — $\frac{1}{2}$" Pitch (light weight)

Revolutions per minute—small sprocket

(Type I, Type II, Type III, Type IV band labels appear diagonally across the table, from low to high rpm.)

No. of Teeth Small Spkt.	50	200	400	600	900	1200	1800	2400	3000	3500	4000	4500	5000	5500	6000	6500	7000	7500	8000	9000
11	0.13	0.44	0.82	1.19	1.71	1.71	0.93	0.60	0.43	0.34	0.28	0.24	0.20	0.17	0.16	0.14	0.12	0.11	0.10	0
12	0.14	0.49	0.91	1.31	1.88	1.95	1.06	0.69	0.49	0.39	0.32	0.27	0.23	0.20	0.17	0.15	0.14	0.12	0.11	0
15	0.18	0.62	1.15	1.66	2.39	2.73	1.49	0.96	0.69	0.55	0.45	0.38	0.32	0.28	0.24	0.22	0.19	0.17	0	
18	0.22	0.75	1.40	2.02	2.91	3.59	1.95	1.27	0.91	0.72	0.59	0.49	0.42	0.37	0.32	0.28	0.25	0.23	0	
21	0.26	0.89	1.66	2.39	3.44	4.46	2.46	1.60	1.14	0.91	0.74	0.62	0.53	0.46	0.40	0.36	0.32	0.29	0	
24	0.29	1.03	1.92	2.76	3.97	5.15	3.01	1.95	1.40	1.11	0.91	0.76	0.65	0.56	0.49	0.44	0.39	0		
30	0.38	1.31	2.44	3.51	5.06	6.55	4.20	2.73	1.95	1.55	1.27	1.06	0.91	0.79	0.69	0				

TABLE 6-11 (Cont.)

ASA No. 50 — 5/8" Pitch

Revolutions per minute — small sprocket

Regions in the table are designated Type I, Type II, Type III, and Type IV.

No. of Teeth Small Spkt.	50	100	300	500	900	1200	1500	1800	2100	2400	2700	3000	3300	3500	4000	4500	5000	5400	5800	6200
11	0.45	0.84	2.25	3.55	6.07	7.86	7.44	5.58	4.42	3.62	3.04	2.59	2.25	2.06	1.68	1.41	1.21	1.07	0.97	0
12	0.50	0.92	2.47	3.91	6.66	8.64	8.50	6.38	5.04	4.13	3.47	2.96	2.57	2.35	1.92	1.61	1.38	1.22	1.10	0
15	0.63	1.17	3.15	4.97	8.48	11.0	11.9	8.89	7.03	5.76	4.83	4.13	3.58	3.28	2.68	2.24	1.92	1.70	0	
18	0.77	1.43	3.83	6.05	10.3	13.4	15.6	11.7	9.24	7.58	6.35	5.42	4.70	4.31	3.52	2.95	2.53	0		
21	0.90	1.69	4.53	7.15	12.2	15.8	19.3	14.7	11.6	9.55	8.01	6.83	5.93	5.44	4.44	3.71	3.19	0		
24	1.04	1.95	5.23	8.26	14.1	18.3	22.3	18.0	14.2	11.7	9.78	8.34	7.24	6.64	5.42	4.54	0			
30	1.33	2.42	6.66	10.5	17.9	23.2	28.4	25.1	19.9	16.3	13.7	11.7	10.1	9.28	7.57	0				

ASA No. 60 — 3/4" Pitch

Revolutions per minute — small sprocket

Regions in the table are designated Type I, Type II, Type III, and Type IV.

No. of Teeth Small Spkt.	50	100	200	500	700	900	1200	1400	1600	1800	2000	2200	2400	2600	2800	3000	3500	3800	4000	4600
11	0.78	1.44	2.69	6.14	8.32	10.5	11.9	9.45	7.70	6.49	5.51	4.78	4.20	3.72	3.33	3.00	2.38	2.10	1.95	0
12	0.85	1.58	2.96	6.73	9.14	11.5	13.6	10.8	8.79	7.42	6.29	5.46	4.79	4.25	3.80	3.42	2.72	2.40	2.23	0
15	1.08	2.01	3.76	8.57	11.6	14.6	18.9	15.0	12.3	10.3	8.77	7.61	6.68	5.93	5.30	4.77	3.79	3.34	3.10	0
18	1.32	2.45	4.58	10.4	14.1	17.8	23.0	19.8	16.1	13.6	11.5	10.0	8.78	7.79	6.97	6.27	4.98	4.39	4.08	0
21	1.56	2.89	5.41	12.3	16.7	21.0	27.2	24.9	20.3	17.1	14.5	12.6	11.1	9.82	8.79	7.90	6.27	5.53	0	
24	1.80	3.34	6.25	14.2	19.3	24.3	31.4	30.4	24.8	20.9	17.7	15.4	13.5	12.0	10.7	9.65	7.66	0		
30	2.29	4.25	7.95	18.1	24.6	30.9	40.0	42.6	34.7	29.2	24.8	21.5	18.9	16.8	15.0	13.5	0			

TABLE 6-11 (Cont.)

ASA No. 80—1" Pitch

Type regions (diagonal): Type I, Type II, Type III, Type IV

No. of Teeth Small Spkt.	Revolutions per minute—small sprocket																			
	25	50	100	200	300	400	500	700	900	1000	1200	1400	1600	1800	2000	2200	2400	2600	2800	3000
11	0.97	1.80	3.36	6.28	9.04	11.7	14.3	19.4	23.0	19.6	14.9	11.8	9.69	8.12	6.94	6.01	5.27	4.68	4.18	0
12	1.06	1.98	3.70	6.89	9.93	12.9	15.7	21.3	26.2	22.4	17.0	13.5	11.1	9.28	7.92	6.86	6.02	5.34	4.78	0
15	1.35	2.52	4.70	8.77	12.6	16.4	20.0	27.1	34.0	31.2	23.8	18.9	15.4	12.9	11.0	9.57	8.40	7.45	0.42	0
18	1.64	3.07	5.72	10.7	15.4	19.9	24.4	33.0	41.3	41.0	31.2	24.8	20.3	17.0	14.5	12.6	11.0	9.79	0	
21	1.94	3.62	6.76	12.6	18.2	23.6	28.8	39.0	48.9	51.7	39.4	31.2	25.6	21.4	18.3	15.9	13.9	0		
24	2.24	4.19	7.81	14.6	21.0	27.2	33.3	45.0	56.4	62.0	48.1	38.1	31.2	26.2	22.3	19.4	17.0	0		
30	2.85	5.33	9.94	18.5	26.7	34.6	42.3	57.3	71.8	78.9	67.2	53.3	43.6	36.6	31.2	24.5	0			

ASA No. 100—1¼" Pitch

Type regions (diagonal): Type I, Type II, Type III, Type IV

No. of Teeth Small Spkt.	Revolutions per minute—small sprocket																			
	10	25	50	100	200	300	400	500	600	700	800	900	1000	1100	1200	1400	1600	1800	2000	2200
11	0.81	1.85	3.45	6.44	12.0	17.3	22.4	27.4	32.3	37.1	32.8	27.5	23.4	20.3	17.8	14.2	11.6	9.71	6.95	0
12	0.89	2.03	3.79	7.07	13.2	19.0	24.6	30.1	35.5	40.8	37.4	31.4	26.8	23.2	20.4	16.2	13.2	11.1	0	
15	1.13	2.59	4.82	9.00	16.8	24.2	31.4	38.3	45.2	51.9	52.2	43.7	37.3	32.4	28.4	22.5	18.5	15.5	0	
18	1.38	3.15	5.87	11.0	20.4	29.5	38.2	46.7	55.0	63.2	68.6	57.5	49.1	42.5	37.3	29.6	24.3	16.6	0	
21	1.63	3.72	6.94	12.9	24.2	34.8	45.1	55.1	64.9	74.6	84.1	72.4	61.8	53.6	47.0	37.3	30.5	0		
24	1.88	4.30	8.02	15.0	27.9	40.3	52.1	63.7	75.0	86.2	97.2	88.4	75.5	65.5	57.4	45.6	37.3	0		
30	2.40	5.47	10.2	19.0	35.5	51.2	66.3	8.10	95.5	110	124	124	106	91.5	80.3	63.7	0			

TABLE 6-11 (Cont.)

ASA No. 120— 1½" Pitch

(Types I, II, III, IV indicated across the table diagonally)

No. of Teeth Small Spkt.	\multicolumn{20}{c}{Revolutions per minute—small sprocket}																			
---	10	25	50	100	150	200	300	400	500	600	700	800	900	1000	1100	1200	1300	1400	1500	1600
11	1.37	3.12	5.82	10.9	15.6	20.3	29.2	37.8	46.3	54.5	46.2	37.8	31.7	27.1	23.5	20.6	18.3	16.3	14.7	0
12	1.50	3.43	6.39	11.9	17.2	22.3	32.1	41.5	50.8	59.8	52.8	43.2	36.2	30.9	26.8	23.5	20.9	18.7	16.8	0
15	1.91	4.36	8.13	15.2	21.9	28.3	40.8	52.9	64.6	76.1	73.6	60.3	50.5	43.1	37.4	32.8	29.1	26.0	4.50	0
18	2.33	5.31	9.90	18.5	26.6	34.5	49.7	64.3	78.6	92.7	96.7	79.2	66.4	56.7	49.1	43.2	38.3	28.6	0	
21	2.75	6.27	11.7	21.8	31.4	40.7	58.7	76.0	93.0	110	122	99.8	83.6	71.4	61.9	54.3	48.2	0		
24	3.18	7.25	13.5	25.2	36.4	47.1	67.8	87.9	107	127	145	122	102	87.2	75.5	66.3	32.6			
30	4.04	9.22	17.2	32.1	46.3	59.9	86.3	112	137	161	185	170	143	122	106	0				

ASA No. 140— 1¾" Pitch

(Types I, II, III, IV indicated across the table diagonally)

No. of Teeth Small Spkt.	\multicolumn{20}{c}{Revolutions per minute—small sprocket}																			
---	10	25	50	100	150	200	250	300	350	400	450	500	550	600	700	800	900	1000	1100	1200
11	2.13	4.86	9.06	16.9	24.4	31.5	38.6	45.5	52.2	58.9	65.5	72.0	75.1	65.8	52.2	42.8	35.8	30.6	26.5	0
12	2.34	5.33	9.94	18.6	26.7	34.6	42.3	49.9	57.3	64.6	71.8	79.0	81.5	71.5	56.8	46.5	38.9	33.2	28.8	0
15	2.98	6.80	12.7	23.7	34.1	44.1	54.0	63.6	73.1	82.4	91.6	101	110	105	83.2	68.1	57.1	48.7	40.0	0
18	3.62	8.26	15.4	28.8	41.4	53.7	65.6	77.3	88.8	100	111	122	133	138	109	90.0	75.0	64.0	22.7	0
21	4.28	9.77	18.2	34.0	49.0	63.4	77.6	91.4	105	118	132	145	158	171	138	113	94.5	69.0	0	
24	4.95	11.3	21.1	39.3	56.6	73.3	89.7	106	121	137	152	167	182	197	168	138	115	46.5	0	
30	6.29	14.4	26.8	50.0	72.0	93.2	114	134	154	174	194	213	232	251	235	175	83.0	0		

ASA No. 160 — 2" Pitch

No. of Teeth Small Spkt.	Revolutions per minute—small sprocket																			
	10	25	50	100	150	200	250	300	350	400	500	550	600	700	750	800	850	900	950	1000
11	3.07	7.03	13.1	24.4	35.1	45.3	55.6	65.5	75.3	84.9	96.7	83.9	73.3	58.3	52.6	47.8	42.0	27.0	13.0	0
12	3.37	7.72	14.4	26.8	38.5	49.9	61.0	71.9	82.7	93.3	110	95.8	83.8	66.7	60.0	54.6	39.4	22.5	0	
15	4.28	9.86	18.3	34.1	49.0	63.5	77.7	91.5	105	119	145	134	117	93.0	83.7	66.3	31.6	9.00	0	
18	5.21	12.0	22.3	41.5	59.6	77.3	94.5	111	128	144	177	176	154	122	90.4	59.4	23.8	0		
21	6.16	14.1	26.3	49.0	70.5	91.4	112	132	151	171	209	214	193	128	87.3	48.6	15.4	0		
24	7.12	16.3	30.4	56.6	81.4	106	129	152	175	197	240	220	203	107	64.2	32.4	0			
30	9.06	20.7	38.7	72.1	104	134	164	193	222	251	270	230	190	65.0	18.0	0				

(Type bands, left to right: Type I, Type II, Type III, Type IV)

ASA No. 180 — 2¼" Pitch

No. of Teeth Small Spkt.	Revolutions per minute—small sprocket																			
	10	25	50	100	150	200	250	300	350	400	450	500	550	600	650	675	700	725	750	800
11	4.24	9.68	18.1	33.7	48.5	62.8	76.8	90.6	104	117	125	106	92.0	81.0	57.0	44.0	28.0	18.0	12.0	0
12	4.66	10.6	19.9	37.0	53.3	69.0	84.3	99.6	114	128	134	110	96.0	81.8	55.0	41.3	25.0	14.4	0	
15	5.92	13.5	25.3	47.1	67.8	87.7	107	127	145	163	161	125	108	84.2	49.0	33.5	16.0	3.60	0	
18	7.21	16.5	30.8	57.4	82.5	107	131	154	177	200	188	152	121	86.6	43.0	25.7	7.00	0		
21	8.52	19.5	36.4	67.8	97.5	126	154	182	209	236	207	169	129	81.9	34.5	17.5	0			
24	9.84	22.5	42.1	78.3	113	146	178	210	241	257	204	166	120	65.1	20.0	0				
30	12.5	28.6	53.6	99.6	143	185	227	268	295	269	200	152	104	32.0	0					

(Type bands, left to right: Type I, Type II, Type III, Type IV)

TABLE 6-11 (Cont.)

346

TABLE 6-11 (Cont.)

ASA No. 200 — 2½" Pitch

No. of Teeth Small Spkt.	Revolutions per minute—small sprocket																			
	5	10	15	20	30	40	50	60	80	100	150	200	250	300	350	400	450	550	600	650
11	3.02	5.63	8.11	10.5	15.1	19.6	24.0	28.2	36.6	44.7	64.5	83.5	102	120	135	122	104	58.2	31.1	0
12	3.32	6.19	8.91	11.5	16.6	21.6	26.3	31.0	40.2	49.1	70.8	91.7	112	132	145	130	110	59.4	29.3	0
15	4.22	7.87	11.3	14.7	21.2	27.4	33.5	39.5	51.2	63.5	90.1	117	143	172	172	151	125	59.0	20.1	0
18	5.14	9.58	13.8	17.9	25.8	33.4	40.8	48.0	62.3	76.1	110	142	174	195	194	168	135	52.5	4.11	0
21	6.07	11.3	16.3	21.1	30.4	39.4	48.2	56.8	73.6	89.9	130	168	205	237	213	180	140	39.6	0	
24	7.01	13.1	18.8	24.4	35.2	45.6	55.7	65.6	85.0	104	150	194	237	258	228	188	140	21.0	0	
30	8.92	16.6	24.0	31.0	44.7	58.0	70.8	83.4	106	132	190	247	301	289	246	191	125	0		

(Lubrication zones across columns: Type I, Type II, Type III, Type IV.)

ASA No. 240 — 3" Pitch

No. of Teeth Small Spkt.	Revolutions per minute—small sprocket																			
	5	10	15	20	25	30	40	50	60	80	100	125	150	175	200	250	300	350	400	450
11	4.88	9.11	13.1	17.0	20.8	24.5	31.7	38.8	45.7	59.2	72.3	88.4	104	120	127	115	94.7	68.2	36.1	0
12	5.36	10.0	14.4	18.7	22.8	26.9	34.8	42.6	50.2	65.0	79.5	97.1	114	131	136	123	101	71.3	35.8	0
15	6.82	12.7	18.3	23.8	29.0	34.2	44.3	54.2	63.8	82.7	101	124	146	168	164	145	116	77.5	31.4	0
18	8.30	15.5	22.3	28.9	35.3	41.6	53.9	65.9	77.7	101	123	150	177	195	189	164	127	78.9	21.7	0
21	9.81	18.3	26.4	34.2	41.7	49.2	63.7	77.9	91.8	119	145	178	209	219	211	180	134	75.7	6.50	
24	11.3	21.1	30.4	39.5	48.2	56.8	73.6	90.0	106	137	168	205	242	241	230	192	137	67.9	0	
30	14.4	26.9	38.7	50.2	61.4	72.3	93.7	115	135	175	214	261	288	278	261	207	131	38.0		

(Lubrication zones across columns: Type I, Type II, Type III, Type IV.)

Type I   Manual Lubrication. Oil applied periodically with brush or spout can. (75 fpm max. chain speed)   Type III   Oil bath or Oil Slinger. Oil level maintained in casing at predetermined height. (800 fpm max.)
Type II   Drip Lubrication. Oil applied between link plate edges from a drip lubricator (220 fpm max.)   Type IV   Oil stream. Oil applied by circulating pump inside chain loop on lower span. (up to max speed shown)

solution

(a) From Table 6-11: single-strand hp $= 31.4$, based on a 15-tooth sprocket turning at 400 rpm. From Eq. (6-25) and Tables 6-9 and 6-10:

$$\text{hp transmitted} = \frac{31.4 \times 3.3}{1.3 \times 1 \times 1} = 79.7 \text{ hp}$$

(b) From Eq. (6-23):

$$D = \frac{p}{\sin\left(\dfrac{180°}{n}\right)}$$

where $p = 1\frac{1}{4}$ in.

For the 15-tooth sprocket:

$$D = \frac{1.25}{\sin\left(\dfrac{180°}{15}\right)} = 6.012 \text{ in.}$$

For the 24-tooth sprocket:

$$D = \frac{1.25}{\sin\left(\dfrac{180°}{24}\right)} = 9.576 \text{ in.}$$

(c) From Eq. (6-28):

$$V = \frac{\pi \times D \times \text{rpm}}{12} = \frac{\pi \times 6.0122 \times 400}{12} = 629.6 \text{ fpm}^2$$

(d) From Eq. (6-26):

$$L = 2C + \frac{N + n}{2} + \frac{(N - n)^2}{4\pi^2 C}$$

$$= (2 \times 30) + \frac{24 + 15}{2} + \frac{(24 - 15)^2}{4\pi^2 \times 30}$$

$$= 60 + 19.5 + 0.07 = 79.57 \text{ pitches}$$

Use 80 pitches:

$$\text{actual length} = 80 \times 1.25 = 100 \text{ in.}$$

(e) Since

$$\text{Working load} = \frac{\text{hp} \times 33,000}{V}$$

assume

$$\text{Maximum power transmitted} = \frac{31.4 \times 3.3 \times 33,000}{629.6}$$

$$= 5430 \text{ lb}$$

From Table 6-8,

$$\text{Ultimate strength} = 24,000 \text{ lb} \times 4 \text{ strands}$$

$$= 96,000 \text{ lb}$$

Therefore,

$$\text{Factor of safety} = \frac{96,000}{5430} = 17.7$$

(f) From Table 6-11: Type 3—bath lubrication.

QUESTIONS AND PROBLEMS

P6-1 To determine the coefficient of friction between a particular belt and pulley, combinations of weights are added to hangers, as shown. The sheave is locked to prevent its rotation. Find μ.

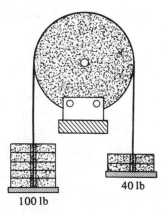

100 lb   40 lb

Figure P6-1

P6-2 Find the torque T required to turn the pulley shown. Assume the coefficient of belt friction to be 0.35.

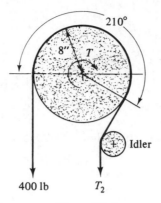

Figure P6-2

P6-3   Determine the centrifugal tension T_c for each of the following:

	BELT SPEED	BELT WEIGHT	BELT DIMENSIONS
(a)	3000 fpm	0.045 lb/cu-in.	10 in. wide by 23/64 in. thick
(b)	40 fps	1.560 lb/ft	—
(c)	5000 fpm	0.003 lb/in.	—

P6-4   A flat belt weighing 3.25 lb per lineal ft is used to transmit 35 hp to an agitator. The belt, which travels at 1500 fpm, has a 220° arc of contact on the driving pulley. If the coefficient of friction between the belt and pulley is 0.40, determine the maximum tight-side and slack-side tensions.

P6-5   The motor shown transmits maximum power at a speed of 1160 rpm. The belt, which is 6 in. wide and $\frac{1}{4}$ in. thick, weighs 0.04 lb

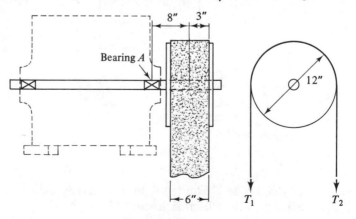

Figure P6-5

per in.³. The maximum allowable tensile stress in the belt is 250 psi, the coefficient of friction between the belt and pulley is 0.4, and the belt strands are vertical. Determine the bending moment and the torque in the motor shaft at bearing A.

P6-6 Plot a curve that will give values of centrifugal tension T_c as a function of belt speed for belt speeds varying between zero and 6000 fpm. Assume a value of $w = 1$ lb/ft, and show how the single curve could be used to give T_c for any value of w.

P6-7 Use Eq. (6-10) to approximate the percentage increase or decrease in belt life for the following conditions:
(a) an increase in small pulley diameter from 18 in. to 20 in.;
(b) a 75% reduction in belt length;
(c) an increase in belt thickness from $\frac{3}{16}$ in. to $\frac{1}{4}$ in.;
(d) a reduction in tight-side tension from 300 lb to 250 lb;
(e) an increase in belt speed from 1000 fpm to 2000 fpm; and
(f) a reduction in small pulley diameter from 14 in. to 12 in., and a 10% increase in tight-side tension.

P6-8 For the open drive shown, determine the contact angles β_1 and β_2 and the approximate belt length.

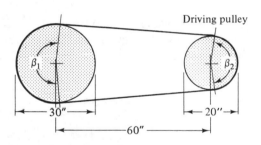

Figure P6-8

P6-9 If the driving pulley of Prob. 6-8 rotates at 900 rpm, determine the belt speed and the angular velocity of the driven pulley.

P6-10 Assume a belt configuration for the pulleys of Prob. 6-8 that causes the driver and driven pulleys to rotate in opposite directions. Determine the required belt length and the contact angles β_1 and β_2.

P6-11 Pulleys B, C, D, E, and F are each driven at 600 rpm by driver A. For the horsepower take-off shown, determine:
(a) the belt length; and
(b) belt tensions T_2, T_3, T_4, T_5, and T_6.

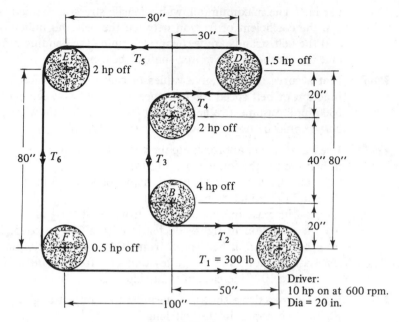

Figure P6-11

P6-12   For the drive shown, $T_1 = 450$ lb, and $T_2 = 200$ lb. By graphical analysis, determine the required idler weight and the bearing forces at shafts A, B, and C.

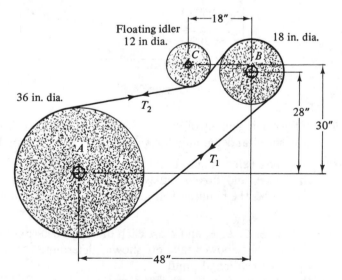

Figure P6-12

P6-13 Determine the design horsepower for each of the following flat belt drives:

	RATED HP	DRIVEN MACHINE	DRIVING MACHINE	SPEED CONDITIONS
(a)	20	flight conveyor	high torque	continuous, wet
(b)	50	positive blower	normal torque synchronous	continuous
(c)	35	2-cyl. reciprocating compressor	line start squirrel cage	frequent start
(d)	25	agitator for semiliquids	wound rotor	frequent start
(e)	100	jaw crusher	diesel engine	continuous

P6-14 For a V-belt-sheave combination, the coefficient of friction between the belt and sheave material is 0.42. Determine the effective coefficient of friction μ_e for groove angles of 34°, 36°, and 38°.

P6-15 A "D" section V-belt weighs approximately 0.40 lb per ft. Determine:
   (a)   the belt density in lb per in.^{3}, assuming a nominal groove angle of 36°; and
   (b)   the tensile stress in the belt for a tensile load of 250 lb.

P6-16 Use Figs. 6-17 through 6-21 to determine the horsepower ratings for the following V-belt drives:

	SECTION SIZE	BELT SPEED	SMALL SHEAVE EQUIVALENT DIAMETER
(a)	B	4000 fpm	5.8 in.
(b)	A	50 fps	3.4 in.
(c)	D	5000 fpm	16.0 in.
(d)	C	4200 fpm	10.5 in.

P6-17 Select, using Fig. 6-16, an appropriate belt section size for the following conditions:

	DESIGN HP	SMALL SHEAVE RPM
(a)	20	500
(b)	30	2500
(c)	25	1750
(d)	100	3000

P6-18 Determine the required pitch length of a V-belt for the following conditions:

> Small sheave O.D. = 12 in. (standard groove)
>
> Large sheave O.D. = 20 in. (standard groove)
>
> Center distance = 62 in.

P6-19 A V-belt has a pitch length of 200 in. and a large and small sheave pitch diameter of 24 in. and 10 in. respectively. Determine the required center distance.

P6-20 A 50-hp, 1200-rpm diesel engine is used to drive a piston compressor at 500 rpm under normal service conditions. The center distance between sheaves is 72 in., and the driven sheave has a pitch diameter of 30 in. Determine:
(a) the required V-belt size;
(b) the belt length; and
(c) the number of belts necessary.

P6-21 A 20-hp, 1800-rpm squirrel cage motor is used to drive a fan at 800 rpm under normal service conditions. Both the driving and driven sheaves have standard pitch diameters; sheave centers must be maintained between 46 in. and 50 in. Select at least two different drive systems that will satisfy the design conditions—answers should include the required belt size, belt length, number of belts required, and the sheave sizes.

P6-22 In the pivoted motor mount shown, T_1 and T_2 are 250 lb and 50 lb, respectively, at full capacity. Determine the proper mounting distance C if the motor weighs 100 lb; neglect the weight of the mounting bracket and centrifugal effects. What are the belt tensions when the motor is sitting idle?

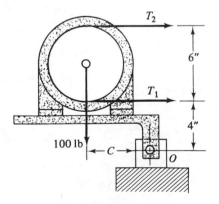

Figure P6-22

P6-23 Solve Prob. 6-22 if the belt pulls are vertically upward. Sketch a bracket design that would accommodate the motor mounting.

P6-24 For the pivoted motor mount shown, assume the following data apply:
1. At maximum load, $T_1 = 300$ lb, and $T_2 = 50$ lb
2. Motor weight $W = 200$ lb
3. Mounting bracket weight $w = 100$ lb at $c_m = 14$ in.
4. Belt pull distances: $d_1 = 10$ in., and $d_2 = 25$ in.
Find:
(a)   the motor position C;
(b)   the belt pulls at $\frac{1}{4}$ load; and
(c)   the belt pulls under zero load conditions.

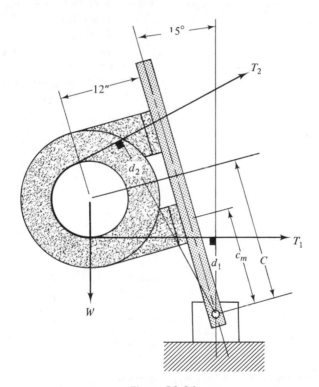

Figure P6-24

P6-25 Determine the pitch diameter and maximum allowable hub diameter of the following standard roller chain sprockets. In each case, estimate the weight per 100 pitches of single-strand chain for each sprocket.

	ASA CHAIN NO.	NO. OF TEETH IN SPROCKET
(a)	35	11
(b)	60	24
(c)	200	20

P6-26   Determine the required chain horsepower for each of the following drives:

	HP TO BE TRANSMITTED	NO. OF STRANDS	OPERATING CONDITIONS
(a)	50	2	Moderate shock and dirt; electric motor drive, continuous operation.
(b)	100	3	I.C. engine with mechanical drive, heavy shock, high temperatures, 18-hour operation.
(c)	20	1	Electric motor, smooth operation, clean conditions, 10-hour operation.

P6-27   The following data apply to a given chain drive:

$$\text{Small sprocket} = 15 \text{ teeth}$$
$$\text{Large sprocket} = 24 \text{ teeth}$$
$$\text{Chain no.} = \text{ASA } 160$$
$$\text{Center distance} = 50 \text{ pitches}$$
$$\text{rpm of small sprocket} = 50$$

Determine the chain length and velocity.

P6-28   Assume the chain length in Prob. 6-27 to be 100 pitches. Determine the center distance and pitch clearance as shown—all other data remains the same.

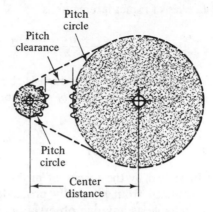

Figure P6-28

P6-29 A three-strand No. 80 roller chain powered by a diesel engine through a mechanical drive operates continuously under heavy shock; the drive also operates under extreme temperature changes. The driving sprocket has 18 teeth and rotates at 700 rpm. The driven sprocket has 30 teeth and the sprockets are separated by 30 pitches. Find:
(a)   the pitch diameter of each sprocket;
(b)   the chain length;
(c)   the horsepower transmitted;
(d)   the working load in the chain; and
(e)   the type of lubrication required.

SUGGESTED DESIGN PROJECTS

DP6-1 Data reliability is, perhaps, the single most important aspect in design. Belt friction, for example, is given as a numerical value that is a function of two mating materials—the belt and the pulley. To say that the coefficient of friction is 0.25 is to mean precisely what? Only after a series of measurements can one establish the degree of accuracy of the *mean* of the series. The difference between any measured observation and the arithmetic average, or mean, of a series of observations is called the *deviation from the mean*. Positive (+) deviation indicates that the measurement is greater than the mean; negative (−) deviation results when a measurement is smaller than the mean. Symbolically, the mean

m of a series of n measurements is

$$m = \frac{a_1 + a_2 + a_3 + \cdots + a_n}{n} = \frac{\bar{Z}(a)}{n}$$

The *average deviation* (a.d.), which is the arithmetic average (a) of all deviations (d) *regardless of sign*, represents the reliability of a single measurement.

$$\text{a.d.} = \frac{\bar{Z}(d)}{n}$$

It can be shown by the theory of probability that an arithmetic mean calculated from n equally probable observations is $\sqrt{n}$ times as reliable as any one observation. To find, therefore, the reliability of the mean of a series of measurements, divide the average deviation by the square root of the number of measurements. The ratio is called the *deviation of the mean* (A.D.):

$$\text{A.D.} = \frac{\text{a.d.}}{\sqrt{n}}$$

The dependability of any measurement can be expressed in terms of the most probable value as

Arithmetic mean $\pm$ A.D.

The belt/pulley testing device shown consists of a means of fastening the belt at A and a pan, also fastened to the belt, on which

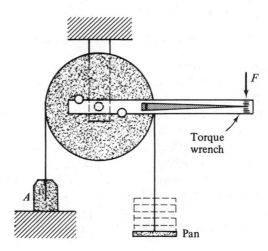

Figure DP6-1

a series of 5-lb weights can be placed. As weights are added to the pan, the couple required to just turn the pulley is measured with a torque wrench. Based on the data that follows, determine the most probable value of μ.

WEIGHT IN PAN (LB)	MEASURED TORQUE	WEIGHT IN PAN(LB)	MEASURED TORQUE
5	29.55 in. lb	30	176.25
10	62.10	35	210.10
15	87.75	40	243.25
20	125.50	45	262.50
25	155.00	50	310.84

DP6-2 The experiment described in Design Problem DP6-1 is to be repeated—this time, however, the belt contact angle is to increase by 10° increments starting with 90° and ending with 120°; the slack-side tension is to remain constant. Design an appropriate device, and describe its use to determine the most probable value of μ.

DP6-3 It is common practice to use a grooved sheave and a flat pulley in combination and driven by one or more V-belts. Two such drives are shown—one with a short center distance and one with a long center distance. Which is the better drive? Give a complete analysis to substantiate your answer.

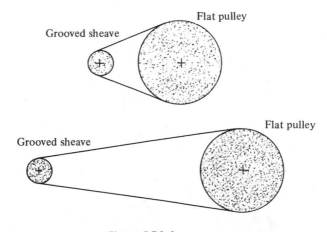

Figure DP6-3

DP6-4 For the automotive accessory drive system shown, determine the minimum required belt tensions at *A*, *B*, and *C* under the conditions that follow. Assume the coefficient of friction between the belt and sheaves is 0.35 and the sheave wedge angle is 36°.

CRANKSHAFT RPM	600	2500	3000	4500
HORSEPOWER LOADS				
FAN AND WATER PUMP	0.20	2.10	5.50	12.00
ALTERNATOR	1.50	2.25	2.00	2.50

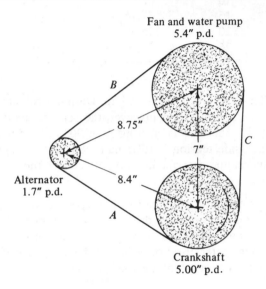

Figure DP6-4

DP6-5 Design the most economical (lightest chain) double reduction drive, of the type shown, for an air compressor running at 140 rpm. The power source is a 100-hp, 900-rpm motor, and the center distances are to be as short as possible. Make your own assumptions regarding service conditions.

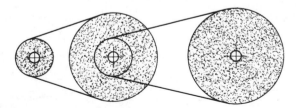

Figure DP6-5

CHAPTER 7

GEARS

"There are about as many rules for computing the power of a gear as there are manufacturers of gears, each foundryman having a rule, the only good one, which he has found in some book, and with which he will figure the power down to so many horses and hundredths of a horse as confidently as he will count the teeth or weigh the casting.

"This variety is very convenient, for it is always possible to fit a desired power to a given gear, and if a badly designed gear should break, it is a simple matter to find a rule to prove that it was just right, and must have met with some accident."

GEORGE B. GRANT
Gearing, 1907

7-1 INTRODUCTION

Gears are among history's oldest mechanical devices; in terms of discovery time, they rank along with the lever, inclined plane, pulley, and screw. While the use of gears probably dates back to 3000 BC, the earliest writings (Aristotle, 330 BC) describe devices such as clocks, grist mills, force multipliers, astronomical models, and perpetual motion machines. Leonardo da Vinci in his first manuscript, known as Codex Madrid I, written about 1495, described an extraordinary device, Fig. 7-1, that converts rotary motion into reciprocating motion. The device makes use of *mutilated* gears on a common shaft—as the crank turns, the wheel reverses its rotation. Da Vinci, seeing the best possible shape for gear teeth to minimize frictional resistance, demonstrated that the cycloidal shape—a shape common

361

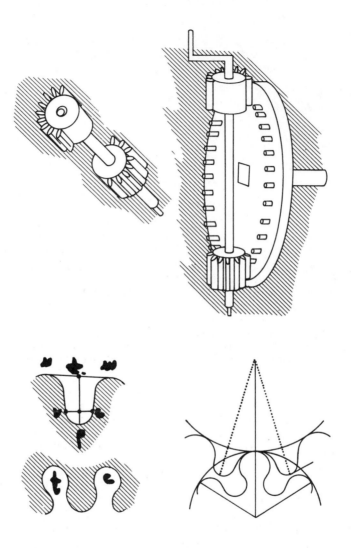

Figure 7-1

even in today's practice—is far superior to pegs. A *cycloid* is the curve generated by the motion of a point on the circumference of a circle rolled in a plane along a straight line, as shown in Fig. 7-2(a). If the circle is rolled on the outside of another circle, Fig. 7-2(b), the curve generated is called an *epicycloid*; if rolled on the inside, Fig. 7-2(c), it is called a *hypocycloid*.

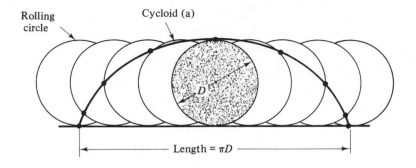

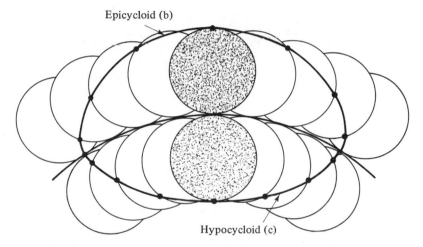

Figure 7-2 / *Cycloidal curves.*

7-2   *INVOLUTE GEOMETRY*

Fundamental to circular gearing is the requirement that the shape of teeth must be such that the ratio of angular velocities of the driving and driven gears must be a constant. Many tooth profiles satisfy this requirement, as history points out. The *involute* curve, however, is the most common for several reasons: it allows meeting teeth to roll in and out of mesh with a minimum of sliding actions; it fulfills the fundamental circular gear velocity requirements; and it is relatively easy to manufacture.

The involute, Fig. 7-3, is a single spirally curved line generated by a fixed point in a straight line that rolls without slipping on a circle. Simply put, it is the curve traced by a point on a taut cord unwinding from

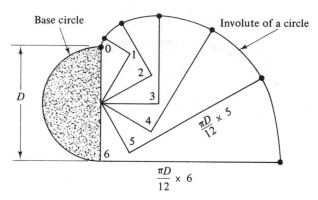

Base circle

Involute of a circle

Figure 7-3 / The involute.

around a circle. Portions of two involutes drawn in opposite directions, Fig. 7-4, are used to generate the spur gear tooth profile.

Basic spur gear nomenclature for a pair of meshed spur gears is shown in Fig. 7-5; note that the smaller of the two gears is generally called the *pinion*.

The definitions that follow are essential for understanding measurements and calculations:

Base Circle. That circle, or cylinder, used to form the involute, as described in Fig. 7-3.

Pitch Circle. When a pinion and gear are in proper mesh, they will come in contact at the pitch circle. The intersection of a tangent to the two base circles and the line-of-centers establishes the diameters of the pitch circles.

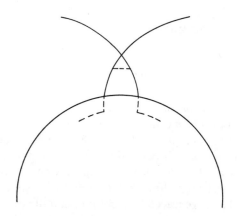

Figure 7-4 / Involute gear tooth profile.

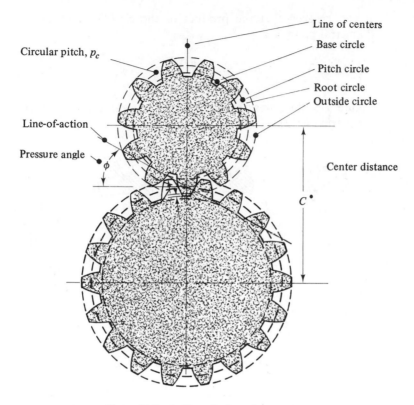

Figure 7-5 / Terminology for spur gears.

Pitch Diameter. The diameter D of the pitch circle.

Pitch Point. The point P of intersection between tangent to the two base circles and the line-of-centers.

Pitch. A measure of the size or spacing of gear teeth around the pitch circle.

Circular Pitch. The distance p_c measured along the pitch circle arc between the same points of adjacent teeth.

$$p_c = \frac{\text{Pitch circle length}}{\text{Number of teeth}} = \frac{\pi D}{N} \qquad (7\text{-}1)$$

Diametral Pitch. The number of teeth per inch of pitch diameter, symbolized by p_d; properly meshing gears must have the same diametral pitch.

$$p_d = \frac{N}{D} \qquad (7\text{-}2)$$

It can be seen that the product of the circular pitch and the diametral pitch is π.

$$p_c \times p_d = \frac{\pi D}{N} \times \frac{N}{D} = \pi$$

Figure 7-6 illustrates the actual sizes of gear teeth of different diametral pitches.

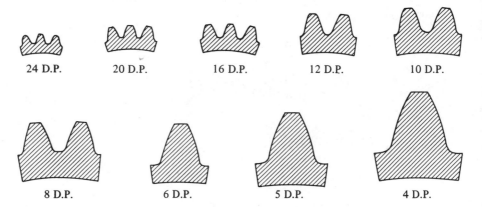

24 D.P.   20 D.P.   16 D.P.   12 D.P.   10 D.P.

8 D.P.   6 D.P.   5 D.P.   4 D.P.

Figure 7-6 | Sizes of gear teeth of different diametral pitches.

Center Distance. The sum C of the radii of the pitch circles—a value that can be expressed in terms of the number of teeth in each gear.

$$C = \frac{D_1 + D_2}{2} = \frac{1}{2} \times \left(\frac{N_1}{p_d} + \frac{N_2}{p_d}\right) = \frac{N_1 + N_2}{2p_d} \qquad (7\text{-}3)$$

Velocity Ratio (Gear Ratio). The ratio Z of angular velocities—ω_1 and ω_2 in radians per sec or n_1 and n_2 in rpm—a value that can also be expressed as the inverse ratio of pitch diameters, pitch radii, or numbers of teeth.

$$Z = \frac{\omega_1}{\omega_2} = \frac{n_1}{n_2} = \frac{D_2}{D_1} = \frac{R_2}{R_1} = \frac{N_2}{N_1} \qquad (7\text{-}4)$$

Pressure Angle. The angle ϕ between the common tangent to the base circles and a perpendicular to the line-of-centers. Two of the most widely used pressure angles are $14\frac{1}{2}°$ and $20°$. The following trigonometric relationship holds:

$$\phi = \cos^{-1}\left(\frac{\text{Base circle diameter}}{\text{Pitch diameter}}\right) = \cos^{-1}\left(\frac{D_b}{D}\right) \qquad (7\text{-}5)$$

Pitch Tooth Thickness. A linear measure T along the pitch circle equal to one-half of the circular pitch. Since circular pitch is related to diametral pitch, tooth thickness can be expressed as

$$T = \frac{p_c}{2} = \frac{\pi}{2p_d} \qquad (7\text{-}6)$$

Addendum. The height A of a tooth as measured between the pitch circle and the outside circle, where

$$A = \frac{1}{p_d} \qquad (7\text{-}7)$$

Dedendum. The depth of a tooth as measured between the pitch circle and the root circle.

Whole Depth. The total height of a tooth—the sum of the addendum and dedendum.

Clearance. The space between the outside circle of one gear and the root circle of the mating gear.

Working Depth. The depth of tooth engagement of two mating gears; the sum of the two addendums.

Backlash. The difference B, as shown in Fig. 7-7, between an excess tooth space and the thickness of the mating tooth—a distance measured at the backside or nondriving side of a pair of gear teeth. Some

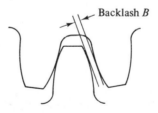

Figure 7-7 / Backlash.

backlash is necessary to allow for tooth deflection, thermal expansion, and lubrication. Backlash in a set of gears that are free to rotate can be felt by holding one gear stationary and moving the mating gear back and forth. The total movement in one direction is the accumulative backlash of the two gears.

$$B = \text{Standard tooth thickness} - \text{actual tooth thickness} \qquad (7\text{-}8)$$

Backlash can be intentionally introduced by slightly increasing the center distance by an amount ΔC. The *linear backlash* B_{LA}, which is the distance as measured with a feeler gage, is given by

$$B_{LA} = 2 \ (\Delta C) \sin \phi \qquad (7\text{-}9)$$

where ϕ is the pressure angle.

The following are commonly used and preferred diametral pitches.

COARSE			FINE		
$\frac{1}{2}$	3	12	20	64	128
1	4	14	24	72	150
2	6	16	32	80	180
2.25	8	18	40	96	200
2.5	10		48	120	

Table 7-1 gives the tooth proportions for the standard systems adopted by the USA Standards Institute and the AGMA (American Gear Manufacturers Association). Molded and sintered gears tend to adopt the 25-degree system. The higher pressure angles run more smoothly and quietly than the $14\frac{1}{2}$-degree system, but they produce somewhat more force on the bearings of the shaft.

TABLE 7-1

STANDARD GEAR SYSTEMS

	$14\frac{1}{2}°$ *AND* $20°$ *FULL DEPTH* (ASA B.6)	$20°$ *STUB* (ASA B.6)	$20°$ *AND* $25°$ *FULL DEPTH COARSE PITCH* 19.99 *AND COARSER* (AGMA 201.02)	$14\frac{1}{2}°$, $20°$, *AND* $25°$ *FULL DEPTH,* 20 *PITCH AND FINER* (AGMA 207.05)
Addendum	$\dfrac{1.000}{P_d}$	$\dfrac{0.800}{P_d}$	$\dfrac{1.000}{P_d}$	$\dfrac{1.000}{P_d}$
Dedendum	$\dfrac{1.157}{P_d}$	$\dfrac{1.000}{P_d}$	$\dfrac{1.250}{P_d}$	$\dfrac{1.200}{P_d} + 0.002$
Clearance	$\dfrac{0.157}{P_d}$	$\dfrac{0.200}{P_d}$	$\dfrac{0.250}{P_d}$	$\dfrac{0.200}{P_d} + 0.002$

The fine pitches are used in instrument systems such as computers and timers.

EXAMPLE 7-1

A 5-DP spur gear pinion with 15 teeth rotates at 1500 rpm driving a gear at 833 rpm. Find: (a) the number of teeth in the gear; (b) pitch diameter of pinion; (c) pitch diameter of gear; (d) O.D. of pinion; and (e) center distance of the two shafts.

solution

(a) $$1500/833 \times 15 \text{ teeth} = 27 \text{ teeth in gear}$$

(b) $$\text{diametral pitch} = N/D$$
$$5 = 15/D \quad \text{and} \quad D = 3.00 \text{ in.}$$

(c) $$27/5 = 5.40 \text{ in.}$$

(d) $$\text{O.D.} = \text{pitch diameter} + 2 \text{ addendums}$$
$$A = \frac{1}{p_d} = \frac{1}{5}$$
$$\text{O.D.} = 3.00 + 0.40 = 3.40 \text{ in.}$$

(e) $$\text{center distance} = \text{sum of pitch diameters}$$
$$= 3.00 + 5.40 = 8.40 \text{ in.}$$

EXAMPLE 7-2

An 18-tooth, 20° pressure angle spur gear is in mesh with a 60-tooth gear. The diametral pitch is 4, and the pinion has an angular velocity of 1200 rpm. Find: (a) the angular velocity of the gear; (b) the center distance; and (c) the linear backlash if the center distance is increased by 0.01 in.

solution

(a) Using Eq. (7-4),

$$Z = \frac{N_2}{N_1} = \frac{18}{60}$$

$$\text{rpm of gear} = \frac{18}{60} \times 1200 = 360 \text{ rpm}$$

(b) Using Eq. (7-3),

$$C = \frac{N_1 + N_2}{2p_d} = \frac{60 + 18}{2 \times 4} = 9.75 \text{ in.}$$

(c)   Using Eq. (7-9)

$$B_{LA} = 2(\Delta C) \sin \phi$$
$$= 2 \times 0.01 \times \sin 20° = 0.00684 \text{ in.}$$

EXAMPLE 7-3

The gear ratio of a pair of spur gears is 4:1, and the desired center distance is 12.600 in. Is it possible to use one of the pitch diameters in the preferred list for this system?

solution

$$N_2 = 4N_1$$

Using Eq. (7-2),

$$N_1 + N_2 = N_1 + 4N_1 = 5N_1 = 2 \times D \times p_d = 2 \times 12.6 p_d$$
$$N_1 = 5.04 p_d$$

Expressing 5.04 as a fraction and reducing to lowest terms gives

$$N_1 = \frac{504}{100} p_d = \frac{126}{25} p_d$$

The denominator 25 falls within Group IV pitches (12 to 50 by increments of 2), but odd-number pitches are not standard. It is, therefore, not possible to use standard gears for this system.

When a pinion with a normal tooth form has fewer than a certain number of teeth, there is interference between the meshing teeth that prevents proper meshing. This interference occurs at the bottom of the pinion teeth, as shown in Fig. 7-8. The minimum number of teeth that provides no interference or required undercutting of the teeth is 32 for a $14\frac{1}{2}°$ pressure angle and is 18 for a 20° pressure angle. For a 20° pressure angle and stub teeth, the minimum number is 14 teeth. If the pinion is produced by a process that generates the teeth, such as hobbing, there is no interference; the process undercuts the base of the tooth. But undercutting is not desirable because it results in a weakened tooth.

If a gear must be keyed to a shaft, the minimum pitch diameter must be twice the shaft diameter. The face width of a gear is usually 3 to 4 times the circular pitch.

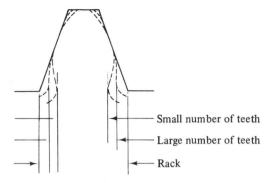

Figure 7-8 / Interference.

In order to mate, two gears must meet the following requirements:

1 / Identical pitch.
2 / Identical pressure angle.
3 / Identical addendum and dedendum.

7-3 FORCES IN GEARS

The forces in gears may be resolved into three components:

1 / A tangential component, tangential to the pitch circle.
2 / A radial force tending to separate the gears.
3 / A thrust force, axial to the shaft. For straight spur gears there is no thrust.

Figure 7-9 shows a mating spur gear and pinion. The reactions between the mating teeth occur at the pitch circle along the pressure line ab, at angle ϕ, the pressure angle. In Fig. 7-10, free-body diagrams of gear and pinion are shown. The action of the shaft is represented by force F. The weight of the gear and pinion are not significant compared with F and W and are not shown. The two forces F and W constitute a couple.

In Fig. 7-11, the forces F and W are resolved into tangential and radial components. Again, W_t and F_t constitute a couple. This couple is the torque which the spur gear transmits:

$$T = W_t R \quad \text{where } R = \text{radius of the pitch circle}$$

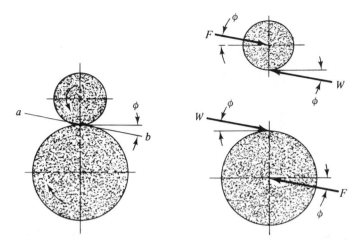

Figure 7-9 / Mating spur gear and   Figure 7-10 / Gear forces.
pinion.

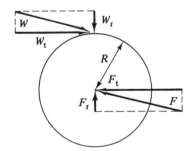

Figure 7-11 / Radial and tangential components of gear forces.

Also from the figure,

$$W_r = W_t \tan \phi$$

and
$$W_t = W \cos \phi \qquad (7\text{-}10)$$

The angle ϕ will be $14\frac{1}{2}°$ or 20° for power transmission, or perhaps 28° for a gear pump. If friction is to be accounted for, a friction angle must be included, making these angles $17\frac{1}{2}°$, 23°, and 31° when calculating radial and tangential components of gear forces.

If V is the pitch line velocity,

$$\text{hp} = W_t V/33{,}000$$

from which W_t may be calculated if the horsepower transmitted is known.

EXAMPLE 7-4

A 16-tooth, 5-pitch, 20-degree spur gear pinion driven at 1725 rpm transmits 5 hp to the gear train shown in Fig. 7-12. Find: (a) the speed of each shaft; (b) the pitch line velocity of the gears; (c) the tooth load at each gear; and (d) the force tending to separate the first two shafts.

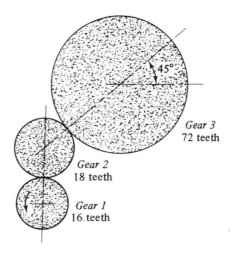

Figure 7-12 / Gear train (Example 7-4).

solution

(a)   Shaft speeds: shaft 1 : 1725 rpm

$$2 : 1725 \times 16/18 = 1533 \text{ rpm}$$

$$3 : 1533 \times 18/72 = 383 \text{ rpm}$$

(b)   The pitch diameters of the gears are found from $\dfrac{\text{number of teeth}}{\text{pitch}}$.

$$\text{gear } 1 : 16/5 = 3.20 \text{ in.}$$

$$2 : 18/5 = 3.60 \text{ in.}$$

$$3 : 72/5 = 14.40 \text{ in.}$$

and $$V = \frac{\pi DN}{12} \text{ fpm} = 0.262DN$$

where $$N = \text{rpm}$$

$$V_1 = 0.262 \times 3.2 \times 1725 = 1445 \text{ rpm}$$

The other pitch line velocities must be the same.

(c) $W_t = 33000\ \text{hp}/V$

for gear 1, $W_t = \dfrac{33000 \times 5}{1445} = 114\ \text{lb}$

If it is assumed that gear 2 is an idler gear, then all the power from gear 1 is transmitted to gear 3. The tangential tooth force then is the same on gear 3, that is, 114 lb.

(d) $W = W_t/\cos 20° = 114/0.94 = 121\ \text{lb}$

$W_r = 121 \tan 20° = 121 \times 0.364 = 44\ \text{lb}$

The free-body diagram for each gear is shown in Fig. 7-13.

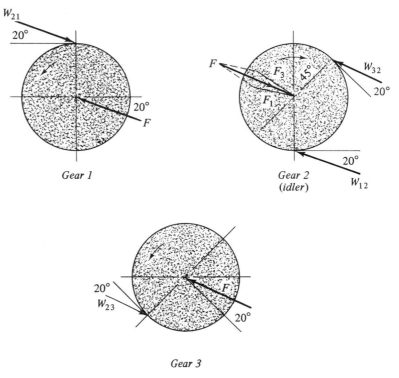

Figure 7-13 / Free-body diagrams for the gear train of Figure 7-11 (Example 7-4).

7-4   THE STRENGTH OF GEAR TEETH

When a gear transmits torque, the gear tooth is loaded as a cantilever beam. If the force exerted on the gear tooth is excessive, the tooth fails at the root, the point of maximum bending moment.

When the tooth of one spur gear first meets a tooth on the mating gear, force is applied at the tip of the tooth. Neglecting any friction force, which will be small except in the case of a worm drive, the applied force W is normal to the tooth profile (Fig. 7-14). This force is assumed to be carried by one tooth only, this being the safest assumption. The force W can be resolved into two components, radial and tangential, W_r and W_t. The radial component exerts a compressive stress on the tooth, tending to separate the mating gears, but presents no design problem, since it is the smaller component and also has little bending effect.

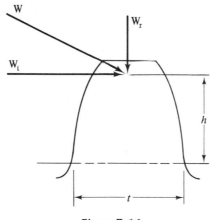

Figure 7-14

The tangential force W_t is taken to be the torque force and is equal to the torque transmitted divided by the pitch radius, as previously discussed. The bending moment on the gear tooth equals

$$M = W_t h$$

See Fig. 7-13. The bending stress equals

$$Mc/I = 6M/bt^2$$
$$= 6W_t h/bt^2$$
$$W_t = \frac{\sigma bt^2}{6h}$$
$$= \frac{\sigma bt^2 p_c}{6h p_c}$$
$$= \sigma b p_c y \qquad (7\text{-}11)$$

where σ = stress in psi, and

$$y = \frac{t^2}{6h p_c}$$

which is a dimensionless quantity called the form factor. The equation $W_t = \sigma b p_c y$ is called the Lewis equation.

A spur gear will not usually have a face width b greater than four times the circular pitch. Let $b = k p_c$, where $k \le 4$. Then the Lewis equation can be expressed as

$$W_t = \sigma p_c^2 k y$$
$$= \sigma \pi^2 k y / p_d^2 \qquad (7\text{-}12)$$

where $\quad \sigma =$ allowable stress.

Form factors (y) may be obtained from Table 7-2.

TABLE 7-2

FORM FACTOR y

NUMBER OF TEETH	$14\frac{1}{2}°$ FULL-DEPTH INVOLUTE OR COMPOSITE	20° FULL-DEPTH INVOLUTE	20° STUB INVOLUTE
12	0.067	0.078	0.099
13	0.071	0.083	0.103
14	0.075	0.088	0.108
15	0.078	0.092	0.111
16	0.081	0.094	0:115
17	0.084	0.096	0.117
18	0.086	0.098	0.120
19	0.088	0.100	0.123
20	0.090	0.102	0.125
21	0.092	0.104	0.127
23	0.094	0.106	0.130
25	0.097	0.108	0.133
27	0.099	0.111	0.136
30	0.101	0.114	0.139
34	0.104	0.118	0.142
38	0.106	0.122	0.145
43	0.108	0.126	0.147
50	0.110	0.130	0.151
60	0.113	0.134	0.154
75	0.115	0.138	0.158
100	0.117	0.142	0.161
150	0.119	0.146	0.165
300	0.122	0.150	0.170
Rack	0.124	0.154	0.175

The allowable stresses for gear materials are given in Table 7-3.

TABLE 7-3
ALLOWABLE STRESSES FOR GEAR MATERIALS

Gray cast iron	6000 psi
High-grade gray cast iron	13,000
Machinery steel, no heat treatment	20,000
Case-hardened steel	50,000
Alloy steel, heat-treated	60,000
Bronze, SAE 20	12,000
Bakelite	7000

These allowable stresses are reduced to allow for the effects of pitch line velocity, in accord with the following formulas if the gears are spur gears:

$$\text{allowable } s = s_o\left(\frac{600}{600 + V}\right)$$

for V (pitch line velocity) less than 2000 fpm, and

$$\text{allowable } s = s_o\left(\frac{1200}{1200 + V}\right)$$

for V between 2000 and 4000 fpm.

The applied force W may have to be increased to allow for dynamic loading, especially at higher velocities. When heavy torques, high horsepower, and high speed are involved, the gear design requires a high degree of sophistication. The design methods discussed here present only the basics of design, which may have to be modified for factors such as fatigue, accuracy of gear teeth, dynamic loading, inertia of moving masses, and any other factors.

In designing a gear train, it is necessary to know all the gear dimensions, including tooth size, before the loads and stresses can be determined. This means that a preliminary estimate must be made of the gear size, and this estimate must be checked out and then altered as the calculations indicate. There is a large element of trial and error. Since the first few trials may not work out correctly, approximations may save time. For example, 3.14 can be taken as 3.

The weaker gear is the one with the smaller sy value, and this governs the design. If the two mating gears are made of the same material, the smaller gear (pinion) will be the weaker because of a higher value of W_t.

In adjusting variables to reach a successful design, two other forms of the Lewis equation may be useful:

$$S = \frac{2TP_d^2}{\pi b N y} \qquad \text{or} \qquad \frac{P_d^2}{y} = \frac{\pi b N s}{2T}$$

The second arrangement of the variables is preferred if the pitch diameter is known because s can be determined and then $\pi s/2T$ is known and fixed.

In making trials, P_d^2/y must approximately equal $\pi b N s/2T$. If P_d^2/y is too small or too large, note that y changes little while P_d^2 changes rapidly with larger or smaller pitches. The other side of the equation can be adjusted by changing b within reasonable limits; b may be as large as $4P_c$.

$T =$ torque (calculated from a known horsepower)
$b =$ tooth width (use an assumed value)
$N =$ number of teeth (from an assumed or known pitch diameter and assumed P_d)
$s =$ allowable stress (calculated from pitch diameter)

EXAMPLE 7-5

A bronze spur gear (pinion) rotated at 600 rpm drives a case-hardened steel gear. The pinion has 16 teeth, 5 diametral pitch, 20° full depth involute, width $1\frac{1}{2}$ inches, and the gear has 50 teeth. Based on allowable stress levels, what maximum horsepower can be transmitted?

	TEETH	s_o	*FORM FACTOR*	$s_o y$
Pinion	16	12,000	0.094	1130
Gear	50	50,000	0.130	6500

solution

The load that the tooth can carry is proportional to the product $s_o y$. Therefore the pinion determines the maximum power.

$$\text{Pitch diameter of pinion} = 16/5 = 3.2 \text{ in.}$$

$$\text{Pitch line velocity} = \pi\left(\frac{3.2}{12}\right)600 = 503 \text{ fpm}$$

This velocity is less than 2000 fpm.

$$\text{Allowable stress} = 12,000 \frac{600}{600 + 503} = 6550 \text{ psi}$$

Then

$$W_t = sbyp_c$$

$$= 6550 \times 1.5 \times 0.094 \times \pi\frac{3.2}{16}$$

$$= 576 \text{ lb}$$

$$\text{Maximum hp} = \frac{W_t V}{33,000}$$

$$= \frac{576 \times 503}{33,000} = 8.8 \text{ hp}$$

7-5 HELICAL GEARS

Like spur gears, helical gears are cut from a cylindrical gear blank, but the teeth are at some helix angle to the shaft. The helix angle is measured on the pitch circle. Helical gears may substitute for spur gears in a parallel-shaft drive but are equally suited for shafts at an angle to each other. For shafts at 90°, they provide only a point contact between the two gears and therefore can transmit only very low-power levels. Particularly in the case of wide gears, a helical gear train transmits force from gear to gear more smoothly, since the whole length of the tooth does not engage all at once but gradually. This action makes them suited to heavy loads, high speeds, and low noise levels. Helical gears use the basic involute shape and the usual pressure angles.

The hand of the helix must be selected correctly. If two mating helical gears are on parallel shafts, one of the pair must have a left-hand helix and the other a right-hand helix, as in Fig. 7-15. The two gears,

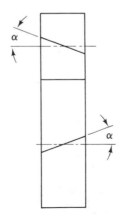

Figure 7-15 / Mating helical gears—one right-hand, one left hand.

however, must have identical helix angles. If the two gears mesh at 90°, both must have the same hand.

A helical gear, like any gear except a straight spur gear, transmits an axial force component to its shaft. A double helical gear or herringbone gear is the equivalent of two helical gears of opposite hand mounted together on the shaft to produce opposing thrusts, which cancel out thrust loading. The axial force in a single helical gear is determined by two different methods, depending on how the pressure angle is defined (see Fig. 7-16). The pressure angle may be measured

1 /   in a plane perpendicular to the axis of the gear

2 /   in a plane normal to the tooth

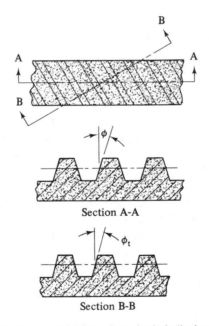

Figure 7-16   /   Axial force in a single helical gear.

If the pressure angle is measured in the plane perpendicular to the axis of the gear, the force components are

$$\text{Tangential force } W_t = \text{torque}/R$$
$$\text{Separating force } W_r = W_t \tan \phi$$
$$\text{Thrust force } W_a = W_t \tan \alpha$$

where R = pitch radius

α = helix angle measured from the axis of the gear (Fig. 7-15)

ϕ = pressure angle measured in a plane perpendicular to the axis of the gear

If the pressure angle is measured in a plane perpendicular to the tooth, the force components are

$$\text{Tangential force } W_t = \text{torque}/R$$

$$\text{Separating force } W_r = \frac{W_t \tan \phi_t}{\cos \alpha}$$

$$\text{Thrust force } W_a = W_t \tan \alpha$$

where ϕ_t is the pressure angle measured in a plane perpendicular to the tooth.

The direction of thrust depends on both the hand of the gear and the direction of rotation. See Fig. 7-17. If the direction of rotation is reversed, the thrust force will be reversed also.

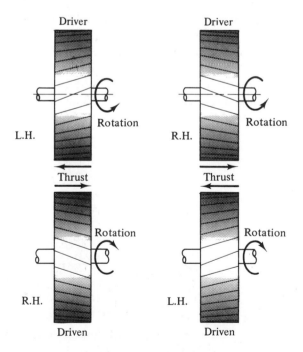

Figure 7-17   /   Direction of thrust force in helical gears.

EXAMPLE 7-6

A right-hand helical pinion drives a left-hand helical gear in a gear pump. The two shafts are parallel. The pressure angle is 28° measured in a plane perpendicular to the tooth, and the helix angle is 20° measured from the axis of the gear. The torque force W_t is 3000 pounds. Find the radial and thrust forces. (See Fig. 7-17.)

solution

$$W_t = 3000 \text{ lb}$$

$$W_r = \frac{W_t \tan \phi_t}{\cos \alpha} = \frac{W_t \tan 28°}{\cos 20°}$$

$$= 1700 \text{ lb}$$

$$W_a = 3000 \tan 20°$$

$$= 1090 \text{ lb}$$

7-6   BEVEL GEARS

Bevel gears resemble a friction cone drive (Fig. 7-18). The two shafts of the bevel gear drive are at an angle to each other, usually 90°. The point of intersection of the two shafts is usually the apex of both pitch cones. The terminology for bevel gears is given in Fig. 7-19.

The tooth section becomes gradually smaller as the apex of the cone is approached. Only a length of cone close to its base is used, actually not more than a third of the full length of the cone. All such dimensions as pitch diameter, diametral pitch, addendum, and dedendum are measured at the large diameter of the bevel gear.

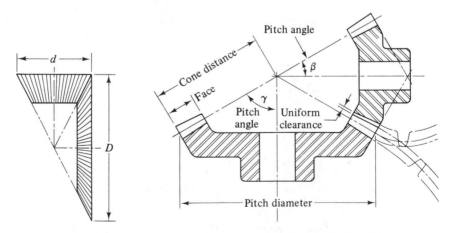

Figure 7-18 / Bevel gears.   Figure 7-19 / Terminology for bevel gears.

Besides the straight-tooth bevel gear, other variants are in use, including helical, hypoid, skew, and zerol types. When two spur tooth bevel gears intersect at right angles and the gears have the same number of teeth, they are termed *miter gears.*

The force components for bevel gears are the following:

1 / tangential force W_t = torque/radius, acting at the average pitch radius (Fig. 7-20).

2 / separating force $W_r = W_t \tan \phi$, where ϕ is the pressure angle.

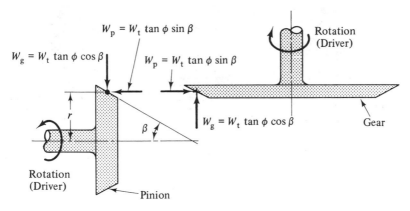

Figure 7-20 / Force components in a bevel gear drive.

The separating force has two components:

1 / the component along the axis of the pinion, W_p.

2 / the component along the axis of the gear, W_g.

As may be seen from Fig. 7-20, the pinion thrust force equals

$$W_p = W_t \tan \phi \sin \beta$$

and the gear thrust force equals

$$W_g = W_t \tan \phi \cos \beta$$

When quiet operation is a requirement, spiral bevel gears are selected. The teeth of a spiral bevel gear are cut as spirals on the face of the pitch cones. See Fig. 7-21.

Hypoid gears resemble spiral bevel gears but are so constructed that the two shafts do not intersect. Thus the two shafts can be extended beyond the gears without interfering with each other. This type of gear was originally developed for automobile rear ends, because the drive shaft could be lowered and therefore also the floor of the vehicle. Hypoid drives are also used for high horsepower, because the hypoid pinion can be larger and stronger than the spiral bevel pinion. Since the hypoid gear slides as well as rolls, it requires an extreme pressure lubricant. See Figs. 7-22 and 7-23.

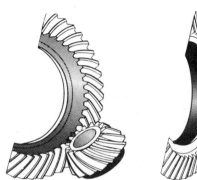

Figure 7-21 / Spiral bevel gears.   *Figure 7-22 / Zerol bevel gears.*   *Figure 7-23 / Hypoid gears.*

7-7   THE SPIROID GEAR

This gear uses for a pinion a tapered worm (Fig. 7-24). It has the advantages of backlash control and great shock resistance, because several teeth are simultaneously in contact. Backlash is removed simply by adjusting either gear or pinion axially. Spiroid gears may be found in many portable power tools, such as hedge trimmers, portable grinders, portable drills, electric can openers, etc.

Figure 7-24 / Spiroid gearing.

7-8   WORM DRIVES

The terminology for worm drives is shown in Fig. 7-25. Worm gears are used to obtain large speed reductions of up to 80: 1 in a small space. Higher reductions, however, are less efficient than smaller reduction ratios.

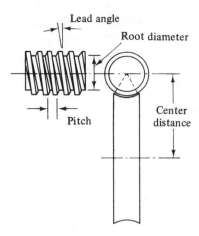

Figure 7-25 / Terminology for a worm drive.

The two shafts are usually at right angles to each other. The worm is a special case of a helical or spiral gear in that the helix wraps completely around the gear like a screw thread. Indeed, the worm teeth are referred to as threads.

There is considerable sliding friction in a worm drive and therefore considerable heat generation. The lubrication problem is then significant in worm drives. While most types of mating gears have efficiencies from 96% to 99%, the efficiency of a worm drive is closer to 50% than 100%. To prevent seizure of worm to gear, the worm is usually a hardened steel and the gear a bronze.

The speed ratio of a worm drive is in part determined by the number of threads on the worm. The number of threads also determines whether the drive is self-locking. A single-thread worm has the smallest helix angle and is self-locking, that is, the worm cannot be turned by the gear. Self-locking worms do not require brakes. Multithread worms, however, are more common than single threads, since they are more efficient.

The American Gear Manufacturers Association (AGMA) recommends the following design equations:

$$p_d \text{ of worm} = C/2.2 = 3p_c \text{ approximately}$$

where C = center distance between the two shafts.

Face width of gear = $0.73 \times$ worm pitch diameter approximately
Axial length of worm = $(4.5 + N_g/50)p_c$

where N_g = number of gear teeth. The number of worm threads + number of gear teeth ≥ 40.

The maximum input horsepower of a worm gear unit is limited by heat dissipation considerations. This horsepower can be estimated from

$$hp = \frac{9.5C^{1.7}}{R + 5}$$

where R = transmission ratio
   C = center distance between the two shafts

The components of force in a worm drive are given by the following equations. The tangential force of the gear will be at right angles to that of the worm.

1 / Tangential force of the worm = $W_{tw} = \dfrac{\text{torque}}{\text{pitch radius}}$

2 / Tangential force of the gear = $W_{tg} = W_{tw}\left(\dfrac{1 - \dfrac{\mu \tan \alpha}{\cos \phi_n}}{\tan \alpha + \dfrac{\mu}{\cos \phi_n}}\right)$

3 / Radial force = $W_{tg}\left(\dfrac{\sin \phi_n}{\cos \phi_n \cos \alpha - \mu \sin \alpha}\right)$

$\qquad\qquad\qquad = W_{tw}\left(\dfrac{\sin \phi_n}{\cos \phi_n \sin \alpha + \mu \cos \alpha}\right)$

where ϕ_n = normal pressure angle, measured in a plane perpendicular to the tooth. This angle is usually $14\frac{1}{2}°$ for single and double threads and 20° for triple and quadruple threads.
    α = lead angle of the worm = helix angle of the gear, where (see Fig. 7-26)

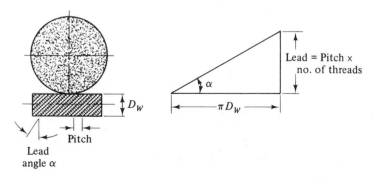

Figure 7-26 / Lead angle relationships.

$$\tan \alpha = \text{lead}/\pi p_d$$
$$= \frac{\text{number of threads} \times \text{linear pitch of worm}}{\pi(\text{pitch diameter of worm})}$$

$\mu = $ coefficient of friction

EXAMPLE 7-7

A double-thread worm has a lead angle of 18°. If the pitch is 1.4 in., what is the worm diameter?

solution

$$\tan \alpha = \frac{\text{lead}}{\pi D}$$

Since the pitch is 1.4 in. and there is a double thread, the lead is 2×1.4 or 2.8 in.

$$\tan 18° = 2.8/\pi D = 0.326$$
$$D = 2.74 \text{ in.}$$

EXAMPLE 7-8

A triple-thread right-handed worm with a pitch diameter of 3.00 in. and pressure angle 20° transmits 10 hp at 1200 rpm to a worm gear. The pitch of the worm is 0.75 in., and the coefficient of friction is 0.1. Determine: (a) the tangential force of the worm; (b) the tangential force on the gear; and (c) the radial force.

solution

$$\text{hp} = TN/63000$$
$$10 \text{ hp} = 1200T/63000$$
$$T = 525 \text{ lb-in.}$$
$$W_t = 525/1.5 = 350 \text{ lb}$$

$$\tan \alpha = \frac{3 \times 0.75}{3\pi} = 0.239$$

$$W_{tg} = 350\left(\frac{1 - \dfrac{0.1 \tan \alpha}{\cos \phi_n}}{\tan \alpha + \dfrac{\mu}{\cos \phi_n}}\right)$$

$$= 350\left(\frac{1 - \dfrac{0.1 \times 0.239}{\cos 20°}}{0.239 + \dfrac{0.1}{0.94}}\right)$$

$$= 990 \text{ lb}$$

$$W_r = W_{tg}\frac{\sin\phi_n}{\cos\phi_n\cos\alpha - \mu\sin\alpha}$$

$$= 990\frac{\sin 20°}{\cos 20°\cos\alpha - 0.1\sin\alpha}$$

$$= 379\ \text{lb}$$

In the calculation for tan α, note that the *pitch* is the axial distance from a point on a thread to the same point on the next thread, while the *lead* is the distance a screw thread advances axially in one turn. For a double thread, the lead is twice the pitch.

EFFICIENCY OF WORM DRIVES

The efficiency of a worm drive may be approximated reasonably closely from the following formula:

$$\text{Eff} = \frac{1 - \mu\tan\alpha}{1 + \dfrac{\mu}{\tan\alpha}}$$

Efficiencies are low, often in the range of 60–70%.

7-9   GEAR LUBRICATION

Wear in gear teeth is produced by five possible mechanisms:

1 / *Seizing*—welding of the teeth due to high local pressure in the absence of lubrication.
2 / *Scuffing*—plastic flow of material near the pitch line.
3 / *Pitting*—due to compressive fatigue stress. A hard tooth surface reduces scuffing.
4 / *Abrasion*—scoring of the teeth by foreign matter.
5 / *Scoring*—due to sharp edges on the teeth. Scoring of this type is not a problem in high-quality gears.

Scuffing and seizing are prevented by proper lubrication. The type of gear and the loading conditions dictate the type of lubricant. A spur gear rolls, but a hypoid or worm gear also slides, and different lubricants are required for these different conditions. Most gears with moderate loads will perform satisfactorily with a rust and oxidation inhibited (R&O) mineral oil. For heavier loads, extreme pressure lubricants are preferred.

Sliding gears such as worm gears require lubricants with special additives to reduce friction and supply resistance to scoring.

Five types of gear lubricant are in common use:

1 | *R&O oil*—rust and oxidation inhibited mineral oil.

2 | *EP oil (extreme pressure)*—contains chemical additives that react with gear materials at high contact temperatures to produce a protective film that reduces metal-to-metal contact.

3 | *Compounded oil*—usually a cylinder oil with a few percent of animal fat to reduce friction.

4 | *Heavy-bodied open gear oils*—these are heavy, sticky oils, applied by hand, dip, or intermittent spray. They are available with or without an EP additive.

5 | *Grease*—a grease is a fluid thickened by an agent to reduce the tendency to flow away from the region being lubricated.

Lubrication requirements for spur, helical, and bevel gearing are identical. R&O oils are used, but EP oils are substituted for heavy and shock loading. For worm gears, EP and compounded oils are preferred. Hypoid gears require EP oils.

7-10 GEAR TRAINS

A power transmission system consisting only of gears is called a *gear train*. There are of several types of gear trains: ordinary, compound, reverted, and planetary (epicyclic). In the planetary train at least one of the gear axes must rotate.

In a *simple gear train* each shaft carries only one gear. The compound gear train contains one or more shafts carrying more than one gear. Consider the simple gear train of Fig. 7-27, where gear A is the driving gear. Speed ratios are determined by the number of teeth; hence

$$\frac{\omega_B}{\omega_A} = \frac{N_A}{N_B} \qquad \frac{\omega_C}{\omega_B} = \frac{N_B}{N_C} \qquad \frac{\omega_D}{\omega_C} = \frac{N_C}{N_D}$$

and
$$\frac{\omega_D}{\omega_A} = \left(\frac{\omega_B}{\omega_A}\right)\left(\frac{\omega_C}{\omega_B}\right)\left(\frac{\omega_D}{\omega_C}\right) = \left(\frac{N_A}{N_B}\right)\left(\frac{N_B}{N_C}\right)\left(\frac{N_C}{N_D}\right) = \frac{N_A}{N_D}$$

Thus the intermediate gears, called *idler gears*, do not influence the overall velocity ratio. Such a simple gear train may be used to fill up a large center distance between driving and driven shafts, to reverse the direction of rotation of the driven shaft, or to take power off the intermediate gears.

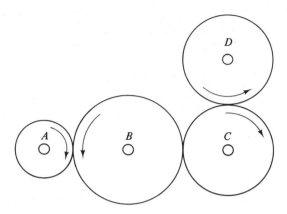

Figure 7-27 | Simple gear train of four gears.

A compound gear train is illustrated in Fig. 7-28. Speed ratios are

$$\frac{\omega_B}{\omega_A} = \frac{N_A}{N_B} \qquad \frac{\omega_D}{\omega_C} = \frac{N_C}{N_D} \qquad \frac{\omega_F}{\omega_E} = \frac{N_E}{N_F}$$

Overall speed ratio $= \dfrac{\omega_F}{\omega_A} = \left(\dfrac{\omega_B}{\omega_A}\right)\left(\dfrac{\omega_D}{\omega_C}\right)\left(\dfrac{\omega_F}{\omega_E}\right)$

where $\omega_B = \omega_C$

$\omega_D = \omega_E$

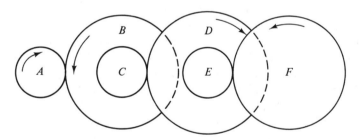

Figure 7-28 | Compound gear train.

Substituting tooth numbers:

$$\frac{\omega_F}{\omega_A} = \left(\frac{N_A}{N_B}\right)\left(\frac{N_C}{N_D}\right)\left(\frac{N_E}{N_F}\right) \qquad\qquad (7\text{-}13)$$

All the gears influence the overall speed ratio. Note that in this speed ratio equation, the numerator is the product of all the driver teeth ($N_A \times N_C \times N_E$), and the denominator is the product of all the driven teeth ($N_B \times$

$N_D \times N_F$). The same equation applies to belt drives and chain drives as well as to gears. For belt drives use pitch diameters.

DESIGN OF A COMPOUND GEAR TRAIN

To design a compound gear train, consider the following examples.

A light precision compound gear train requires a speed reduction of 1:15. Standard 20-degree full depth spur gears with diametral pitch of 24 are selected. A maximum pitch diameter of 3 in. is desirable, and to prevent undercutting, not less than 18 teeth may be used.

Given the requirements, it is obvious that a single pair of gears cannot be used because the number of teeth on the large gear would be 18×15, giving too large a pitch diameter. Therefore to begin the design, try two equal speed reductions:

$$\frac{\omega_{\text{OUT}}}{\omega_{\text{IN}}} = \frac{1}{15} = \frac{1}{3.88} \times \frac{1}{3.88}$$

where $3.88^2 = 15$, and each reduction is to be 1:3.88.

Next change one of the 3.88's to a number which is close to 3.88 but which can be converted into a reasonably simple fraction. Change 3.88 to $3.875 = 31/8$. This alters the speed ratio equation to

$$\frac{1}{15} = \frac{1}{\left(\frac{31}{8}\right)} \times \frac{\left(\frac{31}{8}\right)}{15} = \frac{8}{31}\left[\frac{31}{8 \times 15}\right] = \frac{8}{31}\left(\frac{31}{120}\right)$$

A reduction of 8/31 and then 31/120 (tooth numbers) will give the required ratio of 1:15, but the last gear would have 120 teeth and a pitch diameter of 5 inches. The first gear would have only 8 teeth, an insufficient number. If the number of teeth in the first reduction is doubled to 16:62, 16 teeth are still too few.

Try a larger reduction in the first pair of gears, say 1:4.

$$\frac{1}{15} = \frac{1}{4} \times \frac{4}{15} = \frac{18}{72} \times \frac{20}{75}$$

Eighteen teeth is the choice because this is the smallest allowable number, and four times 18 is the largest possible number, or rather the largest preferred number. The two reduction ratios meet the requirements, though the pitch diameter of the largest gear is slightly over 3 inches.

For a second example, consider a gear reducer which is to reduce an electric motor speed of 1725 rpm to 200 rpm. For compactness no gear is to have less than 20 or more than 50 teeth. Determine the number of gears and the tooth numbers.

$$\text{Speed ratio} = 1725/200 = 8.625$$

Using the smallest and largest gears allowed, 20 and 50 teeth, the maximum speed reduction is $50/20 = 2.50$. Using two pairs it is $50/20 \times 50/20 = 6.25$. This is still less than 8.625. Therefore, three combinations of gears will be required.

Since three combinations are needed, first take the cube root of 8.625, which is 2.05, that is,

$$\frac{2.05}{1} \times \frac{2.05}{1} \times \frac{2.05}{1} = \frac{8.625}{1}$$

Suppose now that we change two of the three ratios to 2.0/1 and then alter the third ratio to suit:

$$2.0 \times 2.0 \times 2.15625 = 8.625$$

For the first two reductions, use tooth numbers of 40 and 20. For the last reduction, if 20 teeth are given to the driving gear, then the final driven gear will have 43 teeth (20×2.15).

THE REVERTED GEAR TRAIN

A *reverted gear train* is a compound train in which both input and output shafts have the same line as axis. See Fig. 7-29. It will be seen that the center distance must be the same for both pairs of gears.

If $C =$ center distance, then $2C = D_A + D_B = D_C + D_D$. But since $D = N/P_d$, $2C = N_A/P_1 + N_B/P_1 = N_C/P_2 + N_D/P_2$ where P_1 and P_2 are two diametral pitches. If a single diametral pitch is used in the reverted gear train, then $2CP_d = N_A + N_B = N_C + N_D$, and thus the same number of teeth must be used in each pair of gears.

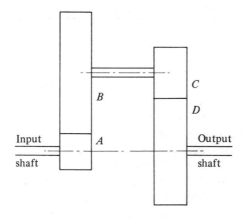

Figure 7-29 / Reverted gear train of four gears.

To design a reverted gear train, consider the following example. The speed ratio between input and output shafts is to be a reduction of 1 : 6. No gear may have less than 18 teeth or more than 96, and a uniform diametral pitch will be used.

The maximum ratio for one pair of gears is 18/96 or 3/16, and for two pairs 9/64. A ratio of 1 : 6 falls between these two numbers, so two reductions are required. Begin as with previous gear train designs by taking the square root of 1/6, which is

$$\frac{1}{2.45} \times \frac{1}{2.45}$$

Try 1/2.5 for the first ratio and alter the second ratio to suit:

$$\frac{1}{6} = \frac{1}{2.5} \times \frac{2.5}{6}$$

But the sums of the numerators $(1 + 2.5)$ and the denominators $(2.5 + 6)$ are not equal, and therefore neither will be the number of teeth in the two combinations.

$$\frac{1}{2.5} \times \frac{2.5}{6} = \frac{2}{5} \times \frac{5}{12}$$

with sums of 7 and 17. Now multiply the first fraction by 17/17 (the sum of $5 + 12$) and the second fraction by 7/7 (the sum of $2 + 5$):

$$\left(\frac{2}{5} \times \frac{17}{17}\right)\left(\frac{5}{12} \times \frac{7}{7}\right) = \frac{34}{85} \times \frac{35}{84}$$

Now the tooth totals are equal. Neither too large nor too small a number of teeth has been used.

7-11 THE PLANETARY OR EPICYCLIC GEAR TRAIN

A simple type of planetary gear train is shown in Fig. 7-30. The distinguishing feature of this type is the rotating arm or planet carrier. If we assume the sun gear to be stationary and the arm to rotate clockwise about axis O, then the planet gear also rotates clockwise about its center C. Actually there are three operating possibilities for this configuration of gears:

1 / Sun gear stationary and arm rotating
2 / Planet gear stationary and both arm and sun gear rotating
3 / Arm stationary and both sun and planet gear rotating.

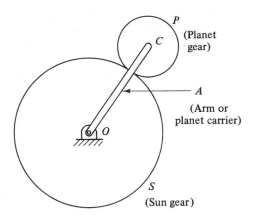

Figure 7-30　/　Simple planetary gear train.

This last possibility reduces the train to a standard train. In an epicyclic train one or more gears rotate about a moving axis.

Figure 7-31 shows a planetary train with two planet pinions. The number of planetary gears has no influence on the speed ratio, but allows more torque to be transmitted through an increased number of planet gears.

The gear speeds of the epicyclic train of Fig. 7-31 can be determined from the equation

$$\omega_S = \left(1 + \frac{N_R}{N_S}\right)\omega_A - \left(\frac{N_R}{N_S}\right)\omega_R \qquad (7\text{-}14)$$

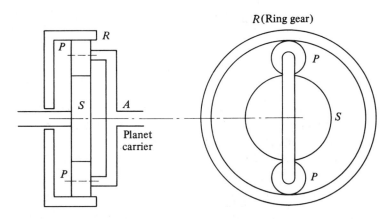

Figure 7-31　/　Epicyclic gear train with two planetary pinions.

where ω_S = rpm of sun gear
         ω_A = rpm of planet carrier arm
         ω_R = rpm of ring gear
         N_R = number of teeth on the ring gear
         N_S = number of teeth on the sun gear

EXAMPLE 7-9

For the train of Fig. 7-30, the following data apply:

$$\omega_S = 1200 \text{ rpm clockwise}$$
$$N_S = 96$$
$$N_P = 24$$
$$N_R = 144$$

Find the planet carrier shaft speed if the sun gear is the input, the ring gear is fixed, and the planet carrier is the output.

solution

$$\omega_S = \left(1 + \frac{N_R}{N_S}\right)\omega_A - \left(\frac{N_R}{N_S}\right)\omega_R$$

$$1200 = \left(1 + \frac{144}{96}\right)\omega_A - 0$$

$$= 2.5\omega_A$$

$$\omega_A = \frac{1200}{2.5} = +480 \text{ rpm}$$

The direction of rotation of the carrier is the same as that of the sun gear since the answer is positive.

QUESTIONS AND PROBLEMS

P7-1   What is the pitch diameter of the following spur gears?
     (a)   8 pitch, 16 teeth;
     (b) $1\frac{1}{2}$ pitch, 40 teeth;
     (c)   12 pitch, 14 teeth;
     (d)   4 pitch, 39 teeth.

P7-2   A small gear of 12 diametral pitch has 11 teeth. For this gear, determine:
     (a)   pitch diameter;
     (b)   addendum; ·
     (c)   outside diameter;

(d)   dedendum, specification ASA B. 6; and

(e)   whole depth of tooth.

P7-3   What is the difference between:

(a)   pitch diameter and diametral pitch?

(b)   diametral pitch and circular pitch?

P7-4   A full-depth spur gear has an O.D. of 3.250 in., a pitch diameter of 3.000 in., and 24 teeth. Find:

(a)   addendum;

(b)   diametral pitch; and

(c)   whole depth of tooth.

P7-5   A full-depth spur gear is 8 diametral pitch, 11 teeth. Determine:

(a)   pitch diameter;

(b)   outside diameter; and

(c)   center-to-center distance of two such mating gears.

P7-6   For a 20° full-depth spur gear, AGMA 201.02, of $\frac{1}{2}$ pitch and 180 teeth, find:

(a)   pitch diameter;

(b)   outside diameter; and

(c)   whole depth of teeth.

P7-7   A 12-diametral-pitch pinion with 36 teeth rotates at 1800 rpm driving a gear at 540 rpm. Find:

(a)   number of teeth on the gear; and

(b)   the center distance.

P7-8   A triple-threaded worm has a pitch diameter of 4.000 in. and an axial pitch of 0.875 in. Determine the lead angle.

P7-9   A worm gear reducer must have a center distance of 12 in. Using the AGMA formula, what should be the diameter of the worm?

P7-10   A worm gear reducer has a center distance of 10.000 in. and a transmission ratio of (a) 20, (b) 40, and (c) 80. Estimate the maximum horsepower input without overheating of the reducer, ignoring strength of the teeth.

P7-11   A cast steel pinion, $s_o = 20{,}000$, is to drive a cast-iron gear, $s_o = 8000$. The speed of the gear is to be one-third that of the pinion. Pitch diameter of the pinion should be approximately 4 in., transmitting 25 hp at 900 rpm. A 20° involute tooth is selected. Design the two gears, preferably for the greatest number of teeth.

P7-12   Solve Prob. 7-11, using instead a bronze spur gear, $s_o = 12{,}000$ psi, and a steel pinion, $s_o = 15{,}000$ psi.

P7-13 A bronze spur gear, $s_o = 12,000$ psi, is to drive a mild steel pinion, $s_o = 15,000$ psi, at 4 times its speed. A pressure angle of 20° is selected. The pinion is to receive 8 hp at 2000 rpm of the pinion. Find a suitably small diameter gearing for the purpose, using not less than 18 teeth.

P7-14 A steel pinion, $s_o = 30,000$, is to transmit 5 hp at 1000 rpm to a steel spur gear, $s_o = 20,000$ psi. Center distance must be approximately 8 in. and pinion diameter approximately 4 in. A pressure angle of 20° is selected. Using the Lewis equation, decide the diametral pitch and face width to provide the largest number of teeth.

P7-15 The accompanying sketch shows two spur gears of $14\frac{1}{2}°$ pressure angle transmitting 20 hp, with the driving gear rotating at 1100 rpm. Determine the force on each of the four bearings.

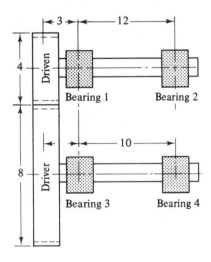

Figure P7-15

P7-16 A helical gear of 16 teeth and 5 diametral pitch has a pressure angle of 20°, measured in a plane perpendicular to the shaft axis, and a 15° helix angle. It transmits $7\frac{1}{2}$ hp to a gear of 36 teeth. The speed of the 16-tooth gear is 1725 rpm. Find the thrust load and radial load.

P7-17 A helical gear of 9.00-in. diameter and 45 teeth has a pressure angle of 20°, measured perpendicular to the tooth, and a helix angle of 30°. It transmits a torque of 1600 lb-in. Determine tangential, radial, and thrust forces.

P7-18   A pair of straight-tooth bevel gears mate in a velocity ratio of
4 to 3. The pinion has a pitch diameter of 6 in. and a face width
of $1\frac{1}{2}$ in., rotating at 540 rpm. Teeth are 5 diametral pitch, with
20° pressure angle, and 16 hp is transmitted. Sketch the gears,
showing dimensions, and find the following:
(a)   mean radius of the pinion;
(b)   pinion torque;
(c)   tangential force at the mean pinion radius;
(d)   pinion thrust force; and
(e)   gear thrust force.

P7-19   A triple-threaded worm has a pitch diameter of 4.00 in. and an
axial pitch of 0.875 in. Determine the lead angle.

P7-20   A worm gear reducer must have a center distance of 12 in. Using
the AGMA formula, what should be the diameter of the worm?

P7-21   A triple-thread right-handed worm, with pitch diameter 2.75 in.
and pressure angle ϕ_n 20°, transmits 13 hp at 1725 rpm to a 60-
tooth worm gear. The pitch of the worm is 0.785 in. The coefficient
of friction is 0.1.
(a)   What is the pitch diameter of the worm gear?
(b)   Determine tangential force of the worm and of the gear.

P7-22   The following are the details of a worm drive.
1. center distance 5.150 in.
2. speed reduction 40:1
3. speed of worm-1800 rpm
4. worm gear-40 teeth
5. 20° pressure angle
6. coefficient of friction-0.1
7. lead of worm 0.6283 (4° 58′)

Determine:
(a)   pitch diameter of worm;
(b)   maximum input horsepower;
(c)   torque of worm;
(d)   tangential force of worm;
(e)   tangential force of gear; and
(f)   radial force.

P7-23   Explain how a compound gear train is designed.

P7-24   A welding turntable is driven by the belts and gearing shown,
with power input supplied by an infintely variable speed motor.
A fillet weld is to be deposited around the circumference of
a 6.000 in. diameter steel tube rotated by the turntable, at a
welding speed of 16 in. per minute. What must be the speed of
the driving motor?

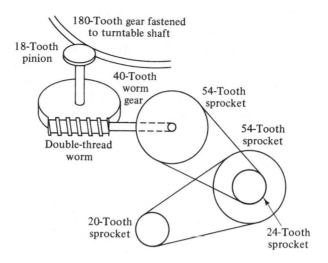

Figure P7-24

P7-25 Design a compound gear train for a speed reduction of 1: 20, using not less than 18 teeth or more than 96 teeth in any gear.
(a) Determine the number of teeth in each gear.
(b) If all gears are 12 pitch, find the center distances.

P7-26 Design a compond gear train for a speed reduction of 1: 11. The maximum number of teeth in any gear must not exceed 54, and the minimum must not be fewer than 18.

P7-27 A pair of mating gears has a diametral pitch of 8. The driving gear rotates at 1800 rpm, and the output speed of the driven gear must be as close to 700 rpm as possible. Center distance must be as close to 4.500 in. as possible. Find the number of teeth in each gear, the output speed, and the center distance.

P7-28 The gear reducer train shown is driven by the first 26-tooth gear on the left at 1750 rpm.

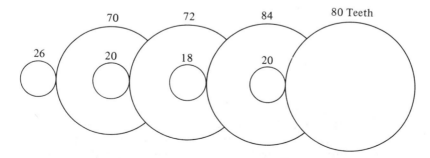

Figure P7-28

(a)   Determine the rpm of the output gear.

(b)   The gears at the output end of the train have a coarser pitch than is used at the input end. Why?

P7-29   A pair of standard 20° full-depth spur gears has a driving gear rotating at 900 rpm and a driven gear at 400 rpm. Center distance must be 3.500 in. closely. A diametral pitch in the range of 10 to 16 inclusive must be used. Decide the diametral pitch, number of teeth, pitch diameters, and speed ratio.

P7-30   A pair of standard 20° fine pitch spur gears mates a driving gear rotating at 5000 rpm with a driven gear to rotate at 1825 rpm as closely as possible. Center distance must be 1.250 in. closely, and the largest gear tooth size allowable corresponds to a diametral pitch of 32. Determine the best possible diametral pitch, pitch diameters, number of teeth, and center distance.

P7-31   Design reverted gear trains to the following specifications. State number of teeth in all gears and give center distances.

(a)   minimum and maximum number of teeth 20 and 100; center distance 3.20 in. closely; speed reduction 1: 10.

(b)   speed reduction 2: 25; minimum and maximum number of teeth 18 and 120; diametral pitch 12.

P7-32   For the planetary train of Fig. 7-30, the sun gear rotates at 500 rpm and the arm at 300 rpm, both counterclockwise. Numbers of teeth are as follows: ring, 108; planet, 18; sun, 72. Determine the rpm and direction of the ring gear.

P7-33   For the planetary train of Fig. 7-30, find the number of teeth in the sun and planet gears, given the following data: ring gear 120 teeth (ring gear is fixed); sun gear rpm 80 clockwise; arm rpm 20 clockwise.

P7-34   Find the speed of rotation of the sun gear in a planetary train similar to Fig. 7-30, given the following data: ring 100 teeth, 25 rpm clockwise; planet 20 teeth; sun 60 teeth; planet carrier rotates at 50 rpm clockwise.

P7-35   Show from the geometry of the planetary train of Fig. 7-30 that $N_R = N_S + 2N_P$.

SUGGESTED DESIGN PROJECT

DP7-1   Besides their uses in power transmission, spur gears are used in gear pumps to pump oil to hydraulic circuits. In a gear pump, oil is captured in the spaces between gear teeth and transported around the gear from input port to discharge port.

(a)   Obtain a gear pump, preferably one whose discharge in gallons per minute you do not know. Use any method available to you to determine the volume of one space between two gear teeth. Determine the delivery of the pump; this is the number of tooth spaces (on both gears of the pump) delivering oil in each revolution times the revolutions per minute. 1 gpm = 231 in³. To allow for leakage losses, use a volumetric efficiency of 90%

(b)   Determine the pressure angle of the gears of the pump. This pressure angle is almost certain to be greater than the standard angles of $14\frac{1}{2}°$ and 20°. Can you think of the reason why a larger pressure angle would be used in a gear pump?

(c)   Determine the load on each bearing in the gear pump due to a pump discharge pressure of 1500 psi. To determine this load, consider the oil pressure (1500 psi) to act against an area equal to the outside diameter of the gear times the width of the gear.

(d)   Determine the area of one pump bearing (length of bearing times bore) and the bearing pressure in psi.

CHAPTER 8

SHAFTS, COUPLINGS, AND CLUTCHES

His first vehicle [Siegfried Marcus's 1864 car] was a cart carrying a two-cycle engine geared to the rear wheels without any intervening clutch. It was started by having a strong man lift the rear end while the wheels were spun.

ENCYCLOPAEDIA BRITANNICA
"Automobile," 1966

8-1 SHAFTS

Power transmission shafting is subject to both torque and bending loads. A gear or a chain drive supplies torque, while the tension in the chain or the contact force between mating gear teeth exerts a bending moment on the shaft. Pronounced bending deflections are not acceptable in shafts, therefore bending stresses are normally quite low. Gears, for example, cannot tolerate misalignment due to shaft deflection. Torsion stresses may be high, however, failures of shafts due to overloading or shock loading are almost always torsion failures. The only other common mode of failure is a fatigue failure of a shaft, usually at a stress concentration such as a keyway or a shoulder.

Loads on the shaft and its bearings are produced by the following:

1 / weight of shaft and its gears, sprockets, rotors, or other attached parts;

2 / centrifugal force of revolving parts, especially if they are unbalanced;

3 / belt, rope, and chain tension;

4 / torque of gears and sprockets.

If horsepower and speed in rpm are known, then torque is

$$T = \frac{63,025 \text{ hp}}{\text{rpm}} \text{ lb-in.}$$

and the driving force at the pitch circle of the gear or sprocket is

$$F = \frac{\text{torque}}{\text{radius}} = \frac{63,025 \text{ hp}}{NR}$$

Loads on the shaft may be applied gradually, or there may be sudden shock loads such as occur in a rock crusher or when a boat propellor strikes a lake bottom. Shock loads on shafts are considered to increase the effective loading by factors of 1.5 to 3.0 times. Shock loads are especially dangerous for shafting of cast iron or other brittle materials. Stepped shafts have shoulders and therefore stress concentrations. The largest possible fillet at the shoulder will reduce such stress concentrations to the minimum possible.

Shafting must have a smooth surface of close tolerances, both to improve fatigue strength and also for the seating of bearings. Smooth finishes are obtained either by cold-rolling or by grinding. Low-carbon cold-finished steel is used for shafting in applications that are not highly stressed, while medium-carbon steels of about 0.4% carbon, either heat-treated or not heat-treated, are used where loads and fatigue conditions are more serious. For severe service conditions, low-alloy steels are used, and for corrosive conditions, as in water pumps, stainless steels of alloys similar to 410, or bronzes, must be used.

8-2 BEARING LOADS

The determination of bearing loads is necessary in order to select suitable bearings for a shaft. As an example, consider the overhung shaft of Fig. 8-1. This shaft transmits 10 hp at 1800 rpm at the overhung pulley, which is 6 in. in diameter.

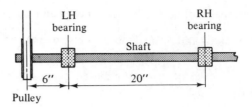

Figure 8-1 / Overhung shaft.

$$\text{Torque} = \text{belt force} \times \text{pulley radius}$$

$$\text{Belt force} = \frac{63{,}000 \times 10}{1800 \times 3} = 116 \text{ lb}$$

$$\text{Force on right-hand bearing} = \frac{116 \times 6}{20} = 35 \text{ lb}$$

$$\text{Force on left-hand bearing} = \frac{116 \times 26}{20} = 151 \text{ lb}$$

The various forces on the shaft may not all be in the same direction, as in Fig. 8-2. The following are then the calculations for bearing loads.

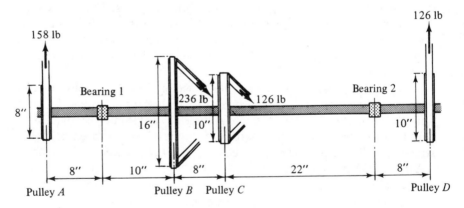

Figure 8-2 / Power transmission shaft. The belt forces on pulleys A and D are vertical, on B and C horizontal.

The forces at each power transmission location:

$$\text{At} \quad A = \frac{63{,}000 \times 10}{4 \times 1000} = 158 \text{ lb}$$

$$\text{At} \cdot B = \frac{63{,}000 \times 30}{8 \times 1000} = 236 \text{ lb}$$

$$\text{At} \quad C = \frac{63,000 \times 10}{5 \times 1000} = 126\,\text{lb}$$

$$\text{At} \quad D = \frac{63,000 \times 10}{5 \times 1000} = 126\,\text{lb}$$

The bearing loads:

PART	LOAD AT BEARING 1	LOAD AT BEARING 2
A	$\dfrac{158 \times 48}{40} = 190$ ↓	$190 - 158 = 32$ ↑
B	$\dfrac{236 \times 30}{40} = 177$ ↓	$236 - 177 = 59$ ↓
C	$\dfrac{126 \times 22}{40} = 69$ ↓	$126 - 69 = 57$ ↓
D	$\dfrac{126 \times 8}{40} = 25$ ↑	$126 + 25 = 141$ ↓

Total radial loads:
Bearing 1: $\sqrt{(190 - 25)^2 + (177 + 69)^2} = 295\,\text{lb}$
Bearing 2: $\sqrt{(141 - 32)^2 + (59 + 57)^2} = 159\,\text{lb}$

8-3 STRESSES IN SHAFTS

The relationship between torque in pound-inches and the maximum shear stress at the surface of a circular shaft is, from Sec. 1-6,

$$\tau = \frac{TR}{2J}$$

and the twist or wind-up of the shaft is

$$\theta = \frac{TL}{GJ}$$

The radial loads, that is, loads applied at right angles to the axis of the shaft, produce bending stresses in the shaft. The shaft can be considered to be a simple beam if supported by bearings at its ends, since the bearings develop only very slight end moments.

A shaft transmitting power through gears, belts, or power chains is subject to combined bending and torsion. Such combined stress was discussed in the first chapter.

From Equation 1-12, the maximum shear stress due to combined loading is

$$\tau_{max} = \sqrt{\tau^2 + \frac{s_t^2}{4}}$$

and for a solid cylindrical shaft, from this equation can be derived the following

$$\tau_{max} = \frac{5.1}{D^3}\sqrt{T^2 + M^2}$$

where D = shaft diameter
T = torque, inch-pounds
M = bending moment, pound-inches

The maximum tensile stress due to combined loading is

$$\sigma_p = \frac{\sigma}{2} + \sqrt{\frac{\sigma^2}{4} + \tau^2}$$

For a solid cylindrical shaft this equation may be modified to

$$\sigma_p = \frac{5.1}{D^3}(M + \sqrt{M^2 + T^2})$$

EXAMPLE 8-1

A rotating shaft with 1.25-in. diameter and 20 in. long between bearings carries a 720-lb load at its center. Torque is 2500 lb-in. applied with light shock loading (use a constant of 1.5 for both torque and bending moment). What is the maximum shear stress due to combined loading?

solution

$$M = \frac{PL}{4} = 3600 \text{ in.-lb}$$

$$\tau = \frac{5.1}{D^3}\sqrt{(1.5 \times 2500)^2 + (1.5 \times 3600)^2}$$

$$= 17,200 \text{ psi}$$

The ASME code defines an allowable shear stress, which is the smaller of the two following values:

1 /   0.3 times the tensile yield stress of the material, or
2 /   0.18 times the ultimate tensile stress of the material.

The code recommends that the bending moment and the torsional moment be multiplied by shock and fatigue factors, C_m and C_t, respectively. Hence,

$$\tau = \frac{5.1}{D^3} \sqrt{(C_m M)^2 + (C_t T)^2} \qquad (8\text{-}1)$$

If stress concentrations such as keyways are present, the code recommends a 25% reduction in allowable stress, or if the consequences of failure are serious, a further 25% reduction.

TABLE 8-1
VALUES OF BENDING MOMENT FACTOR C_m AND TORSION-MOMENT FACTOR C_t FOR ROTATING SHAFTS

TYPE OF LOADING	C_m	C_t
Load applied gradually	1.5	1.0
Steady load	1.5	1.0
Load applied suddenly, minor shock	1.5–2.0	1.0–1.5
Load applied suddenly, heavy shock	2.0–3.0	1.5–3.0

8-4 CRITICAL SPEEDS OF SHAFTS

If the center of gravity of a shaft lies on its axis of rotation and the shaft does not deflect under load, the shaft will not vibrate. These two conditions almost never hold, however. Only a large shaft under light loads supported by bearings not too far apart would have virtually zero deflection. Further, manufacturing tolerances often indicate that the center of gravity of a shaft cannot lie on its axis of rotation. Shafts for agricultural equipment, for example, are allowed a total indicator runout (TIR) of 0.005 in. maximum. The method of measuring TIR is shown in Fig. 8-3, the runout being the total movement on the dial indicator as the shaft is rotated 360°. The actual deformation is of course only one-half of this figure, or 0.0025 in. maximum.

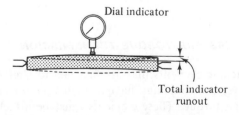

Dial indicator

Total indicator runout

Figure 8-3 / Measurement of total indicator runout.

Because of such imperfections in shafts, contrifugal forces are set up tending to increase the deflection of the shaft, and as the rotational speed increases, a speed is reached at which vibration (*whip*) is strong enough to be noticed. This speed is the critical speed of the shaft. If the speed is still further increased beyond the critical speed, the vibration becomes reduced, increasing again at multiples of the critical speed. As a general rule, shafts should not operate at speeds exceeding 80% of their critical speed. The critical speed is very nearly the natural frequency of the shaft in bending.

If the shaft has a concentrated load and is supported by two end bearings, the critical speed can be approximated from this formula:

$$\text{critical speed, rpm} = N_{cr} = \frac{188}{\sqrt{\delta}}$$

where δ = deflection of the shaft as determined by bending formula.

If the shaft is of solid steel, cylindrical, then the critical speed is

$$N_{cr} = 387 \times 10^3 \frac{D^2}{ab}\sqrt{\frac{L}{P}}$$

where D = shaft diameter
a, b = distances from the load to the bearings, inches
L = center-to-center distance between bearings, inches
P = concentrated load on the shaft, pounds

EXAMPLE 8-2

A steel shaft 1.25 in. in diameter and 24 in. long between bearing centers carries a load of 100 lb centrally between the bearings. Determine the critical speed.

solution
Here $a = b = 12$

$$N_{cr} = 387 \times 10^3 \frac{(1.25)^2}{12 \times 12}\sqrt{\frac{24}{100}}$$

$$= 2057 \text{ rpm}$$

8-5   FASTENERS FOR TORQUE TRANSMISSION

Most of the machine elements that are fastened to shafts, such as gears, sprockets, pulleys, and couplings, have as their functions the transmission of power through torque. These elements must be attached to the shaft in some manner that prevents relative motion between them and the shaft

but which allows the element to be removed from the shaft for mainte-
nance.

8-6   FASTENERS WITH LOW TORQUE CAPACITY

The *setscrew* (Fig. 8-4) is used only in fraction horsepower applications.
This is a headless screw with a hexagon socket head and a conical tip that
bears on the shaft. The setscrew is threaded into the hub of the pulley or
gear, and should be short enough that it does not project above the hub
to act as a possible hazard.

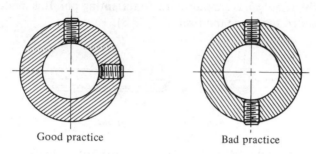

Good practice   Bad practice

Figure 8-4 | Use of setscrews.

If two setscrews are used in one hub, they should be at 90° to
each other. If installed opposite each other, the shaft is supported on the
two setscrews and not against the hub. With the right-angle arrangement,
the shaft must lie against the hub, giving an improved frictional resistance
to prevent relative movement between shaft and hub.

Taper pins (Fig. 8-5) have a standard taper of $\frac{1}{2}$ inch per foot. The
seating hole for the pin is finished with a matching taper reamer, and the
pin is driven into place.

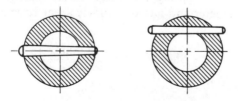

Figure 8-5 | Concentric and eccentric installation of taper pins.

The *groove pin* (Fig. 8-6) has one or more longitudinal grooves
over a part of the length, so that the farther it is driven into the hole, the
tighter it locks.

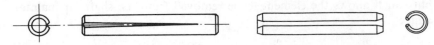

Figure 8-6   /   Groove pin.   *Figure 8-7   /   Spring pin.*

The *spring pin* or *roll pin* (Fig. 8-7) is a hollow tube with a longitudinal slot down the whole length and tapered ends to facilitate driving
the pin. The slot allows the diameter to reduce slightly when the pin is
driven into its hole. The torque resistance can be improved by driving a
spring pin of the next smaller size into the first pin. This is an economical
pin to install, because a drilled hole is sufficient, without reaming.

The *spiral pin* is similar in use to the spring pin. It is made of sheet
metal wrapped twice around itself (Fig. 8-8).

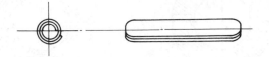

Figure 8-8   /   Spiral pin.

A *shear pin* is used as a weak torque-transmitting element stressed
in shear. It is designed to fail at a predetermined stress that is less than the
stress required to damage more costly transmission elements in the power
train such as gears or sprockets. The shear pin thus protects a power
transmission drive against damage from overloads. If the pin shears, the
hub in which it is located will idle or stop until the drive is shut down and
the shear pin replaced.

EXAMPLE 8-3

Find the diameter of a shear pin to the nearest 16th of an inch, if the
pin is to shear at 15 hp and 1000 rpm. The pin attaches a hub to a shaft
1.25 in. in diameter and has an ultimate shear strength of 40,000 psi.

solution

$$\text{Shear force} = \frac{63,000 \times 15}{0.625 \times 1000}$$

$$= 1512$$

$$\text{Shear pin cross section} = \frac{1512}{40,000} = 0.0375 \text{ in.}^2$$

$$\text{Pin diameter} = 0.222 \text{ in.}$$

Use a diameter of 1/4 in.

8-7 KEYS AND KEYWAYS

The key is a common torque-transmitting fastener. The simplest key shape is the square key. Half the height of the key lies in the shaft and the other half in the hub of the machine element. In addition to the square key, the round end key, the taper key, and the Woodruff key of Fig. 8-9 are also used. All these shapes of course result in stress concentrations because of the required keyway.

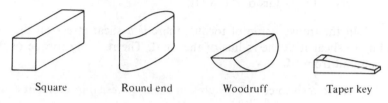

Square   Round end   Woodruff   Taper key

Figure 8-9 / Types of keys.

The rules for sizing plain and round end keys are:

1 /   key width should be $\frac{1}{4}$ of shaft diameter;

2 /   keyway depth should be $\frac{1}{6}$ of shaft diameter;

3 /   minimum key length should be $1\frac{1}{2}$ shaft diameters; and

4 /   the depth of a square keyway in the shaft or hub should be one-half the width of the key.

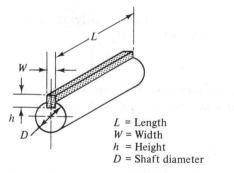

L = Length
W = Width
h = Height
D = Shaft diameter

Figure 8-10 / Key dimensions.

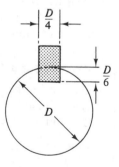

Figure 8-11 / Sizing of square keys.

Key thickness is measured radially, key width tangentially.

The semicircular side of the Woodruff key is seated in the shaft, the keyway being cut with a milling cutter of the diameter and width of the key. The size of a Woodruff key is given by a system of digits. The last two digits of the key number indicate the key diameter in 8ths of an inch, and the digit or digits preceding the last two give the width of the key in 32nds an inch. For example, the 1012 size has a diameter of $\frac{12}{8}$ or $1\frac{1}{2}$ in. and a width of $\frac{10}{32}$ or $\frac{5}{16}$ in.

In determining the stresses in keys and keyways, two simplifying assumptions are made:

1 / The force on the bearing face of the key is uniform over the area.

2 / The force on the side of the key and on the side of the keyway acts at the radius of the shaft.

In the transmission of torque, there is a shear stress on the key tending to shear it at the radius of the shaft. The resisting torque of the key is found from

$$\text{Torque resistance} = T = \text{area in shear} \times \text{resisting moment}$$

$$= WL \times \tau \frac{D}{2}$$

See Fig. 8-10. Hence, $\tau = 2T/WLD$.

The resisting torque due to the bearing strength of the key surface or of the keyway surface may be expressed as

$$T = \text{area in bearing} \times \text{resisting moment}$$

$$= \left(\frac{h}{2} \times L\right) \times \frac{s_b D}{2}$$

and bearing stress $= s_b = 4T/hLD$.

The key is an inexpensive item of hardware and therefore should be designed to fail before the shaft or hub fails.

EXAMPLE 8-4

Determine the shear and the compressive stress on a square key 0.50×0.50 in. in section and 4 in. long, used in a 2-inch shaft to transmit 100 hp at 1000 rpm.

solution

$$T = \frac{63{,}000 \times 100}{1000} = 6300 \text{ lb-in.}$$

$$\tau = \frac{2T}{WLD} = \frac{2 \times 6300}{0.5 \times 4 \times 2} = 3150 \text{ psi}$$

$$s_c = \frac{4T}{hLD} = \frac{4 \times 6300}{0.5 \times 4 \times 2} = 6300 \text{ psi}$$

8-8 SPLIT TAPER BUSHINGS

A flanged type of split taper bushing used with a square key is shown in Fig. 8-12; flangeless types are also used. These bushings are used to fasten any shaft-mounting part that is equipped with a hub.

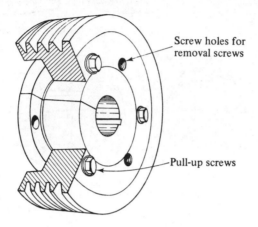

Screw holes for removal screws

Pull-up screws

Figure 8-12 / Flanged split taper bushing attaching a pulley to a shaft.

The split taper bushing is flexible and is drawn into the mating internal taper by three pull-up screws. It seats with very high friction and therefore requires removal screws for dismounting.

8-9   SPLINES

Splines may be considered as a series of teeth cut into the periphery of a shaft that mate with a series of keyways cut into the hub of the mounted part. They are used instead of keys when a sliding connection is required and when heavy torque loads, particularly reversing loads, are present, as in power take-offs in farm machinery. To ensure sliding under load, the bearing stress against the sides of the splines is often limited to 100 psi.

Splines of many shapes are in use, perhaps the more common being the parallel-side spline and the involute spline. Spline shapes are shown in Fig. 8-13. The involute spline resembles the shape of a gear tooth but is modified from the standard gear tooth profile. The involute shape gives greater strength and ease of manufacture.

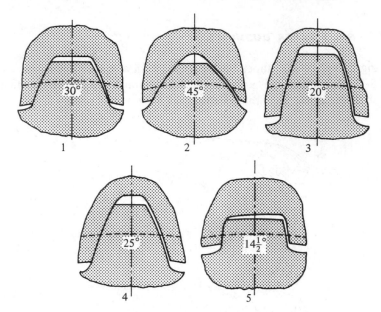

Figure 8-13 / Spline shapes.

8-10 SHAFT COUPLINGS

A common requirement in the design of machines is the connecting of two shafts axially. For example, an electric motor may be direct-connected to a pump. The pump manufacturer will buy a standard electric motor and join the motor and pump shafts by some type of rigid coupling.

Shaft couplings are grouped into two broad classes: rigid and flexible. A rigid coupling has no provision for misalignment of the two shafts joined, nor will it reduce shock or vibration across it from one shaft to the other.

Shafts are frequently subject to the various kinds of misalignment shown in Fig. 8-14. Where such misalignments occur, a flexible coupling must be employed. The misalignment may be:

1 / radial, such that the axes of the two shafts are parallel but offset;

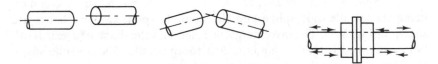

Figure 8-14 / Types of shaft misalignment: radial, angular, and float.

2 / *angular*, such that the two shafts make a slight angle with each other; or

3 / *float*, such that there may be a small axial movement of one or both shafts.

Some misalignment may be due to deflection from loads on the shafts. Severe misalignment must be corrected; slight misalignments can be accommodated by flexible couplings, thus preventing fatigue failure or destruction of the bearings.

8-11 FLANGED COUPLINGS

The usual rigid coupling is the flanged shaft coupling of Fig. 8-15. The flange at the outside diameter is a safety feature to reduce the hazard of rotating bolt heads catching in clothing. Torque is transmitted from one shaft to the other by compressive stress in the hub and finally by shear stress in the bolts of the two-piece coupling. The analysis of stresses in keyed and bolted connections has been previously explained. The load is assumed to be divided equally among the bolts, a safe assumption only if the bolt holes are reamed after drilling.

The diameter of the hub of the coupling is usually sized to be 1.75 times shaft diameter plus 0.25 in.

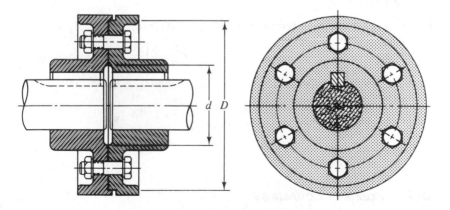

Figure 8-15 / Flanged coupling.

EXAMPLE 8-5

A flanged coupling has a bore of 2.002 in. The flanges of the coupling are joined by 4 bolts in reamed holes on a bolt circle of diameter 5.25 in. Both bolts and shaft are of SAE 1040 steel, ultimate strength

90,000 psi, yield strength 55,000 psi, both in tension. Find the required bolt diameter such that the bolts supply the same torque capacity as the shaft. Use the ASME shaft code, noting that the shaft has a keyway.

solution

The ASME shafting formula is

$$D^3 = \frac{16}{\pi S_s} K_t M_t$$

where D = shaft diameter
M_t = torque
K_t = constant for service conditions (assume $K_t = 1.0$)
S_s = allowable shear stress = 0.18(UTS) or 0.3(Y.S.)
= 16,200 psi (0.18 × 90,000)

The code requires that this allowable stress be multiplied by 0.75 to allow for the effect of the keyway. Therefore, the allowable stress of the shaft in shear is reduced to 12,150 psi.

$$2^3 = \frac{16}{12,150\pi} M_t$$

and $M_t = 19,200$ lb-in.

The bolts must provide the same torque in shear, with each bolt carrying one-fourth of the total torque, since the bolts are fitted into reamed holes.

Torque requirement per bolt = 4800 lb-in.

$$= \left(\frac{\pi}{4} d^2\right)\frac{5.25}{2}\tau$$

$$= 0.785d^2(2.625)(16,200)$$

where d = bolt diameter.

$$d = 0.38$$

Use $\frac{3}{8}$-in. bolts.

8-12 *FLEXIBLE COUPLINGS*

One class of flexible coupling contains a nonmetallic flexing insert. A flat rubber membrane or cushion may be included in the coupling to compensate for angular and axial misalignment. Since rubbers develop heat when flexed, this type of coupling is limited to low-speed applications and therefore low horsepower. The shear type of rubber coupling (Fig. 8-16) can be used for higher speeds and horsepowers.

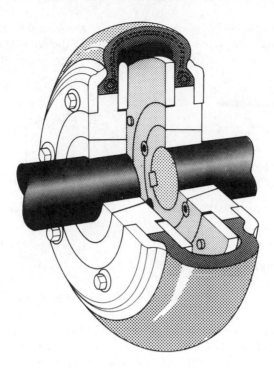

Figure 8-16 / Shear type of flexible coupling.

The *chain coupling* consists of two identical sprockets coupled by roller chain (Fig. 8-17). Disconnection is easy, since only one pin in the chain is removed. Misalignment of as much as 5° can be accommodated by this coupling. However, it transmits shock loads and generates some noise. The gear type of flexible coupling is similar to the sprocket but substitutes gears of special tooth profile and a housing sleeve with internal splines.

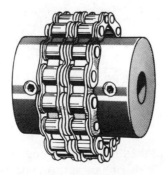

Figure 8-17 / Chain coupling.

The *Falk Steelflex coupling* includes a continuous steel strip that weaves through serrations in the two flanges of the coupling (Fig. 8-18).

The spider type of coupling uses a spider inside the jaws of the two coupling hubs (Fig. 8-19).

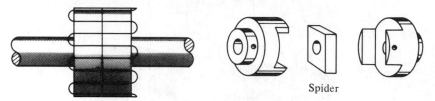

Spider

Figure 8-18 / Falk Steelflex coupling.   Figure 8-19 / Spider type of flexible coupling.

8-13 UNIVERSAL JOINTS

For very large angular or offset misalignment between two shafts, a universal joint must be employed. Figure 8-20 shows the simplest type of universal joint, consisting of a yoke on each shaft connected by a central bar. Universal joints are almost always used in pairs as in the figure, with equal angles in the two joints. The use of equal angles provides uniform angular velocity in the driven shaft.

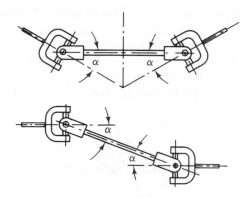

Figure 8-20 / Universal joint.

8-14 CLUTCHES

A clutch is a mechanism used for repeated connection and disconnection of a torsion load from the driving power. A clutch is designed to transmit a certain maximum torque, hence its horsepower rating depends on rpm.

8-15 POSITIVE ENGAGEMENT CLUTCHES

In a positive engagement clutch, the engaging clutch surfaces interlock to produce a rigid joint. Most positive clutches are of the jaw type. One of the jaws must be splined or keyed so that it can move axially to engage. The jaws may be square (Fig. 8-21) or spiral (Fig. 8-22). The spiral clutch is an overrunning clutch; that is, it transmits power in one direction only.

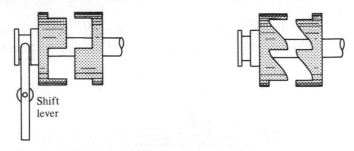

Figure 8-21 / Square-jaw clutch.   Figure 8-22 / Spiral-jaw clutch.

Positive clutches are inexpensive and generate no heat. But their applications are limited to small torques and low speeds. Engagement of the jaws produces considerable shock. The design of a jaw clutch is a relatively simple matter: there must be sufficient bearing area and shear area. The clutch surfaces must be wear-resistant.

The slip clutch of Fig. 8-23 is a positive-engagement clutch provided that the torque is below a value set by the force of the spring. The triangular teeth on the clutch faces give a component of the transmitted torque in the axial direction, which cannot exceed the spring force.

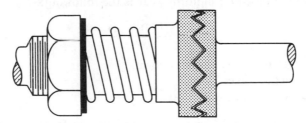

Figure 8-23 / Spring-loaded slip clutch.

8-16 FRICTION CLUTCHES

Friction clutches are adapted to the smooth engagement of shafts with a large difference in relative speeds. Two mating friction surfaces are forced into contact and torque is transmitted by means of friction. The driven shaft can be accelerated up to the speed of the driver shaft, because the clutch can slip. This type of clutch can be used as an overload protective clutch by designing for slip when a certain torque is reached, and by slipping whenever a shock application occurs. Most of the friction clutches are either of the *disk* or *rim* type.

The friction element of the disk type consists of one or more friction disks that engage axially. The single disk is usually run dry, while multiple-disk clutches usually run in oil for smoother engagement and heat dissipation.

The basic disk or plate clutch is shown in Fig. 8-24. The inner diameter of the friction surface is designated *d* and the outer diameter *D*.

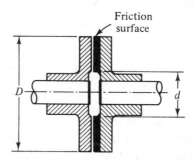

Figure 8-24 / Single-disk friction clutch.

Since the friction torque that can be transmitted is proportional to diameter, *d* is usually at least one-half of *D*. For these clutches, it is assumed either that clutch pressure is uniformly distributed or that surface wear is uniform, though neither assumption is strictly true. A reasonably useful design equation based on uniform wear is the following:

$$T = \frac{\mu P(D + d)}{4}$$

where T = transmitted torque in lb-in.
P = axial force
μ = coefficient of friction

The coefficient of friction depends on the facing materials and their condition and temperature. A worn facing has a reduced coefficient; for design purposes, it is assumed that a worn condition exists. For gray cast iron against gray cast iron, both dry, the friction coefficient is 0.20, and the allowable pressure is 150 psi maximum. For gray cast iron against steel, both dry, the corresponding values are 0.30 and 150 psi maximum. To allow for slipping and inertia of the mass to be rotated, the nominal required torque is increased by a factor of 1.5 to 2.0. A shock factor is also included. If the clutch will be subject to frequently repeated operation, the pressure is reduced to increase the life of the friction facing.

The capacity of this clutch may be increased by increasing the number of friction plates; two plates of the same design supply double the torque capacity. The previous equation (for transmitted torque) applies to each of the several plates. The number of plates cannot be indefinitely increased because of increasing heat generation and increasing difficulty of complete disengagement. One kind of multiple-disk clutch is shown in Fig. 8-25. When air is forced into the air tube, this tube forces the two floating plates with their friction faces against the backplate. Thus, four friction disks transmit torque. The torque is directly proportional to air pressure. To disengage, the air pressure is removed, and springs recenter the plates.

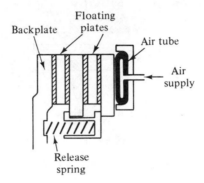

Figure 8-25 / Air-operated multiple-disk clutch.

The *cone clutch* of Fig. 8-26 is another friction clutch. In this design, a small axial force results in a large force between the friction surfaces. The clutch angle is in the range of 8° to 15°, smaller angles giving greater contact forces. The transmitted torque may be found from the following equation, which is for a plate clutch modified to allow for the conical angle:

$$T = \frac{\mu P(D+d)}{4 \sin \alpha}$$

The cone clutch is used less in this country than it formerly was. Its disadvantages are a tendency to grab and some reluctance in disengagement.

The *radial clutch*, also called a rim, shoe, band, or ring clutch, has a cylindrical friction surface (Fig. 8-27). This type can be built into pulleys and flywheels. One widely used type is the Fawick Airflex Clutch, with two concentric friction rims which are engaged by air pressure in an an-

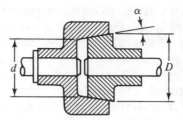

Figure 8-26 / Cone clutch.

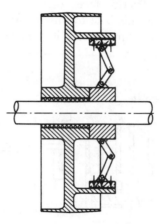

Figure 8-27 / Radial friction clutch built into a belt pulley.

nular tube fastened to the outer rim in the same manner as described for a multiple-disk clutch. Disengagement occurs by exhausting the air in the tube and is assisted by centrifugal force on the tube. The air tube compensates automatically for wear.

8-17 CENTRIFUGAL CLUTCHES

The centrifugal clutch engages by centrifugal force after a certain driven rpm is reached. This clutch is recommended for connecting equipment of high inertia to the commonly used alternating-current induction motor, since this motor can then pick up its load at a speed giving a higher torque than it can deliver at standstill. The electric motor can then be of smaller capacity than would be required by direct connection to the load. Another advantage is automatic disengagement at overload, when the electric motor decelerates. This clutch is also used with internal combustion engines, which cannot be started under load. Such clutches use shoes acting against a drum by centrifugal action. Applications include fans, chain saws, compressors, presses, and conveyor belts.

This type of clutch is not readily adaptable to variable-speed drives, because its horsepower rating varies with the cube of the rpm. Nor is it suited to low speeds, which would require a large diameter to obtain sufficient centrifugal effect.

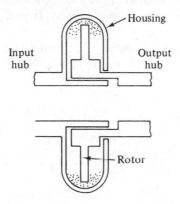

Figure 8-28 / Dry fluid clutch, a type of centrifugal clutch. Centrifugal force cause fine steel shot to wedge between the rotor and the housing.

Centrifugal force is proportional to the square of the speed. Torque is directly proportional to centrifugal force and also varies as the square of the speed. Power is the product of speed and torque, hence it varies as the cube of the speed.

8-18 ELECTRIC CLUTCHES

A rotating field clutch is shown in Fig. 8-29. This is a single-disk friction clutch that is operated electrically. The rotating electromagnet on the driving shaft pulls in the armature on the driven shaft. The friction surfaces

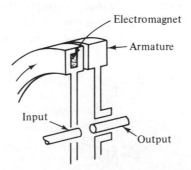

Figure 8-29 / Rotating field clutch. The clutch engages when the armature on the output shaft is pulled to the input electromagnet.

on the electromagnet and armature transmit the required torque. Usually the electromagnet, being heavier than the armature, is mounted on the driver shaft so that its greater weight does not have to be accelerated when the clutch is operated.

An alternative construction is the stationary field clutch, shown in Fig. 8-30. Here the magnetic coil is fixed. It pulls the output armature in to the input rotor.

The *magnetic fluid clutch*, which may also be used as a brake, uses two parallel magnetic plates, as in Fig. 8-31. One of these contains

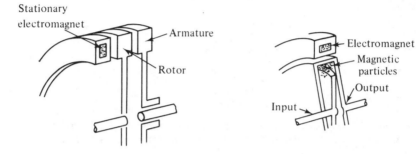

Figure 8-30 / Stationary field clutch.   **Figure 8-31 / Magnetic fluid clutch.**

an electromagnet—the other is an input housing containing a driving disk. The magnetic particles are carried in an oil. By varying the excitation of the magnetic coil with a direct current, the shear strength of the magnetic fluid suspension can be controlled, so that any speed of the driven shaft is possible, from idling to full speed. A disadvantage is the considerable heat generated during slip of the driven shaft.

The *magnetic particle clutch* is similar but uses a dry magnetic powder.

The *eddy current clutch* transmits torque by magnetic drag from eddy currents induced in a driven member by a rotating magnetic field (Fig. 8-32). The driven shaft always slips (lags) behind the speed of the

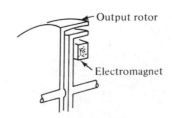

Figure 8-32 / Eddy current clutch.

driven shaft. This clutch delivers a constant torque once it reaches rated slip speed. At zero slip (that is, both shafts at the same speed), there is no torque transmission.

Magnetic flux links the input drum and the output rotor. The eddy currents in the input drum interact with the magnetic field induced in the output rotor to produce a coupling torque that is proportional to the exciting electric current.

The eddy current clutch, because of its high slip, is best employed as an adjustable speed drive. There is no wear in this clutch, so its service life is long. Like the magnetic clutch, this clutch generates considerable heat and may require cooling.

8-19 THE HYDRAULIC COUPLING

The hydraulic or fluid coupling uses a fluid for transmitting torque. The input part of the coupling (the *impeller*) has radial vanes to provide kenetic energy to the fluid—this energy being absorbed by the output part of the coupling (the *runner*), except for losses due to fluid friction. The losses appear as heat, which is dissipated to the surrounding air by cooling ribs on the housing.

Coupling of the two shafts can occur only when the driver shaft is rotating. There is some slip of the driven shaft behind the driving shaft—this slip being in the range of 3–6%. Heat generation increases with slip. Maximum torque occurs when the driven shaft stalls.

The fluid coupling is not adaptable to shaft misalignment but has the ability to absorb both shock and vibration. It will stall if overloaded. Power output varies with the cube of speed. If a stator is included between the two halves of the coupling, then the coupling becomes a torque converter, which may be used as a variable speed drive.

If the fluid coupling connects a torque load to an electric motor, it picks up the load gradually and smoothly, greatly reducing peak current demand on the motor. Because this coupling will stall, it provides overload protection, although a prolonged stall may overheat the hydraulic oil.

The transmission of power through a fluid coupling may be understood by reference to Fig. 8-33. In the primary half of the coupling, fluid moves radially outward, then axially to the more slowly rotating secondary half. In the secondary, it must decelerate, because the secondary is moving more slowly and because the fluid also moves toward the center of rotation. There is no transfer of energy unless the secondary rotates at a slower speed than the primary. The difference of speeds is called *slip*, which is usually expressed as a percent:

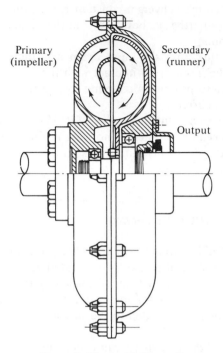

Figure 8-33 / Fluid coupling.

$$\text{Slip} = \frac{\text{primary speed} - \text{secondary speed}}{\text{primary speed}} \times 100\%$$

Output torque is always equal to input torque, and power output is less than power input by the percent slip:

$$\text{Efficiency} = \frac{\text{power output}}{\text{power input}} \times 100\%$$

$$= 100\% - \text{slip}$$

8-20 SELECTION OF CLUTCHES

Many service factors must be considered for the specification of a clutch. After torque and power requirements, perhaps the first consideration is heat generation. Most catalog ratings for clutches are based on only a few engagements of the clutch per hour, but if engagements are frequent, then the heat-dissipating capacity of the clutch becomes a governing factor, or special cooling arrangements may be required. Heat dissipation improves at higher speeds. Heat however is generated only during clutch slip: heat in horsepower equals delivered horsepower multiplied by percent slip.

Both the power source and the type of load influence the performance of a clutch. An electric motor or a steam turbine run at a uniform speed, but an internal combustion engine produces a speed pulsation as each cylinder fires. A single-cylinder engine presents the worst case of speed pulsation. On the load side of the clutch, the moving masses may present a high-inertia or a low-inertia load. Further, the time required to engage or disengage the load will influence the rapidity of temperature rise in the friction areas of the clutch.

Rating factors are therefore applied to the torque or horsepower capacity of the clutch, as in Table 8-2. These rating factors for load and prime mover should be multiplied together. Selection of load factors however requires some experience, and the following factors do not take into account the frequency of engagement. It must be noted too that not all manufacturers of clutches use the same rating factors.

TABLE 8-2
CLUTCH AND BRAKE RATING FACTORS

PRIME MOVER	
Electric motor	1.00
Steam turbine	1.00
Internal combustion engine	1.50
Single-cylinder engine	2.00
LOAD	
Fan, belt conveyor, generator	1.00
Line shaft	1.50
Crusher	2.00
Low-inertia load	0.75–1.00
Average inertia	1.00–1.25
High inertia	1.50–2.00

Since higher speeds give lower torque requirements, the clutch would normally be located in the power train at the place of highest rotational speed.

The consequences of a wrong selection of clutch must be considered. If the clutch capacity should be too small, the clutch will not pick up the load but instead will overheat and gradually destroy itself. But too large a clutch may exceed the overload capacity of the prime mover that drives it or may hold engine or motor speeds to low levels. If the primer mover is an electric motor, the clutch should have a torque capacity slightly below the motor breakdown torque.

AN EXAMPLE OF BASIC CLUTCH DESIGN

A multiplate friction clutch is to be designed to transmit a torque of 350 lb-in. under normal operation and must slip at 200% of rated torque to protect a gear transmission from damage. Hardened steel plates with a friction coefficient of 0.1 will be used as the friction surfaces. Space limitations in the machine dictate a maximum diameter of 4.00 in. for the friction disks.

The basic design equation is, from Sec. 8-16,

$$T = \mu N P \frac{D + d}{4} \qquad (8\text{-}2)$$

where N is the number of friction plates. It is not usual for d to be smaller than $0.5D$. Choose $d = 2.00$ in. P can be as high as 150 psi. Suppose for long life that P is reduced to 100 psi.

$$P = 100 \times \text{disk area} = 100\pi \left(\frac{4^2 - 2^2}{4}\right) = 942 \text{ lb}$$

$$T = 350 \text{ lb-in.} \times 200\% = 700 \text{ lb-in.}$$

Solving for N gives $N = 5$ disks (6 plates).

8-21   HEAT GENERATION IN BRAKES AND CLUTCHES

Brakes are often similar in construction to clutches, since a clutch becomes a brake if it can decelerate the driven member. The kinetic energy of rotation must be dissipated in braking, and this energy is transformed by friction into heat. The heat dissipation problem becomes a major consideration in a brake.

The torque required to accelerate or decelerate a rotating mass is

$$T = J\alpha$$

where J is the polar moment of inertia of the mass, and α is the angular acceleration in radians/sec². If the acceleration or deceleration is constant, the acceleration for N rpm is

$$\frac{2\pi N}{60t} \text{ radians per sec}^2$$

where t = braking time in seconds. Then,

$$T = 12J\frac{2\pi N}{60t} \quad \text{or} \quad t = \frac{2\pi NJ}{5T}$$

where T is in lb-in.

The energy required for a braking or clutching operation is

$$E_b = \frac{2\pi T}{12}\left(\frac{1}{2}\frac{N}{60}\right)t = \frac{2\pi TNt}{1440}$$

Here,

$\frac{2\pi T}{12}$ = the work done per revolution

$\frac{Nt}{2 \times 60}$ = number of revolutions for a speed change from zero to n rpm

Substituting $t = 2\pi NJ/5T$ in $2\pi TNt/1440$ gives

$$E_b = \frac{\pi^2 N^2 J}{1800}$$

or in Btu per braking operation, $Q = E_b/778$ Btu.

The temperature rise of the clutch plates at the end of a braking operation will be approximately

$$\Delta t = \frac{Q}{Wc}$$

where Δt = temperature rise
   $Q = E_b/778$ Btu
   W = weight of the clutch plates
   c = specific heat of the clutch plates (0.12 for cast iron or steel)

The rate of heat generation may be determined also from the rate of frictional work:

$$Q = \frac{pA\mu V}{778}\text{ Btu/min}$$

where Q = heat generation in Btu/min
   p = average contact pressure, psi
   A = friction area, sq. in.
   μ = coefficient of friction
   V = peripheral velocity, fpm

If there is sufficient time between braking operations, this braking heat will be dissipated by conduction, radiation, and convection of the heat. If not, some heat is retained for the next braking operation, until an elevated equilibrium temperature is reached. This equilibrium temperature must not be so high as to be destructive to the brake material.

EXAMPLE 8-6

A small, solid-steel flywheel 12 in. in diameter and 2 inches thick is rotated at 1720 rpm by an induction motor. Determine the kinetic energy of the flywheel at this speed and the braking torque required to stop it in 1 second.

solution

$$\text{Weight of the flywheel} = \frac{\pi 12^2 \times 2 \times 0.285}{4} = 64.5 \text{ lb}$$

The weight of steel is 0.285 lb/sq. in.

$$J = \frac{1}{2}\frac{W}{g}r^2$$

$$= \frac{1}{2}\frac{64.5}{32.2}\left(\frac{1}{2}\right)^2 = \frac{1}{4}$$

The kinetic energy of the flywheel is

$$\frac{\pi^2 n^2 J}{1800} = \frac{\pi^2 (1720)^2 (0.25)}{1800}$$

$$= 4052 \text{ ft-lb}$$

$$T = J\alpha = J\frac{\omega}{t} = 12J\frac{2\pi n}{60t}$$

$$= \frac{12 \times 0.25 \times 2\pi \times 1720}{60 \times 1}$$

$$= 540 \text{ lb-in.}$$

8-22 *THE SHOE BRAKE*

The external shoe or block brake has a shoe that presses against a rotating brake drum. The shoe may either be rigidly mounted on a lever or pivoted to the lever (Fig. 8-34).

The free-body diagram of such a brake is shown in Fig. 8-35. If the total shoe angle does not exceed 60°, the normal force N and friction

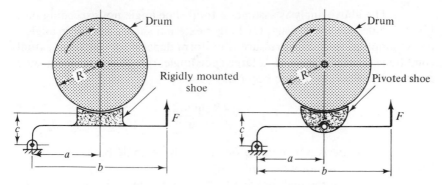

Figure 8-34 | Shoe brakes, rigid and pivoted.

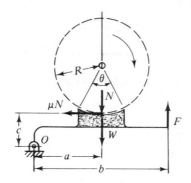

Figure 8-35 | Free-body diagram of a shoe brake.

force μN may be assumed to act at the midpoint of contact. Taking moments about the fixed pivot:

$$(N + W)a - \mu Nc - Fb = 0$$

$$F = \frac{(N + W)a - \mu Nc}{b} \qquad (8\text{-}3)$$

If the drum is rotating clockwise, the friction force μN assists the force F in braking the drum; that is, the brake is self-actuating or self-locking. If the brake is wholly self-locking, then F must be zero or negative in the above equation. For such a case, and supposing W to be not significant,

$$F = \frac{Na - \mu Nc}{b} \leq 0 \qquad \text{or} \qquad \frac{a}{c} \leq \mu$$

Note that a self-locking brake may tend to "grab" however. For a brake that is not self-locking, the braking torque is μNR.

The above analysis assumes a total shoe angle not exceeding 60°. Pivoted shoes are more usual for longer external shoes. For large angles, the assumption that shoe pressure is uniform departs too far from actual conditions. For a brake with a large shoe angle, due to nonuniform shoe pressure the braking torque becomes

$$T = \mu N \frac{4R \sin \theta/2}{\theta + \sin \theta}$$

Since the pressure is not uniform on such a wide-angle brake, the brake friction surface will not wear uniformly.

To reduce the high bearing loads imposed by a single shoe, or to obtain greater capacity, a double-shoe brake may be used, as in Fig. 8-36.

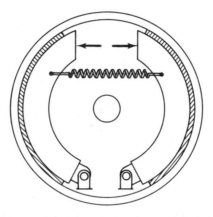

Figure 8-36 / Internal shoe brake.

Here the assumption must not be made that the normal forces are equal on both sides. For a shoe angle of less than 60°,

$$T = \mu(N_L + N_R)R$$

or for a large shoe angle,

$$T = \mu(N_L + N_R)\frac{4R \sin \theta/2}{\theta + \sin \theta}$$

EXAMPLE 8-7

Determine the braking torque of a single block brake under a closing force of 100 lb. The shoe angle is 70°, the drum is 20 in. in diameter, and the coefficient of friction is 0.40 (asbestos fabric against dry gray cast iron).

solution

Note that 70° is 1.22 radians.

$$T = \frac{4\mu NR \sin \theta/2}{\theta + \sin \theta}$$

$$= \frac{4 \times 0.40 \times 100 \times 10 \times 0.574}{1.22 + 0.940}$$

The internal shoe brake of Fig. 8-36 is familiar in automobiles. Two long pivoted shoes are used, one providing self-actuation for forward motion and the other for reverse motion. The pivot points of the shoes are usually adjustable to compensate for brake wear. Wear and normal pressure in this type of brake are proportional to the vertical distance from the pivot point.

8-23 THE BAND BRAKE

This brake uses a steel band lined on the inside with a friction material and wrapped against a rotating drum (Fig. 8-37). The difference in the tensions F_1 and F_2 at the ends of the belt is the friction force (similarly for

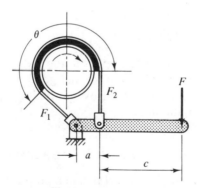

Figure 8-37 / Band brake.

a driving belt the difference in tension is the driving force—see Chapter 6). Therefore, the braking torque is

$$T = R(F_1 - F_2)$$

where R = brake drum radius. The tension forces can be determined from the following relationships.

$$\frac{F_1}{F_2} = e^{\mu\theta}$$

and
$$F_1 = Rbp$$

where p = maximum allowable pressure, psi
  b = width of brake band

EXAMPLE 8-8

From the following information, determine the braking torque of a band brake 24 in. in diameter, width 5 in.

1 / $\mu = 0.15$
2 / $\theta = 3.67$ radians
3 /   maximum allowable pressure = 10 psi

solution

$$e^{\mu\theta} = 2.718^{0.15 \times 3.67} = 1.73$$

$$F_1 = Rbp = 12 \times 5 \times 10 = 600\,\text{lb}$$

$$F_2 = \frac{F_1}{e^{\mu\theta}} = \frac{600}{1.73} = 346\,\text{lb}$$

Braking torque = $R(F_1 - F_2) = 12(600 - 346) = 3050$ lb-in.

8-24 THE DISK BRAKE

The discussion of the disk clutch applies to the disk brake. This brake has the advantages of large braking surface and good heat dissipation; grabbing, chatter, squeal, and brake fade (loss of braking capacity with increasing temperature) are minimized.

QUESTIONS AND PROBLEMS

P8-1 Where would you locate a clutch in a power train, at a high-speed or at a low-speed location?

P8-2 Which of the types of clutches discussed can be adjusted for torque?

P8-3 What advantages do splines offer over keys?

P8-4 Select two types of shaft couplings that can compensate for the following varieties of misalignment between shafts:
  (a) radial;
  (b) angular;

(c)   float; and

(d)   shock.

P8-5   A countershaft carries two pulleys as shown in the accompanying figure. The shaft rotates at 900 rpm, receiving 12 hp from the small pulley and transmitting it to the other pulley. Both belt tensions are horizontal but in opposite directions. The ultimate tensile strength of the steel shaft is 90,000 psi. The pulleys are mounted with keys, and shock factors C_m and C_t are both taken as 1.5. Determine:

(a)   maximum torque;

(b)   maximum bending moment;

(c)   shaft diameter; and

(d)   angular deflection of the shaft in degrees. Modulus of rigidity is 12×10^6 psi.

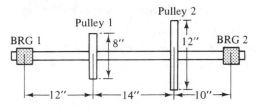

Figure P8-5

P8-6   Determine the shaft diameter for the case of Prob. 8-5, supposing however that the 8-in. pulley is removed, the 12-in. pulley remains, and that the torque load is a sprocket 10 in. in diameter with a horizontal chain force, mounted 3 in. outboard of the left-hand bearing.

P8-7   A steel shaft transmits 100 hp at 3600 rpm from an A.C. motor to a D.C. generator. There is no bending moment on the shaft, shock factor is unity, and allowable stress is 6000 psi in shear. Size the shaft diameter.

P8-8   Determine the diameter of the steel shaft shown in the accompanying figure. The shaft is a 1040 steel with an ultimate tensile stress

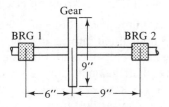

Figure P8-8

of 90,000 psi, and the 9-in. gear is keyed to the shaft. The gear forces at the pitch diameter are 150 pounds tangential (vertical) and 65 pounds radial. Shock factors of 1.5 are applied to both moment and torque. The shaft can be considered as a simple beam between the two bearings.

P8-9 A steel shaft 2.000 in. in diameter and 36 in. between bearings has a load of 250 lb at 16 in. from one of the bearings. Find the critical speed.

P8-10 Find the diameter of a shear pin to the nearest 16th inch, if it is to protect at 25 hp and 900 rpm. The pin connects a hub to a shaft 2.00 in. in diameter and has an ultimate shear strength of 36,000 psi.

P8-11 Size a square key for a 3-in. shaft.

P8-12 The key of Prob. 8-11 transmits 50 hp at 1200 rpm. Determine the shear and compressive stress in the key.

P8-13 A square key connects a $1\frac{1}{2}$-in. shaft to a hub $2\frac{1}{2}$-in. long. Shaft and key are of the same material, with an allowable shear stress of 8000 psi. What percent torque overload is possible without exceeding the allowable shear stress?

P8-14 A shaft 2.000 in. in diameter has 10 splines with a height of 0.214 in. A hub 1.87-in. long slides on the splines, and to permit sliding, the allowable normal pressure on the sides of the splines is limited to 100 psi. Determine the horsepower that may be transmitted to the hub by the splines at 3000 rpm.

P8-15 A square key $\frac{1}{2} \times \frac{1}{2}$ in. connects a 2-in. shaft to the hub of a gear. The gear transmits a torque of 12,000 lb-in., to which is applied a shock factor of 1.5. Determine the required length of the key, using a medium-carbon steel with a yield stress of 60,000 psi and a factor of safety on yield stress of 4.0.

P8-16 Completely dimension a fixed coupling for the shaft of a 25 hp alternating-current induction motor rotating at 1725 rpm. The coupling is to have 4 bolts of 75,000 psi ultimate tensile stress fitted into reamed holes. Use a shock factor of 1.5 for torque. Obtain the shaft diameter from an electric motor catalog.

P8-17 A flanged coupling has a bore of 2.002 in., and the flanges are joined by 5 bolts fitted into reamed holes on a bolt circle 5.00 in. in diameter. The bolts have an ultimate tensile strength of 80,000 psi, and the shaft has an ultimate tensile strength of 75,000 psi. The shaft has a keyway. Use $K_t = 1.5$. Find the required bolt diameter such that the bolt torque capacity in shear approximates but does not exceed the torque capacity of the shaft, in accord with the ASME method.

P8-18 The three-jaw clutch of the accompanying sketch is to transmit 6 hp at 1000 rpm. Determine the bearing stress and shear stress in the key and in the jaws.

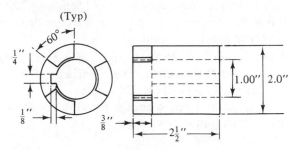

Figure P8-18

P8-19 A coupling of the Falk Steelflex type (Fig. 8-18) joins two 1.25-in. shafts rotating at 1000 rpm. Allowable stress in the shafting is 6000 psi. The coupling strip is 0.010 × 0.120 in. in cross section, with an allowable shear stress of 27,000 psi and is located at an average radius of 2.00 in. from the shaft axis. If the coupling is to have 90% of the strength of the shaft, how many passes of the strip through the coupling are needed?

P8-20 A flanged coupling joins two shafts 2.00 in. in diameter, the shafting material being allowed 6000 psi in shear. Four bolts are used in the coupling, the bolt circle being 5.000 in. in diameter. Bolt holes are reamed. Determine a standard bolt diameter to transmit almost the torque that the shaft can transmit. Assume a sufficient hub thickness that the coupling cannot fail in bearing stress and an allowable shear stress in the bolts of 12,000 psi.

P8-21 Determine the torque capacity of a single-plate clutch with friction surfaces having an inside radius of 2.00 in. and an outside radius of 4.00 in. The axial force is 1000 lb and the coefficient of friction is 0.25.

P8-22 A pneumatically operated radial clutch uses air pressure of 60 psi to engage the friction surfaces. The friction rims are 2 in. wide, with an I.D. of 2.5 in. The coefficient of friction is 0.20. What is the torque capacity, assuming that wear is uniform and therefore so is pressure?

P8-23 A multiple-disk plate clutch has 4 steel disks and 3 bronze disks which are engaged by a 90-lb axial force. The coefficient of friction is 0.20. The effective disk radius is 3.00 in. Uniform wear is assumed. Determine torque and horsepower capacity at 600 rpm.

P8-24 A multiple-disk clutch must transmit 12 hp at 800 rpm. The inside radius of the disks is 1.50 in., and the outside radius is 3.00 in. The coefficient of friction is 0.15. Average pressure must not exceed 100 psi. How many pairs of friction surfaces are required? When the number of pairs of surfaces are determined, calculate the actual pressure required.

P8-25 A multiple-disk clutch has 7 plates (6 contact surfaces) with I.D. 2.25 in. and O.D. 4.00 in. The coefficient of friction is 0.2. The average pressure is 60 psi. Determine:
(a)   operating force to engage the clutch;
(b)   clutch torque capacity; and
(c)   hp capacity at 750 rpm.

P8-26 For a slip clutch of the type shown in Fig. 8-23 with 45° tooth angle and an effective clutch radius of 6.000 in., at what torque will the clutch slip against a 1000-lb. spring force?

P8-27 The slope of a cone clutch is 8°, the large diameter is 13.00 in., and the length of the face of the cone is 2.50 in. The coefficient of friction is 0.15. What axial force is required to produce a torque capacity of 1600 lb-in.?

P8-28 How much heat in Btu must be dissipated to bring a 3000-lb vehicle to a stop from 60 mph?

P8-29 A clutch engages a bulk conveyor. The steady load is a torque of 10,000 lb-in. at the surface of the 24-in. diameter conveyor drive pulley, plus the force required to lift the maximum conveyor load of 67,200 pounds up a slope of 5°. On start-up, the conveyor must reach its steady speed of 10 fpm in 1 second. Determine:
(a)   the operating torque measured at the drive pulley.
(b)   the accelerating torque, $T = J\alpha$, measured at the drive pulley. The polar moment of inertia, $J = W/g \times$ (radius of gyration)^2, where $W = 67,200$ pounds, and the radius of gyration may be taken as equal to the radius of the drive pulley.
(c)   The conveyor drive uses a speed reducer of 700:1. The clutch will therefore be located at the high-speed end of the drive. What torque and horsepower capacity are required in the clutch to handle the sum of start-up and steady loads? Assume 50% efficiency in the drive train.

P8-30 Estimate the heat in Btu/min generated in a friction brake with a friction area of 10 sq. in., a contact pressure of 120 psi, coefficient of friction 0.25, and average velocity of 100 fpm between full speed and stop.

P8-31 A disk flywheel 24 in. in diameter and 3-in. thick is revolved at 800 rpm. Determine the kinetic energy of the flywheel at this speed and the braking torque required to stop it in 3 seconds.

P8-32 The accompanying figure gives the dimensions of a shoe brake with a torque capacity of 2000 lb-in. at 600 rpm and 0.25 coefficient of friction. Determine:
(a) total normal force N between shoe and brake wheel;
(b) total frictional force;
(c) the force required to brake against counterclockwise rotation;
(d) the force required to brake against clockwise rotation;
(e) heat generation in Btu/min; and
(f) a suitable dimension for d to make the brake self-locking. Self-locking is possible only for clockwise rotation.

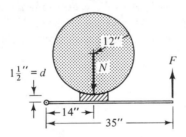

Figure P8-32

P8-33 For the block brake of the accompanying figure, using a coefficient of friction of 0.3, determine the torque capacity and the heat generated in Btu/min if braking at 500 rpm.

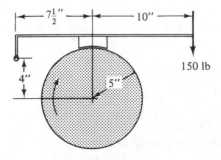

Figure P8-33

P8-34   The band brake of the accompanying figure is engaged with a force of 60 lb as shown. The coefficient of friction is 0.35. Determine the maximum and minimum force in the band and the braking torque.

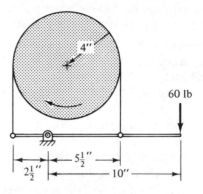

Figure P8-34

P8-35   In the band brake of the accompanying figure, the band wraps over the brake wheel an angle of 225°. Braking torque is 1350 lb-in. Determine the belt tensions, if the coefficient of friction is 0.2.

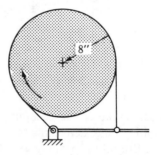

Figure P8-35

P8-36   Determine the torque capacity of a single-plate clutch with a friction surface having an inside radius of 50 mm and an outside radius of 100 mm. Axial force is 400 newtons, and the coefficient of friction is 0.25.

P8-37   The slope of a cone clutch is 10°, the large diameter is 300 mm, and the length of the face of the cone is 65 mm. The coefficient of friction is 0.15. What axial force in newtons is required to provide a torque capacity of 200 newton-meters?

CHAPTER 9

BEARINGS

The use of steel balls for reducing the friction of bearings is being rapidly extended, and new applications are daily being made. The fact that one company alone has a monthly output of from 15 million to 20 million balls, ranging in size from 1/16 inch to 4 inches, is a sufficient index of the importance of this relatively new branch of machine design. . . .

The first name that comes into one's mind when the subject of ball bearings is mentioned is that of John. J. Grant. [The Grant Axle and Wheel Company made ball-bearing axles for bicycles and other equipment.] Mr. Grant gave the rather unexpected and decidedly depressing information that he himself did not know how to proportion ball bearings, and that he believed no one else did.

C. H. BENJAMIN
*"The Design and Construction
of Ball Bearings",*
The Practical Engineer,
December 1898.

9-1 INTRODUCTION

All bearings are grouped into two types: those in which the mating parts slide, called *plain bearings* or *sleeve bearings,* and those in which the moving parts in contact roll on one another, called *antifriction bearings.* Ball bearings and roller bearings are antifriction bearings, characterized by very low friction.

Bearings may be required to resist either radial loads or thrust loads or both. Radial loads are forces at right angles to the axis of the shaft, such as the loads imposed by straight spur gears, drive chain, or V-belts. Thrust loads are applied parallel to the axis of the shaft, such as the loads imposed by turntables on their vertical shafts, or the loads on the crane hook of Fig. 9-1.

441

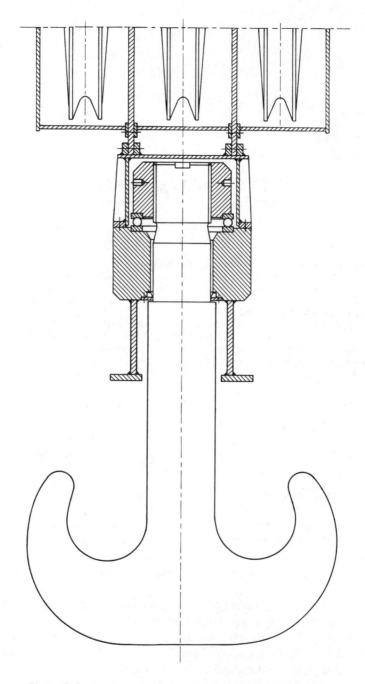

Figure 9-1 / Crane hook supported by a ball thrust bearing.

9-2   DECIDING ON THE TYPE OF BEARING

The choice between a sleeve bearing and an antifriction bearing may be governed by any of the following considerations.

1 /   Space limitations. If the bearing must have the smallest possible outside diameter, then a sleeve bearing with its thin cylindrical shape must be the choice. But if the bearing must be as short as possible, then an antifriction bearing is selected, since these bearings have a length-to-diameter ratio of much less than 1.

2 /   If starting torque is high, the antifriction bearing has the advantage, since it rolls.

3 /   If the bearing must resist forces parallel to the shaft, then the usual types of sleeve bearings cannot be used.

4 /   If shafting and machine elements such as gears or grinding wheels must be accurately located, then the antifriction bearing is the choice. Clearances are much smaller in antifriction bearings.

5 /   If electric currents must not be allowed to flow across the bearing, then a sleeve bearing with its insulating oil film must be selected.

6 /   In the case of fractional horsepower drives, where power losses must be closely controlled, the antifriction bearing develops less frictional horsepower. The friction torque of an antifriction bearing is given approximately by

$$0.0015 \times \text{bearing load} \times \text{shaft radius}$$

7 /   The sleeve bearing is preferred for high speed, heavy loadings, and long life. Antifriction bearings produce more noise at high rotational speeds.

9-3   FRICTION IN BEARINGS

In the transmission of power, friction is an essential element. Without friction, belt drives, clutches, and brakes could not function. But in a bearing, friction is undesirable, because it produces heating effects, power loss, and wear.

Friction may be either sliding or rolling friction. Sliding friction occurs between surfaces either with or without lubrication. Rolling friction, as in a ball bearing, reduces friction losses to about one-tenth those of sliding friction. Sliding friction is a more complex effect, since such friction is very high when movement is begun but is reduced greatly when a steady velocity is reached. High break-away friction is apparent in the piston of a pneumatic or hydraulic cylinder, which is difficult to start but slides readily once in motion. Similarly, it is more difficult to start an automobile rolling than to keep it rolling.

Sliding friction between two surfaces may be reduced by lubrication. If there is no lubrication, the relationship between the surfaces is as shown in Fig. 9-2; due to some surface roughness, the two surfaces are in

Figure 9-2 / Microscopic surface roughness of two apparently smooth surfaces in contact.

contact only through small areas. If sliding occurs, there will be a tendency for the irregularities of the two surfaces to interlock. If a considerable force pushes the two surfaces together, the small contact areas will weld together or *seize*. The force required to break these small welds will result in a high friction effect as the surfaces move relative to one another. A sleeve bearing normally is made of a different material from that of the shaft since dissimilar materials are less likely to seize should the lubricant fail to separate them.

9-4   LUBRICATION

A lubricant is a substance, solid, liquid, or gas, that will produce a friction-reducing film between two sliding surfaces by preventing close contact of the two surfaces. Such lubricants serve three important functions in bearings:

1 /   reduction in friction;

2 /   reduction in wear, thus increasing the life of the bearing;

3 /   removal of frictional heat from the bearing.

Lubricant systems are often designed for a fourth function: the prevention of foreign particles from entry to the bearing.

The most important property of a lubricant is its viscosity. Viscosity is resistance to flow. A liquid with a high viscosity, such as a heavy oil or grease, flows reluctantly; a liquid with a low viscosity, such as water, flows or pours readily. Viscosity is influenced by temperature; if the temperature is elevated, then the viscosity of any liquid is reduced. Hence the necessity for using different lubricants for high- and low-temperature applications. Lubricants of higher viscosity must be used for supporting heavy loads on bearings. Therefore, the viscosity of an oil must be selected to suit the pressure and the temperature of the application.

The units of measurement of viscosity are a source of great confusion, and a great many of these units are in use. Any thorough discussion of this subject would serve little purpose here. The most commonly used units of measurement heretofore in use have been the poise and the SUS (Saybolt universal second). The metric unit of the poise is more commonly given as centipoises, where 100 centipoise $=$ 1 poise. The SUS is the time in seconds required for a certain quantity of oil to run out of the cup of a Saybolt viscosimeter. The higher the number of poises or SUS, the higher the viscosity.

The variation of viscosity with temperature is shown in Fig. 9-3.

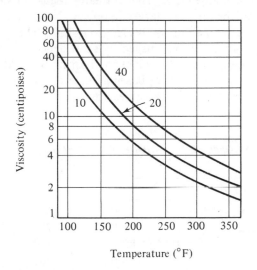

Figure 9-3 / Variation of viscosity with temperature of three SAE lubricating oils.

Plastic bearings such as nylon and Delrin are able to operate unlubricated under light loads and slow speeds.

9-5 SLEEVE BEARING CHARACTERISTICS

For a sleeve bearing, the most important operating characteristics are the bearing pressure P in pounds per square inch and the speed of a point on the rotating member V in feet or inches per minute. The most important characteristic of the lubricant is its viscosity Z. Bearings are therefore characterized by the quantities PV and ZN/P (where N = rpm). For example, consider a bearing $\frac{3}{4}$-in. long with a $\frac{3}{4}$-in. bore and a bearing load of 400 lb rotating at 500 rpm. To determine the bearing pressure P, consider the bearing load to be distributed over the shaft diameter ($\frac{3}{4}$ in.) and the bearing length ($\frac{3}{4}$ in.), so that an area of 0.56 sq. in. carries this load. Therefore

$$\text{Pressure} = P = 400/0.56 = 700 \text{ psi}$$

To determine V in inches per minute, consider that a point on the shaft surface rotates on a $\frac{3}{4}$-in. diameter 500 times each minute.

$$V = \tfrac{3}{4}\pi \times 500 = 1185 \text{ ipm,} \quad \text{or almost 100 fpm}$$

The severity of the service increases with PV, that is, with an increase in load, diameter, or rpm. Greater loads must be carried by larger shafts and bearings of larger diameter. Note that a heavier load cannot be carried by making the bearing very long. The ends of an excessively long bearing would be crushed by shaft deflection. Sleeve bearings therefore are usually about as long as the bearing bore. Maximum values of P, V, and PV for a number of bearing materials are given in Table 9-1. Note that P values are in psi; V values are in fpm.

TABLE 9-1
MAXIMUM PRESSURE AND VELOCITY FOR BEARING MATERIALS

MATERIAL	P	V	PV
Sintered leaded bronze	800	1500	80,000
Cast bronze	3000	750	75,000
Tin babbitt	1500	1200	30,000
Lead babbitt	1300	1400	18,000
Cast nylon	3000	450	17,000
Delrin plastic	1000	1000	2000
Teflon	300	100	500

9-6 OPERATING CONDITIONS FOR SLEEVE BEARINGS

To understand how sleeve bearings operate, consider the graph of the coefficient of friction versus ZN/P in Fig. 9-4. If we suppose this graph to apply to a specific bearing, then P and Z are fixed quantities, and only the

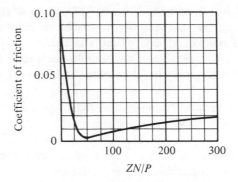

Figure 9-4 / Variation of coefficient of friction with ZN/P.

rpm can vary. When the machine is stopped, N and ZN/P are zero. Assuming that oil is not pumped into the bearing, when the shaft is at rest it is lying against the bearing and is unlubricated. At start-up, the friction is very high. As the shaft gains speed, lubricant begins to smear the mating surfaces and friction falls to a minimum. As the shaft draws oil around its periphery, it becomes self-lubricating. Finally, if the shaft accelerates to a high speed, there is a slight increase in friction resulting from the power required to churn the lubricant in the bearing.

Since at speed a shaft draws lubricant around itself to produce separation from the bearing, in theory a shaft that is never stopped would have an infinite bearing life. This of course does not happen because of corrosion, contamination, shaft deflection, and other factors. Nevertheless, while it is difficult to obtain a service life of 100,000 miles from an automobile engine used in frequent starts and stops, the life of the engine of a Greyhound bus is expected to be 2,000,000 miles.

When the shaft is up to operating speed, the fully lubricated condition on the right-hand side of the graph of Fig. 9-4 prevails. Here the rotating shaft serves as its own oil pump, supposing that forced lubrication from an oil pump is not used. Figure 9-5 shows the relationship of shaft

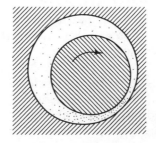

Figure 9-5 / Hydrodynamic lubrication in a sleeve bearing.

and bearing for this condition. The shaft floats to a slightly eccentric position in the bearing. Because of the clockwise rotation, more oil is pulled under the right-hand side of the shaft than escapes from under the left-hand side. This self-lubricating condition is called hydrodynamic lubrication. For hydrodynamic lubrication, the oil hole introducing the oil must be located in the minimum pressure area of the bearing.

If a shaft must continually operate under the conditions of the left side of the graph of Fig. 9-4, there is no hydrodynamic lubrication, and journal and bearing must be protected from wearing each other away. A lube oil or grease is of assistance, but often solid lubricants such as graphite and molybdenum disulfide are employed.

9-7   BEARING MATERIALS

Several characteristics are required of an ideal bearing material:

   1 /   *Hardness.* Greater hardness is accompanied by greater strength. A harder bearing material will require a harder shaft. Soft bearing materials will generally have a greater tendency to weld to the shaft than hard materials.

   2 /   *Compatibility.* The bearing material must not tend to weld or seize to the shaft when metal-to-metal contact occurs. A hard steel shaft will not tend to weld to a gray iron bearing material, but an aluminum shaft in an aluminum bearing may seize readily. In this respect, the bearing material should be incompatible rather than compatible with the shaft material.

   3 /   *Conformability.* Conformability in the bearing may be needed to compensate for slight misalignment and deflection of the shaft.

   4 /   *Embeddability.* This characteristic will permit abrasive particles to embed and sink into the bearing material, thus protecting the journal. A bearing material sufficiently soft for embedment will also provide the necessary conformability.

   5 /   *Corrosion resistance.* The bearing material must not be attacked by corrodents in the atmosphere around it, by water, or by corrosive agents in the lubricant.

The softer types of bearing materials include the babbitts. These are either lead- or tin-bearing alloys. The lead babbitts contain about 80% lead, with antimony and tin; the tin babbitts, used in high-grade bearing applications, contain about 85% tin, with antimony and copper. Because of their softness, babbitt bearings are not suited to high loadings or high temperatures, but their conformability and embeddability are excellent.

Thin babbitt on a steel backing will support heavier loads, but the thin overlay must not be allowed to wear completely away.

Copper alloys such as brasses and bronzes are suited to heavy loads and more severe wear conditions than babbitts can sustain. Leaded bronzes and tin bronzes are excellent bearing materials. The lead or the tin in the bronze alloy provides some of the bearing properties of the babbitts, while the copper enables such an alloy to support a higher bearing load.

Gray cast iron is a suitable bearing material for slow speeds and moderate loads. This material contains graphite flakes which act as a solid lubricant to prevent seizing.

Sintered metal bearings are usually bronze. These bearings are porous and can retain a supply of lubricant in the pores, usually sufficient for about two years of lubrication. A *PV* factor of about 25,000 can be used with these bearings, though higher values have been used where a shorter bearing life is acceptable.

For these metallic bearing materials, the bearing clearance is about 0.001 in. per inch of shaft diameter. If the bearing is a press fit into the housing, allowance for the reduced diameter due to pressing must be made. This reduction will approximate the interference of the fit but cannot be more than this.

Rubber bearings are used where, as in marine service, the presence of water makes oil lubrication difficult. Because of the excellent embeddability of rubber, sand in the water has not too damaging an effect. Low *PV* factors, below 15,000, must be used with rubber and low pressures below 50 psi.

Nylon and acetal (Delrin) bearings are used for low-pressure and low-speed applications. Graphite makes a self-lubricating bearing like these plastics but cannot sustain high pressures. Sintered bronze, plastics, and graphite bearings should be used when speeds are too low for hydrodynamic lubrication (less than about 25 fpm) and when there is considerable possibility of seizure.

9-8 BALL AND ROLLER BEARINGS

Those bearings in which rolling elements are included are referred to as *antifriction bearings*. The rolling components are either balls or rollers, and the rollers may be cylindrical or tapered. In most types of antifriction bearings, the rolling components may also slide as well as roll, as we shall see. The parts of an antifriction bearing are named in Fig. 9-6.

The load-carrying capacity of a ball or roller increases with the square of the diameter. Thus a $\frac{1}{4}$-in. ball can support only one quarter the load of a $\frac{1}{2}$-in. ball.

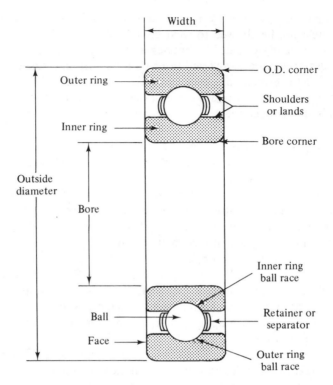

Figure 9-6 / Ball bearing parts.

This load capacity of a rolling bearing depends on whether the load is static (nonrotating) or dynamic (rotating). The basic static load is defined as the static radial load that will produce a permanent deformation of the rolling element and raceway of 0.0001 of the diameter of the rolling element at the most heavily stressed contact. Clearly, the static load rating would govern such an installation as the crane hook of Fig. 9-1, which experiences little rotation. These static load ratings are given in manufacturers' catalogs.

An antifriction bearing properly sized and selected for the application and suitably maintained will fail as a result of fatigue. Once each revolution, each ball or roller receives a high contact stress of hundreds of thousands of pounds per square inch. The fatigue failure appears as a spalling or separation of flakes of metal from the rolling elements or from the inner race. When this fatigue fracture occurs, the bearing must be replaced, even though it will still rotate. Harder materials have longer fatigue lives, hence the necessity for a very hard ball or roller. The life of a rotating antifriction bearing, therefore, is a certain number of revolutions at a specified load. The calculation of this life is discussed presently.

The required hardness in the rolling elements is provided by using high-carbon steel alloys. The most common ball and roller alloy is 52100, containing about 1% carbon and 1.3–1.6% chromium. Rolling elements that must resist corrosive attack are usually made of stainless steel alloy 440C, which contains about 1% carbon and 16–18% chromium. The chromium provides corrosion resistance except against seawater. Stainless 440C is also used for bearings in high-temperature applications.

9-9   LUBRICATION FOR ANTIFRICTION BEARINGS

The lubrication requirements for an antifriction bearing are different from the case of a journal bearing, because the bearing load is carried on the rolling elements and not by a film of oil. Nevertheless, lubricant is required and serves the same three functions in both types of bearings:

1 /   reduction of sliding wear;

2 /   removal of heat from the bearing;

3 /   protection of the bearing from dirt, water, and other contaminants.

Since an antifriction bearing rolls, it might be supposed that lubrication is unnecessary for the first function above, reduction of sliding wear. Actually, most ball and roller bearings involve some sliding. This is explained for a ball bearing by Fig. 9-7. The surface speed of a point near the outside diameter of the inner race is $\pi D_1 \times$ rpm. The speed at the bottom of the groove in this race is less, because the diameter D_2 is less. In the case of the ball, the maximum surface speed is the ball rpm $\times \pi d_2$, but at a point at d_1, the diameter is less, and the surface speed is lower. Thus, a slower part of the ball is in contact with the fastest moving part of the race, and therefore, there must be sliding. The manufacturer makes the groove in the race with a radius about 4% larger than the ball radius; this partially prevents slippage between them. Hence, the roll path is a narrow path of width w instead of the full groove width W. Sliding friction and wear are slight if the bearing is lightly loaded but become more severe under heavy loads. If there is end thrust on the ball bearing, there is a displacement of races and ball as in Fig. 9-8, because the grooves have a larger curvature than the balls. Under this condition of end thrust, sliding wear is more severe.

Normally these bearings are lubricated with grease, which in addition to lubrication prevents the entry of contaminating particles. High-speed or high-temperature bearings, however, are lubricated with oil. Greases are mixtures of an oil with a soap base for a thickener. Calcium soaps have good resistance to water but are not suited to high speeds or

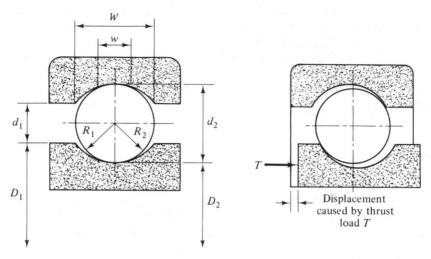

Figure 9-7

Figure 9-8 / Displacement in a rolling bearing due to thrust.

elevated temperatures. Sodium greases are less resistant to water but have a broader speed and temperature range. Lithium greases, sometimes called multipurpose greases, combine the virtues of calcium and sodium greases.

The viscosity of a grease changes with the rate of shearing. When the bearing is not rotating, the grease is quite stiff, but in motion the viscosity approaches that of its constituent oil. The soap in the grease acts as a sponge to hold the oil and release it at a slow and controlled rate. It is assumed that the useful life of a grease is reduced by one-half for every 25°F rise in temperature.

A sealed bearing should be packed one-third full of grease.

9-10   BEARING SEALS AND SHIELDS

Dirt must be prevented from entering any kind of bearing, because it will act as a grinding abrasive within the bearing. Seals and shields are required to prevent the ingress of such contamination and also to retain oil or grease lubricants applied to the bearing. Felt, rubber, leather, or other materials are used as seals in a variety of configurations to seal lubricants in and contaminants out (Fig. 9-9).

Felt seals should be soaked in oil of about SAE #20 before installation. This soaking reduces the coefficient of rubbing friction by about one-third. Other seal materials are oiled after mounting in place.

Since the seal rubs on the shaft, a smooth shaft surface is required to prevent abrasion of the seal.

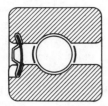

Figure 9-9 / Ball bearing with seal.

Figure 9-10 / Ball bearing with shield.

Shields (Fig. 9-10) are made of strip steel fastened to the outer race of the bearing. Shields do not rub on the shaft or inner race as a seal does. There is just enough clearance between the shield and the inner race to allow excess lubricant to leak but still prevent entrance of foreign matter.

9-11 TYPES OF ANTIFRICTION BEARINGS

The common types of antifriction bearings are illustrated in Figs. 9-11 to 9-21. The following brief remarks explain their characteristics and functions.

1 / The *self-aligning ball bearing* (Fig. 9-11) has two rows of balls which roll on the spherical surface of the outer race. This spherical surface allows for angular misalignment in the shaft arising from errors in mounting, deflection of the shaft, or frame distortion in the machine. This bearing type is suited to the carrying of radial loads or moderate thrust loads.

2 / *Single-row deep-groove ball bearings.* This type of bearing (Fig. 9-12) will support radial or thrust loads and is the type of bearing most commonly selected.

3 / *Angular contact ball bearings.* This bearing (Fig. 9-13) has deep grooves in the races and one shoulder of the outer race partially removed to allow assembly. The design allows the bearing to carry

Figure 9-11 / Self-aligning ball bearing.

Figure 9-12 / Single-row deep-groove ball bearing.

Figure 9-13 / Angular contact ball bearing.

a heavier thrust load or a combined thrust and radial load because of the deeper groove and the large number of balls used. The thrust can be taken in one direction only. To carry a radial load only, such a bearing must be mounted in pairs as in (a) or (b) of Fig. 9-14. The figure shows a double-row angular contact bearing in a single unit. Fig. 9-14(a) is an externally converging type and (b) is internally converging (angles converge inside the bearing).

4 / The *double-row deep-groove ball bearing* (Fig. 9-15) resembles the single-row deep-groove bearing in its design. The load lines through the balls may be outwardly converging or inwardly converging. This bearing can sustain high radial and thrust loading.

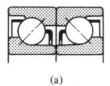
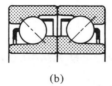

(a)   (b)

Figure 9-14 / Duplex mounting, angular contact bearings: (a) externally converging, (b) internally converging.

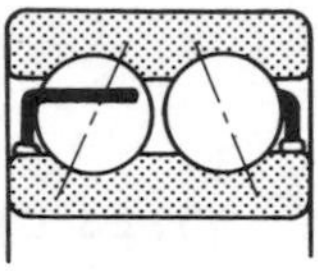

Figure 9-15 / Double-row deep-groove ball bearing.

5 / *The spherical roller bearing* (Fig. 9-16) has a spherical surface on the outer race and thus is self-aligning. It too is adaptable to radial or thrust loads or combinations of both.

6 / *The cylindrical roller bearing* (Fig. 9-17) is designed for very high radial load capacity but not thrust. There is little sliding effect in this bearing so that friction is unusually low and high speeds are possible.

7 / *The ball thrust bearing* (Fig. 9-18) can sustain only axial loads.

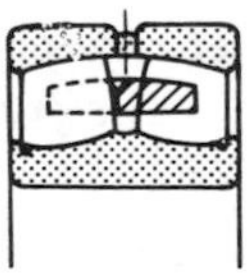

Figure 9-16 / Spherical roller bearing.

Figure 9-17 / Cylindrical roller bearing.

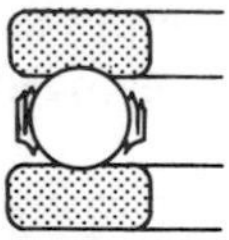

Figure 9-18 / Ball thrust bearing.

8 / *Tapered roller bearings* (Fig. 9-19) can carry heavy radial and axial loads. Such bearings are mounted with another tapered roller bearing elsewhere on the shaft, the two bearings opposing each other's thrust. The steeper the cone angle, the greater the thrust load that can be carried—but at a reduction in radial capacity. These bearings are unsuited to high speeds.

9 | *The spherical roller thrust bearing* (Fig. 9-20) carries large thrust loads but has limited radial capacity. It is self-aligning.

Figure 9-19 | Tapered roller bearing.

Figure 9-20 | Spherical roller thrust bearing.

10 | *Precision bearings.* Many of the above bearing types are available as either standard or precision bearings. The use of the phrase "precision bearing" seems to infer that the standard antifriction bearing is not made to precise tolerances. Actually all antifriction bearings are precision bearings, being made to tolerances of a few ten-thousandths of an inch. A precision bearing is simply a standard bearing for which the tolerances on dimensions are about one-half those allowed in standard bearings. Thus, if the bore tolerance allowed in a standard bearing is 0.0000-in. oversize and −0.0005-in. undersize, for a precision bearing the same tolerances might be 0.0000-in. oversize and −0.0002-in. undersize.

Precision bearings are required for high-speed applications and for precision machinery such as engine lathes, automatic lathes, and grinding machines.

11 | *Bearings with larger clearances.* Sometimes operating conditions require bearings with clearances larger than normal. A hot inner race could expand and cause binding in the bearing. Bearings with special clearances are available for such equipment as fans handling hot gases.

12 | *Needle bearings.* These are bearings with rollers of very small diameter and no inner race (Fig. 9-21). The needles roll on the

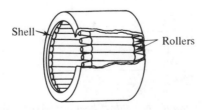

Shell — Rollers

Figure 9-21 | Needle roller bearing.

shaft itself. This type of bearing is used when a bearing of smallest possible diameter must be used.

9-12 MOUNTING ARRANGEMENTS

If bearing loads are too heavy for ball bearings, an arrangement such as that shown in Fig. 9-22 may be used. The roller bearings in the figure support the radial loads, while the ball bearing carries any thrust load and also fixes the shaft axially.

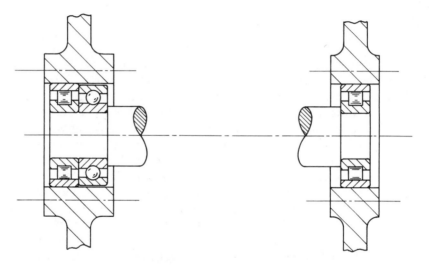

Figure 9-22 / Roller bearings for the radial loading—ball location bearing for the axial "fix" and the thrust load.

The arrangement of two angular contact bearings, as in Fig. 9-23, is used where axial adjustment is required to remove slackness. This is done by controlled clamping pressure on the outer ring of one of the bearings.

A bearing design suited to grease lubrication is shown in Fig. 9-24. Grease is charged through the grease nipple at the top of the housing,

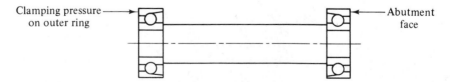

Clamping pressure on outer ring ——→   ←—— Abutment face

Figure 9-23 / Two angular contact bearings. Axial adjustment is effected by a sliding fit of the outer ring of the left-hand bearing in the housing, with controlled pressure against this ring.

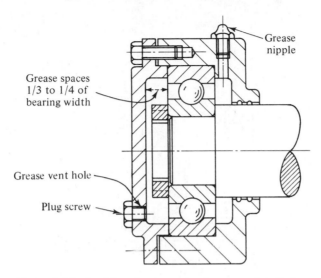

Figure 9-24 | A bearing mounting for grease lubrication.

while the existing grease discharges through the grease vent at the bottom of the housing by removing the grease vent plug. The grease space on each side of the bearing should not be unnecessarily wide; about one-third of the bearing width is sufficient. Note that the grease vent is on the opposite side of the housing to the grease nipple.

Other design details should be noted. Two grease grooves are machined in the housing where it bears against the shaft. These serve as a sufficient grease seal at the shaft. The clearance between housing and shaft at this seal should be about 0.010 in. on radius (0.020 in. on diameter). Note also that the shaft is reduced in diameter where it enters the bearing. The shaft shoulder must have a fillet with a radius slightly smaller than the corner radius of the bearing. A minimum of three bolts are required in the cover plate on the left-hand side of the mounting.

For oil lubrication, a design such as in Fig. 9-25 is used, with a sealing ring to seal the oil at the shaft. An oil vent hole at the proper oil level prevents charging too much oil to the bearing.

A mounting design for a fixed and an axially floating bearing is shown in Fig. 9-26. The right-hand bearing can be adjusted axially. As temperature of the shaft changes, the shaft can expand at the left-hand bearing.

If the bearing is held to the shaft by a clamping nut as in Fig. 9-27, the nut will be wider than the inner race which it must clamp, and this is undesirable. Therefore, the nut is chamferred to reduce the width of the nut against the bearing. An alternate clamping arrangement is given in Fig. 9-28, using a minimum of three bolts. Clamping of the outer race is shown in Fig. 9-29.

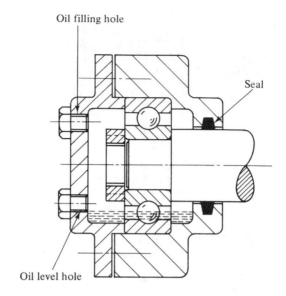

Oil filling hole

Seal

Oil level hole

Figure 9-25   /   A bearing mounting for oil lubrication.

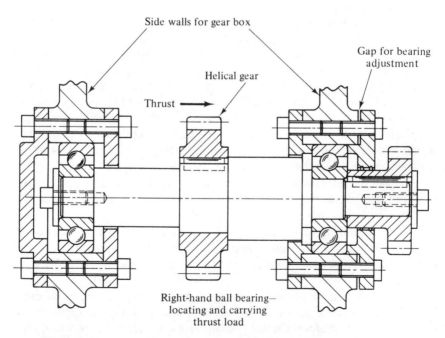

Side walls for gear box

Gap for bearing adjustment

Helical gear

Thrust

Right-hand ball bearing— locating and carrying thrust load

Figure 9-26   /   Fixed and floating bearing to accommodate temperature change in a gearbox.

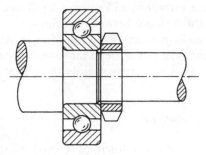

Figure 9-27 / Clamping nut chamferred to avoid pressure on outer race.

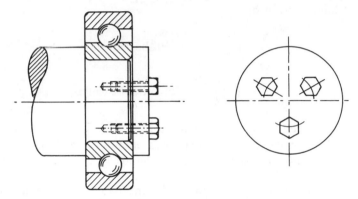

Figure 9-28 / A clamping plate.

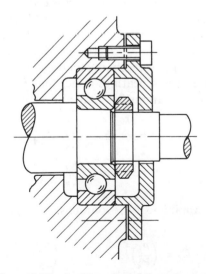

Figure 9-29 / Clamping of the outer race.

Temperature variations will cause a shaft to expand, resulting in small variations in the distance between bearings. One of the bearings on a shaft should be fixed in place axially in the housing. All the other bearings should have sufficient axial clearance to move with expansion and contraction. The "held" or fixed bearing is usually the one at the drive end.

9-13 BEARING CAPACITY

Antifriction bearing capacity is influenced by load and speed, both of which determine the fatigue life of the bearing. If the load is doubled, the life of the bearing is reduced to one-eighth. The relationship between life, capacity, and load is given by

$$L_n = \left(\frac{C}{P}\right)^3 \quad \text{for ball bearings}$$

or
$$L_n = \left(\frac{C}{P}\right)^{10/3} \quad \text{for roller bearings} \qquad (9\text{-}1)$$

where L_n = life in millions of revolutions
   C = specific dynamic capacity in pounds (see below)
   P = radial or equivalent load, pounds

The specific dynamic capacity C is the load that 90% of a group of identical bearings can carry for one million revolutions of the inner ring before evidence of fatigue appears. This factor C is obtained from manufacturers' catalogs. The life L_n is the minimum or "B-10" life of the bearing, defined as the number of revolutions that will be reached or exceeded before fatigue failure by 90% of a group of identical bearings operated under identical conditions. Some manufacturers use the average life, which is that of 50% of a group of bearings rather than 90%.

The following relationships may also be used:

For ball bearings: $\dfrac{L_A}{L_B} = \left(\dfrac{P_B}{P_A}\right)^3$

For roller bearings: $\dfrac{L_A}{L_B} = \left(\dfrac{P_B}{P_A}\right)^{10/3}$

where L is the life in millions of revolutions. If L_B is taken as 10^6 revolutions, then

$$L_A = \left(\frac{P_B}{P_A}\right)^3 \quad \text{or} \quad \left(\frac{P_B}{P_A}\right)^{10/3}$$

If the life in hours is required,

$$L = \frac{16{,}667}{\text{rpm}}\left(\frac{C}{P}\right)^3 \quad \text{or} \quad \frac{16{,}667}{\text{rpm}}\left(\frac{C}{P}\right)^{10/3}$$

EXAMPLE 9-1

A ball bearing has a catalog value of C of 118,000. What is its rating life under a radial load of 10,000 pounds at 1000 rpm?

solution

$$C/P = \frac{118{,}000}{10{,}000} = 11.8$$

$$L = \frac{16{,}667}{1000}(11.8)^3 = 27{,}340 \text{ hours}$$

EXAMPLE 9-2

Select a bearing to carry a 3000-lb radial load for a life of 1.5×10^6 revolutions, using the following data. Select a bearing bore of approximately 2 inches.

BORE NO.	BORE, in.	C	DIAMETER	WIDTH
05	0.9842	1650	1.8504	0.4724
06	1.1811	2200	2.1654	0.5118
07	1.3780	2650	2.4409	0.5512
08	1.5748	2850	2.6772	0.5906
09	1.7716	3550	2.9528	0.6299
10	1.9685	3750	3.1496	0.6299
11	2.1654	4800	3.5433	0.7087
12	2.3622	5100	3.7402	0.7087
13	2.5590	5300	3.9370	0.7087

solution

$$C = P\sqrt[3]{L_n}$$
$$= 3000\sqrt[3]{1.5}$$
$$= 3430$$

A bearing with C not less than 3430 is selected. Bore nos. 09 or 10 are suitable.

Referring back to the bearing life equation, P represented the radial load or an equivalent radial load in pounds. An equivalent load means some calculated combination of radial and thrust load. The equivalent load is determined in different ways by different manufacturers but may be approximated closely enough and conservatively for single-row bearings by the formula

$$P_e = 0.5R + 1.7T$$

where P_e = equivalent radial load, pounds
        R = radial load, pounds
        T = axial thrust load, pounds

If this equivalent load should be less than the radial load, the radial load should be used in the life equation.

EXAMPLE 9-3

Select a 10 series bearing (see the table in Example 9-2) to carry a radial load of 2800 lb and a thrust load of 1000 lb, with a life of 100 hours at 800 rpm.

solution

The total number of revolutions is 4.8×10^6.
The equivalent radial load is

$$P_e = 0.5R + 1.7T$$
$$= 1400 + 1700 = 3100 \text{ lb}$$

This is greater than the radial load.

$$C = P\sqrt[3]{L_n}$$
$$= 3100\sqrt[3]{4.8} = 5280$$

From the previous table, a bore no. 13 is selected.

Because of centrifugal forces in the bearing and frictional heat, there is a limit imposed on the maximum speed at which any bearing may be operated. This limit is expressed by the factor $25.4DN$, where D is the bore diameter in inches, and N is the rpm. If the bore diameter is given in millimeters, the factor is DN. For radial ball bearings, DN is 500,000; for double-row bearings and radial roller bearings, DN is less (300,000).

Thus for the above bore no. 10 bearing, with a bore of about 50 mm, the maximum allowable speed is about 10,000 rpm. Manufacturers' tables are available for rating the fatigue life of bearings operated at high speeds. Higher speeds are possible with oil lubrication than with grease.

In most antifriction bearing applications, the inner race rotates. If it is the outer race that rotates, the circumstances are not the same. The inner race has a sharper curvature, and this curvature is opposite to the curvature of the ball or roller. Hence, there is a smaller contact area between the rolling element and the inner race than between the rolling element and the outer race and therefore a higher contact stress. When the inner race rotates, the same point in the outer race sustains the radial load, but because of the lower stress in the outer race, this is not a critical matter. But if the outer race rotates, the radial load is carried on the same point on the inner race, at a higher stress level than for the outer race. To allow for this difference in stress levels, if the outer race rotates, then the radial load is multiplied by 1.2.

While the above methods of selecting bearings have general application, each manufacturer has a different method of design, selection, and rating of bearings. Since the ratings are taken from the manufacturer's catalog, it is necessary to use that manufacturer's design and selection methods.

9-14 BEARING CLEARANCE AND PRELOADING

The manufactured bearing clearance may be different from the operating clearance, due to such factors as increased temperature during operation, the pressing of the inner race onto the shaft, and the pressing of the outer race in the bearing housing. All these factors reduce the clearance. At the operating temperature, a mounted ball bearing should have a radial clearance ranging from zero to 0.0001 in., or for a roller bearing, slightly more than this range.

Normally, antifriction bearings, like journal bearings, will "float" axially and radially to the limit of clearance in both directions. Such float, however, is not acceptable in some types of equipment such as rotating radar antennas, machine tool spindles, precision gearing, and precision instruments and tools. To eliminate this float, the bearings are preloaded, that is, an initial load is applied to remove end play or radial clearance.

To understand the method of preloading, see Fig. 9-8, where a thrust load is applied to the bearing, thus displacing one race with respect to the other. A typical graph of load against displacement is given in Fig. 9-30. Notice that the first increments of load remove most of the clearance:

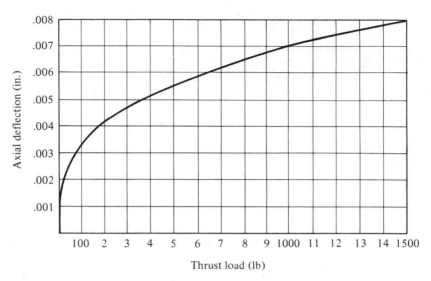

Figure 9-30   /   Typical deflection curve for thrust loading on a single-row ball bearing.

a 100-lb thrust removes 0.0035-in. clearance, while the next 100 lb reduces the clearance only by another 0.001 in.

Suppose this bearing to be preloaded with a 650-lb load during installation, and that the working load on the bearing is an additional 300 lb. The working load produces an additional deflection of only 0.001 in. beyond the preload deflection of 0.006 in. A higher preload would reduce the deflection by the working load still further but might possibly reduce the life of the bearing. Note that without preload, a working load of 300 lb would deflect the bearing almost 0.005 in.

9-15 HANDLING OF ANTIFRICTION BEARINGS

Because all antifriction bearings are precisely manufactured, they require care in handling so that they are protected from dirt and deformation. The bearing should be removed from its package only when conditions are ready for installation. Clean working areas and clean hands and gloves will prevent dirt from entering the bearing. When pressing a bearing on a shaft, no force must be used on the outer race; similarly, when pressing a bearing into a housing, no force must be used on the inner race. See the installation diagrams of Fig. 9-31.

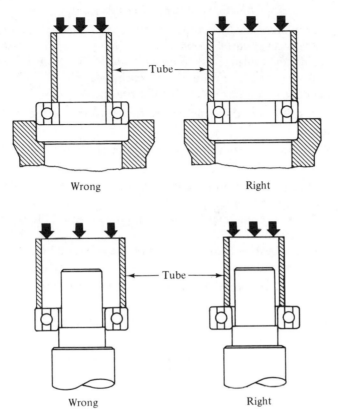

Figure 9-31 / Installation of rolling bearings.

QUESTIONS AND PROBLEMS

P9-1 State what type of rolling bearing you would select for the following conditions:
(a) radial load only, no thrust load;
(b) shaft must be held by the bearings against axial movement;
(c) very high radial load on one bearing, small radial load on the other bearing;
(d) absolute minimum bearing diameter required; and
(e) very high speed, small radial load.

P9-2 In welding turntables, the welding current usually must pass through the turntable bearing. Explain why an antifriction bearing cannot be used in such an application.

P9-3   List the four functions of a bearing lubricant.

P9-4   A gear pump shaft rotating at 1500 rpm is carried symmetrically in two sleeve bearings each $\frac{3}{4} \times \frac{3}{4}$-in. long. The force of the shaft against the bearings due to oil pressure is 3370 lb. Determine the bearing pressure in psi and the bearing velocity V in fpm. What material would be selected for the bearings?

P9-5   In Fig. 9-4, why does friction loss in a bearing increase with higher speeds?

P9-6   Why are sleeve bearings not made of the same material as the shaft?

P9-7   From the table of bore numbers on p. 461, select a bearing for the following applications:
(a)   the bearing load is 900 lb, and the expected life is 10×10^6 revolutions.
(b)   the bearing load is 900 lb, and the expected life is 100×10^6 revolutions.

P9-8   Determine the equivalent radial load for a bearing carrying 2400 lb radially and 600 lb thrust load.

P9-9   Explain how preloading reduces slack in a bearing.

P9-10   The accompanying figure shows two spur gears mounted on a shaft. What is the resultant radial load on each bearing?

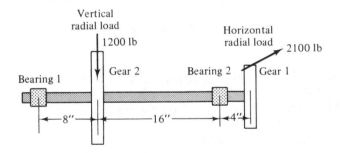

Figure P9-10

P9-11   The accompanying drawing of a rolling bearing mounting gives certain basic dimensions in millimeters. Convert these to inches and thousandths of inches, and supply all missing dimensions for this mounting, particularly for the cap over the bearing. (Note that you do not have information to specify the fillet on the shoulder of the shaft. Omit this shoulder dimension.)

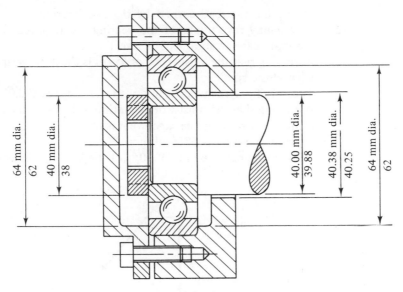

Figure P9-11

P9-12 A deep-groove ball bearing has a specific dynamic capacity C of 10,400 lb. At 1500 rpm and a radial load of 1800 lb, what is the life of this bearing in revolutions and in hours?

P9-13 A bearing has a specific dynamic capacity C of 5400 lb. What radial load can the bearing sustain at 1500 rpm for a life of 2000 hours?

P9-14 The specific dynamic capacity C of a rolling bearing is 7500 lb. If subjected to an equivalent radial load of 1000 lb, what is the expected life in millions of revolutions?

P9-15 A bearing with a specific dynamic capacity C of 9000 is under consideration. What radial load can it carry at 400 rpm if the desired life of 90% of the bearings is 5000 hours?

SUGGESTED DESIGN PROJECT

DP9-1 The accompanying sketch shows the sheave of a mine hoist, hoisting mine cages weighing 100,000 lb. Ten percent is added to this static load to account for acceleration force, since the cage must be accelerated to 65 fps. The main sheave shown has a pitch diameter of 248 in. or 20.65 ft. Spherical roller bearings are used on the shaft, and the bearing load is the resultant of the two cable forces shown plus 10%.

(a) Determine the load carried by the two bearings.
(b) How many revolutions of the bearing are required to lift the mine cage from a 3000-ft level?
(c) Assuming that the hoist speed of 65 fps is the average speed, what is the rpm?
(d) The specific dynamic capacity of the bearings is 590,000 lb. Assuming that the hoist works continuously 24 hours a day lifting and dropping the cage, what is the expected life of the bearings in millions of revolutions?
(e) What is the expected life of 90% of the bearings in hours?

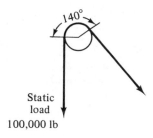

Static
load
100,000 lb

Figure DP9-1

CHAPTER 10

MACHINERY ALLOYS AND THEIR PERFORMANCE

Some less reverent comments on the practice of Machine Design:
In any given computation, the figure that is most obviously correct will be the source of error.
The probability of a dimension being omitted from a set of drawings is directly proportional to its importance.
All constants are variables.
The cause of problems is solutions.
The amount of work yet to be done on a project increases in proportion to the amount of work already completed.
Identical parts tested under identical conditions will not perform identically in the field.
Interchangeable parts won't.

10-1 *MACHINE DESIGN AND THE DESIGNER*

There are four steps in the design of a machine:

1 / the design of individual machine elements such as shafts, bearings, or gears;

2 / the selection of materials for these components;

3 / the combining of the individual machine elements into an integrated machine, for example, aligning two bearings that support a shaft, or ensuring a correct center-to-center distance for two mating gears;

4 / the building, testing, and evaluation of prototype machines.

When the machine is designed, the designer has no assurance that his design is a successful one. Prototype tests must be undertaken to provide information and insight that is not otherwise obtainable, and it is a rare machine that is not modified after such testing. Finally the machine goes into use, and throughout the life of the design, users continue to feed back information about performance, this information leading to further modifications. Suppliers of materials and parts provide advice from which further modifications are developed. Machine design is a continuing process of redesign that ends only when production of the machine ceases.

The machine designer, therefore, is a coordinator of a vast amount of information. He must be able to assess information—is the information useful, useless, false, or irrelevant? Especially, he must be able to make compromises, for if he delays until he has the perfect design, it will be too late to build the machine. Much art, imagination, judgment, and compromise must go into his work.

This book deals largely with the design of machine components. This step is only the first step. Coordinating the parts into a complete and functioning machine is an art rather than a science and is learned with practice. Selecting materials is equally an art and an exercise in engineering judgment, since it requires that the designer balance properties such as strength, ductility, wear resistance, machinability, and weldability against cost.

To illustrate the complex problem of selecting materials for a machine and also some of the unusual design problems that can arise, we consider as an example a portable trigger-operated pneumatic perforator.

10-2 DEVELOPING THE DESIGN

The essentials of the portable gun perforator are shown in Fig. 10-1. It consists of a piston with an integral pin, fired by compressed air, the pin punching a hole in the material to be perforated. The pin must be corrosion-resistant.

Perhaps the first problem to be solved is "What air pressure is required for perforating the materials that the designer has in mind?". Most air compressors can supply 80 psi at the tool. Suppose we tentatively select a 2-in. piston. This has an area of about 3 sq. in., and 80 psi will give a force of 240 lb approximately. This size and pressure are tried in a few tests but are not adequate. The designer can go to a 3-in. cylinder or can use a higher air pressure.

Tests with a 3-in. ϕ pneumatic cylinder show that two things are wrong with it. First, this larger size of cylinder makes the perforator exces-

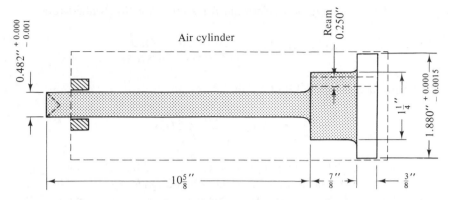

Figure 10-1 / Gun perforator.

sively heavy for use as a portable tool. Second, when using the perforator on brittle board material, instead of a neat hole, the whole board is smashed into pieces.

So here is another design problem. How do you punch a neat hole in brittle materials? The designer recalls that if you throw a baseball at a window, the whole pane of glass is smashed, but if a bullet is fired into the glass, the result is a neat hole. Impact is needed. Clearly 80 psi of air does not supply impact. After further experiment, let us suppose that the designer settles tentatively for a 2-in. ϕ piston and 200 psi of air. Researching the problem of impact also discloses that the piston requires neither too little nor too much mass for the best impact effect; in particular, too much mass has too high an inertia. So different weights of pistons must be tried in order to find the best weight for the purpose.

In pursuing this design, almost none of the information needed can be found in handbooks. Like so much data required in mechanical design, the designer must find it from the results of trials.

10-3 AIR CYLINDER AND PISTON

The air cylinder will be made of seamless hydraulic annealed tubing, with an inside diameter of about 2 in. This type of tubing has an ultimate tensile strength of 65,000 psi. The designer is concerned with the factor of safety, since both impact and fatigue are involved, and the explosion of an air cylinder is a serious matter. The cylinder is not yet designed, and there may be stress concentrations to lower fatigue strength drastically. For example, welds have failed in fatigue at stress levels as low as 10,000 psi. So the designer selects an allowable stress of 5000 psi.

The following is the formula for stress in a thin-walled tube:

$$\text{Allowable stress} = \frac{\text{pressure} \times \text{O.D.}}{2 \times \text{wall thickness}}$$

Assuming a cylinder O.D. of 2.5 in.,

$$5000 \text{ psi} = \frac{200 \times 2\tfrac{1}{2}}{2t}$$

and $t = 0.05$ in. The tubing supplier offers either 2.000 O.D. by 16-gage wall (0.0598) or 2.000 I.D., 2.250 O.D. The designer selects 2.000 × 16-gage. The I.D. then is 1.880 in. With this size, the wall stress is below 5000 psi.

Since we are primarily concerned with materials selection in machine design, the detailed design of the cylinder will not be discussed.

The O.D. of the sliding piston can next be decided. The piston will not have O-rings or seals, since it is suspected that the high impact speeds will wreck any such seals. A seal must be made by a tight clearance between cylinder and piston that is still sufficient for sliding. A clearance of less than 0.001-in. per inch of diameter will make the assembly too tight and result in excessive wear. A diametral clearance of 0.002 in. is chosen. The piston must be ground to nominal dimensions of 1.880–0.002 or 1.878. But some tubing may be slightly undersize, and a piston diameter of 1.877 + 0.0000 − 0.001 is the decision. The tolerance of 0.001 in. is not a difficult one for a grinder operator to meet.

Final dimensions for the piston are shown in Fig. 10-1. This seems to provide the proper mass for impact. The 0.250-in. reamed hole fits a hardened pin that prevents the piston from spinning in the cylinder.

10-4   *DESIGN OF THE PERFORATOR*

Piston and perforator could be machined out of one piece of 2-in. ϕ steel, but this would be expensive in material and machining cost. The perforator pin will be a separate component provided that the designer can develop a joining method that will not fail in impact or fatigue. The following are the requirements in the two components:

> *1 /   The piston.* The sliding surface of the piston against the cylinder must be wear-resistant. The piston alloy must be weldable if we assume that the two components are to be welded. Steel alloys too high in carbon will harden and become brittle if welded;

nevertheless, hardness increases with carbon content, and hardness is needed for wear resistance.

2 / *The perforator.* The end of the perforator must be hard if it is to perforate—say a Rockwell hardness on the C scale of 50. The end of the perforator must also be corrosion-resistant. Chromium-plating would supply corrosion resistance, but impacts will surely cause the plating to flake off.

The piston must have enough carbon for hardness and wear resistance but not so much carbon that it will become brittle if welded. A medium-carbon steel is a suitable compromise, either 1035 (0.35% carbon) or 1040 (0.4% carbon). A 1040 steel will give a maximum hardness of Rc 55.

The steel alloys considered for both hardness and corrosion resistance in the pin are the martensitic steels 410 and 420. These have equal corrosion resistance. Stainless 410 will furnace-harden to a maximum of Rc 40, which may not be quite hard enough, while 420 will harden to Rc 50, which is adequate. But 420 is not as readily obtainable as stainless 410 and is more expensive. The designer would prefer to use 410, but more hardness is needed. A processing trick comes to the rescue. By means of the induction-hardening process, an electrical method of hardening, a Rockwell C hardness of 45 can be produced in the stainless 410. This is judged to be adequate. Only the end of the pin will be hardened by induction. After hardening, the pin will be centerless-ground to size.

The piston will be machined; then the sliding surface will be induction-hardened and ground. It is not desirable to harden the piston throughout, as it will then be too brittle.

10-5 JOINING PIN AND PISTON

The pin could be threaded into the piston. This method will not be considered. There is impact and fatigue to contend with, and threads provide a stress concentration that invites failure under such conditions.

Consider welding, after the pin is pressed into a reamed hole in the piston. A fillet weld at the pin end of the piston would again be a stress concentration and cannot be trusted. It would be better to weld at the back end of the piston. To provide space for depositing weld metal, the piston would be grooved-out. Both the alloys to be welded, 410 and 1040, are hardenable, so that the heat of welding can be expected to produce a hardening effect in both, which would be dangerous. The designer may not know how to specify a welding procedure to avoid hardening, but he should be aware of the problem. He should also specify a sufficient weld area to sustain the impact stresses.

10-6 *DESIGN DECISIONS*

The above example is typical in its design problems. Not all of the decisions were based on stress calculations.

In machine design, there is never a single solution. The designer is concerned with finding the best in a range of solutions. Some of the decisions were based on price and availability, including the selection of tubing size and the alloy for the pin. The design problem began with almost no available data and was pursued even without any ordered procedure. A certain degree of fudge, fumble, and finagle was required to find an acceptable design. Factor of safety was a major concern, especially for the air cylinder, which could explode if underdesigned. Many of the design decisions required an extensive and shrewd knowledge of metals, their costs and availability, and their processing characteristics.

We turn our attention now to the more commonly used machinery materials.

10-7 *PLAIN CARBON STEELS*

Plain carbon steels are alloys of iron containing usually less than 1% carbon with fractional percentages of silicon and manganese. The function of silicon is to deoxidize the steel. Dissolved oxygen in the molten steel combines with silicon to form submicroscopic inclusions of silicon dioxide. If not thus converted, dissolved oxygen makes steels brittle at low temperatures. Manganese combines with sulfur in steel, which forms manganese sulfide inclusions. Otherwise, sulfur makes steel brittle at red heat, making it difficult to hot-roll, weld, or heat-treat. Manganese also improves the hardenability of machinery steels.

The principal function of carbon is to provide strength, hardness, and wear resistance. Iron without carbon has an ultimate tensile strength of 42,000 psi. The addition of 0.2% carbon increases this strength to about 60,000 psi; the addition of 0.4% carbon gives a strength of about 90,000 psi, though even higher-strength levels are possible with heat-treating. However, these strength levels produced by higher carbon are accompanied by lower ductility and toughness, since it is a general rule that a gain in strength is paid for in loss of ductility, and vice versa. Thus in the case of the perforator, a high level of carbon would have been unsuitable for the pin. Low carbon was required for shock resistance.

Plain carbon steels have some limitations. They tend to warp when heat-treated, and they harden to a depth of only about $\frac{1}{8}$-in. from the surface. If a heat-treated part must not deform, or must be deep-hardened, then an alloy steel must be selected.

The plain carbon steels fall into three broad groups:

TYPE	CARBON CONTENT	AREA OF USE
Low carbon	0.03–0.25%	Plate, sheet, structurals
Medium carbon	0.30–0.55%	Machine elements
High carbon	0.60–1.4%	Tools and tooling

Low carbon is a requirement in sheet, plate, and structurals for two principal reasons:

1 / such materials must be ductile and therefore readily formable; and

2 / they must be easy to weld.

Low carbon or "mild" steels are readily bent and shaped without cracking. If welded, they do not become brittle. A machinery steel requires a higher carbon level so that it may be hardened by heat treatment to gain improved strength, hardness, and wear resistance. A minimum of 0.30% carbon is necessary to make a heat-treatable steel. But too much carbon would result in steels too brittle for machine parts, so that the high carbon steels find their applications in punches, dies, and cutting tools. A good general rule is to use the lowest carbon content that will meet the service requirement of the part.

The AISI–SAE steel alloy numbering system used for machinery steels is discussed in the following section. This numbering system designates the plain carbon steels by a four-digit number beginning with 10xx. The first two digits, 10--, designate a plain carbon steel—that is, one with negligible alloy content. The last two digits indicate the carbon content. Thus, 1040 is a plain carbon steel containing 0.40% carbon.

Wear resistance may be assumed to be roughly proportional to the carbon content, though there are other factors and alloying elements that also improve resistance to wear. The lowest Rc hardness that will provide satisfactory wear resistance seems to be Rc 50–55. This hardness can be obtained with 0.40% carbon. The maximum hardness obtainable in heat-treated steels is Rc 65–68, but to obtain such a level of hardness, a minimum of 0.60% carbon is required. Still higher carbon contents do not increase the hardness but have a favorable influence on wear resistance. See Fig. 10-2.

Steel bars are obtainable in either the hot-rolled, cold-finished, or centerless-ground condition. Hot-rolled bars have a coating of black mill scale that must be machined away. If the bar surface is to be used as

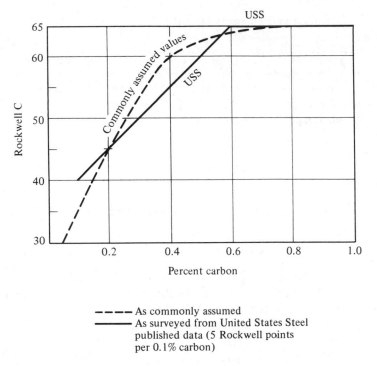

──── As commonly assumed
━━━━ As surveyed from United States Steel
published data (5 Rockwell points
per 0.1% carbon)

Figure 10-2   /   Maximum hardness in steels.

received, then either cold-finished or centerless-ground bar is used. The diameters of such finished bars are produced to closer tolerances than hot-rolled bars.

10-8 *THE AISI–SAE MACHINERY STEELS*

The abbreviations refer to the American Iron and Steel Institute and the Society of Automotive Engineers. These two institutions have established a numbering system for the machinery steels using four digits. A few machine alloys belong to one system but not the other. The last two digits give the carbon content:

$$
\begin{array}{ll}
1040 & 0.40\%\ \text{carbon} \\
8620 & 0.20\%\ \text{carbon} \\
52100 & 1.00\%\ \text{carbon}
\end{array}
$$

The first two digits are not quite so systematic and will not be discussed. An explanation of the complete system may be found in any alloy steel handbook. The 10xx steels are plain carbon steels, and most of the others are low chrome-moly-nickel steels. Only a handful of these machinery steels are in wide use, chiefly the following:

1015	12L14
1020	4130
1025	4140
1035	4340
1040	6150
1045	8620
1060	

The 10xx carbon steels are the lowest in cost. In machine parts where their limitations cannot be tolerated, alloy machinery steels must be substituted. These limitations are chiefly hardening only to shallow depths, a tendency to warp when hardened, lower strength, lower fatigue resistance, and lower wear resistance. The addition of alloy ingredients, however, does not alter the hardness levels obtainable. For example, a ball bearing made of 10100 steel (1% carbon) is not harder than 52100, the actual alloy used in ball bearings, but 52100, containing about 1% chromium, has a better fatigue life and hardens deeper, thus producing a bearing with much longer life.

Machine parts that require considerable machining are often made of 12L14. This is a low-carbon steel containing lead, the lead acting as a lubricant for the cutting tool and thus permitting much higher machining speeds.

The 4xxx steels, principally 4130, 4140, and 4340, are used for more exacting machinery applications, such as aircraft undercarriages, balls for trailer hitches, and wrench sockets. These are deep-hardening steels that can be hardened to high strength levels.

Alloy 8620 is a preferred alloy for parts to be case-hardened.

10-9 STAINLESS MACHINERY STEELS

Most stainless steels, particularly the 300 alloy numbers such as 304, are not generally used for machine parts. The 300 series also are not heat-treatable. A few of the 400 series of stainless alloys are hardenable and, therefore, are used where corrosion resistance or resistance to higher temperatures is required in a machine part. Such hardenable stainless steels are referred to as martensitic stainless steels. Pump shafts and the heads of golf clubs, for example, may need resistance to corrosion by water, as well as needing hardness, and such products are made of 410 stainless steel. Stainless knives must hold a cutting edge and are made of 420 alloy, which has more carbon than 410.

Stainless 410, 420, and 440 may be considered as basically the same alloy but with the carbon content varied to give a range of levels of maximum hardness:

ALLOY	% CR	% C	MAX RC
410	12	0.1	40
420	12	0.2	50
440	16	0.4+	60

(Stainless 430 is not included: it is not a martensitic stainless steel.)

A few other 400 alloys are martensitic, such as 431. Alloy 416 is a free-machining grade of 410 containing sulfur.

10-10 EFFECT OF ALLOY ADDITIONS TO STEEL

While the effects of any alloy addition to a basic carbon steel may be numerous, only the basic effects are discussed here.

Manganese. In amounts up to 1% or slightly more, manganese improves strength, hardenability, and low-temperature toughness. Its effect on toughness is not as pronounced as that of nickel.

Nickel. The principal effect of nickel is to greatly improve the toughness, fatigue, and impact resistance of steels, especially at low temperatures. Pure nickel has a remarkable ability to resist impact fractures, even when stress concentrations are present. Usually $3\frac{1}{2}\%$ or less is used in machinery steels.

Chromium, Molybdenum, and Vanadium. These elements are added in small amounts to most machinery alloys. They bring significant increases in strength and wear resistance at the expense of some loss in ductility. In large amounts (12% or more), chromium provides remarkable corrosion resistance, as in the stainless steels.

A chrome-plated surface provides low friction as well as hardness and wear resistance. Chrome-plating, therefore, is frequently applied to parts that must slide against another surface.

10-11 ALUMINUM ALLOYS

Aluminum alloys are used for machine parts because of their light weight, corrosion resistance, and easy formability. Pure aluminum is of marginal interest to the machine designer; however, it is remarkably soft and weak, with a tensile strength of about 12,000 psi. It can be alloyed to give strength levels of about four times as much, and heat-treating of suitable alloys will result in strength levels of 70,000–88,000 psi. The higher-strength alloys are difficult to form and to weld and have limited corrosion resistance.

The aluminum casting alloys are designated by three-digit numbers such as 356. The wrought alloys, shaped by rolling or extruding, use four-digit numbers such as 2024 or 6061. Silicon alloys are preferred for castings but are not easy to machine. The most commonly used alloys for rolled and extruded products are copper or silicon-magnesium. Alloys 6061 and 6063 are typical silicon-magnesium aluminums, with strengths of about 40,000 psi. Such alloys are selected for welded products. Alloy 2024 is the most popular copper aluminum. The copper alloys are selected when hardness, strength, and machinability are required, but they are more difficult to form.

10-12 BRASSES AND BRONZES

The brasses are alloys of copper and zinc. While copper, with an elongation of 46% in a tensile test, is readily formable, the addition of zinc improves the elongation and, therefore, the formability still further, as shown in Table 10-1. It is a general rule that if the strength of a metal is increased, then ductility is lost, or if ductility is increased, then strength is lost. This general principle does not apply to the zinc brasses. Note from the table that a 30% zinc brass, called *cartridge brass*, will elongate by two-thirds of its original length before failure, a remarkable figure.

TABLE 10-1

STRENGTH AND ELONGATION OF BRASSES

% ZN	TENSILE STRENGTH (PSI)	ELONGATION (% IN 2 IN.)	BRINELL HARDNESS
0	32,000	46	38
5	36,000	49	49
10	41,000	52	54
15	42,000	56	58
20	43,000	59	56
25	45,000	62	54
30	46,000	65	55
35	46,000	60	55
40	54,000	45	75

A bronze is a copper alloyed with elements other than zinc. The *phosphor bronzes* are erroneously named, being tin bronzes. The small phosphorus content serves not as an alloying element but as a deoxidizing ingredient in copper alloys. These bronzes have high strength, toughness, and a low coefficient of friction. The *silicon bronzes* are even stronger. Other bronzes have a more restricted range of uses.

10-13 THE PLASTICS

While all the metals have certain generally similar characteristics such as hardness and stiffness, there are almost no common characteristics for the plastics, other than a high coefficient of thermal expansion. Indeed, many plastics have no plasticity—that is, they are entirely elastic. All the thermosetting plastics are elastic in a stress–strain test rather than plastic.

The most serious weakness of the plastics is their limited dimensional stability. Under moderate loads continuously applied, plastics will deform and warp after a period of time, even though they can sustain such loads for shorter periods. Other factors contributing to poor dimensional stability are a high thermal expansion, approximately ten times that of the metals, and a modulus of elasticity approximately one-tenth that of the metals. Dimensional stability is greatly improved, sometimes by a factor of ten, by reinforcing with fiberglass.

All the plastics fall into two broad groups: thermosetting plastics and thermoplastics. The thermoplastics can be softened by heat, while the thermosets have no melting or softening point, like wood, though they may be damaged by heat. The thermosets are brittle, hard, and strong plastics; the thermoplastics are generally ductile, low in strength, resistant to impact, and limited in their resistance to temperature effects.

THERMOSETS	*THERMOPLASTICS*
phenol-formaldehyde (Bakelite)	polyethylene
urea-formaldehyde	polypropylene
allyls	polystyrene
epoxies	polyvinyl chloride (vinyl)
polyesters	Teflon (trade name)
	polycarbonate
	Delrin (trade name)
	nylon (trade name)
	Plexiglas, Lucite (trade names)
	polysulfene
	ABS

It is not possible to discuss the properties of individual plastic materials in brief, because each of the 60 types of plastics is offered in many grades and formulations. Polyethylene for example is available in at least the following types: low-density, medium-density, high-density, ultra-high molecular weight, cross-linked, and even as a rubber (chlorosulfonated polyethylene). Any of these types may be foamed. Each type has a unique combination of characteristics. Besides foam formulations, any plastic may be reinforced, and any foamed plastic may be reinforced also. Reinforcements and fillers may be used to increase stiffness, hardness, or strength, to reduce shrinkage, or to increase resistance to environmental factors. If a certain type of plastic should be a candidate for a machine part, such as nylon for a roller or a gear, the supplier's technical literature and recommendations must be carefully assessed. Because of the huge range of plastic materials available, more knowledge, experience, judgment, and imagination are required for designing in plastics than are required for designing in metals. Certainly they must never be considered as straight substitutes for metals, because they are entirely different materials with different responses to stress and environment. For example, a nylon sleeve bearing must have a larger clearance than a metal bearing; moisture absorption by nylon will reduce the nominal clearance.

Table 10-2 gives properties of the plastics that are most commonly used for machine parts. The table includes both unreinforced and reinforced plastics and the structural polyurethane foams. Some general characteristics should be noted.

Thermosetting polyester, epoxies, and phenol-formaldehyde are thermosetting; the other unreinforced plastics listed are thermoplastic. The large elongations of the thermoplastics (except acrylics) contrast with the very limited elongations of the thermosets. E-values are usually in the range of 0.3 to 0.4×10^6, or much less than one-tenth of the E-value of metals. Ultimate tensile strength averages about 8000 psi; the coefficient of thermal expansion is about 10 times that of the metals. But the light weight of the plastics is advantageous; most of them have a specific gravity of 1.1 to 1.2, or about 70 pounds per cubic foot.

Since the stiffness of a section in bending is proportional to EI, when designing in plastics the low E-value is compensated by increasing the depth of the plastic section, the moment of inertia being proportional to depth cubed. For the same stiffness, EI, the plastic part will usually be lighter in weight than the corresponding metal part.

Any plastic can be reinforced with glass fiber. The effect of such reinforcement is to increase the E-value to about 10^6 or approximately that of wood. Tensile strength will also be increased. The effect is proportional to the amount of reinforcing.

Most of the plastic foams are in a stage of rapid development,

TABLE 10-2
TYPICAL PROPERTIES OF PLASTICS FOR MACHINE PARTS

	SPECIFIC GRAVITY	COEFFICIENT OF THERMAL EXPANSION (IN./IN./°F)	ULTIMATE TENSILE STRENGTH (PSI)	ELONGATION (%)	E-VALUE TENSION	MAXIMUM OPERATING TEMP	WATER ABSORPTION (%)	FLAMMABILITY
UNREINFORCED PLASTICS								
Acetal	1.4	0.0000014	8.8×10^3	12–75	0.4×10^6	185	0.25	Slow burning
Acrylic	1.2	0.00003	8.7–11×10^3	3–6	0.35–0.45×10^6	130–195	0.25	Slow burning
Teflon	2.15	0.00003	2–5×10^3	75–400	0.05–0.1×10^6	500	0.01	No
Nylon 6/6	1.15	0.00003	11–13×10^3	60–300	0.4–0.48×10^6	250	1.3	Self-extinguishing
Polycarbonate	1.2	0.00004	8–9.5×10^3	60–125	0.345×10^6	250	0.15	Self-extinguishing
Polysulfone	1.25	0.000016	10.2×10^3	30–80	0.35×10^6	300–390	0.22	Self-extinguishing
Polyester (thermoplastic)	1.3	0.000053	8×10^3	300	0.34×10^6	300	0.3	Slow burning
Epoxy	1.1–1.4	0.000025–40	4–13×10^3	3–6	0.35×10^6	250–550	0.1	Slow burning
Phenol-formaldehyde	1.3	0.00004	5–9×10^3	1.5–2	0.4–0.7×10^6	160	0.1	Very slow
Polyester (thermosetting)	1.1–1.5	0.000036–60	6–13×10^3	4	0.3–0.64×10^6	250	0.15–0.6	Slow burning
REINFORCED PLASTICS								
Acetal, glass-filled	1.6	0.00002	18.5×10^3	3	1.25×10^6	220	0.2	Slow burning
Polycarbonate, glass-filled	1.24–1.5	0.00001–2	12–25×10^3	1–5	0.5–1.7×10^6	275	0.15	Self-extinguishing
Thermoplastic polyester, reinforced	1.5	0.000033	17,000	5	1.2×10^6	350	0.06	Slow burning
Thermosetting polyester, glass-reinforced	1.35–2.1	0.000015–30	15–50×10^3	0.5	0.8–4.5×10^6	300–350	0.01–1.0	Slow burning
Phenol-formaldehyde, glass-filled	1.7–2.0	0.000005–10	5–18×10^3	0.2	0.190–0.33×10^6	350–550	0.03–1.2	Very slow
Epoxy, glass-filled	1.6–2.0	0.00001–2	10–20×10^3	4	0.3×10^6	300–500	0.1	Slow burning

Rigid urethane foams: Ultimate tensile strength $= 23\ D^{1.1}$
Ultimate compressive strength $= 12.8\ D^{1.4}$ where D = density, lb/cu ft

and some have not yet entered the machine designer's repertory of materials. Of the older and more established foams, only rigid polyurethane has sufficient dimensional stability to be considered for machine parts. Strength and E-value of the urethane foams increase as some power of the density.

10-14 FATIGUE FAILURES

The machine designer must always be aware that the commonest type of mechanical failure probably is a fatigue failure. The appearance of a typical fatigue fracture is shown in Fig. 10-3. The steady propagation of the fracture is shown by the small ridges or beach marks, whose curvature encloses the initiating crack. Fatigue has been discussed in Sec. 1-13; in this section, some observations on the management of fatigue are presented.

Figure 10-3 / Fatigue failure in a shaft. Failure originated at the upper right-hand end of the milled flat.

Materials of higher strength will have correspondingly higher endurance limits, since the endurance limit for steels is 40–50% of the ultimate tensile strength. Therefore, strong alloys are selected for fatigue resistance, such as 4130, 4140, and 4340. However, material strength determines endurance limit only in the absence of stress concentrations. Any stress concentration results in a drastic reduction in fatigue strength. Residual stresses, which are particularly present in welds, are equally harmful to the fatigue performance of a part. Figure 10-4 shows the Goodman chart for a structural steel. For 100% reversing load, the as-rolled material has an endurance limit of 50% of ultimate tensile strength at 10^5 cycles. But a butt weld in the material has an endurance limit of only 33% at 10^5 cycles and as little as 20% for 10^6 cycles. The reason for the low

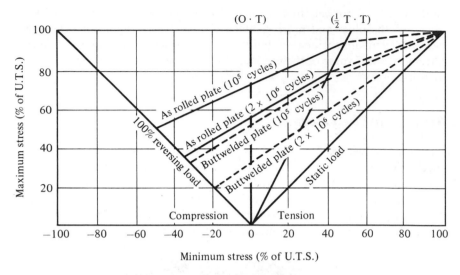

Figure 10-4 / Goodman fatigue diagram for a low-carbon steel.

values for butt welds is the presence both of stress concentration and residual stress. For fillet welds, not shown in the figure, a safe endurance limit at 10^6 cycles would be not more than 10,000 psi for this steel or any other steel, assuming reversed loading. Figure 8-23 shows a slip clutch fillet-welded to a shaft. Rotation of the shaft represents a case of 100% reversing load, and the stress on the throat of the fillet should not exceed 10,000 psi if fatigue failure is to be prevented. Even at 10,000 psi, the factor of safety against fatigue is little better than 1.0.

The residual stress in a fillet weld and therefore the fatigue strength may be improved by a proper shape of weld. All welds shrink, and this shrinkage results in a residual tensile stress in the weld. Suppose the usual flat fillet weld is made. As the weld freezes, it shrinks. Figure 10-5 shows

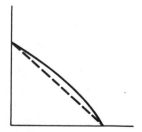

Figure 10-5 / Flat or concave fillet shrinks to a longer length, increasing residual stress.

Figure 10-6 / Convex fillet shrinks to a shorter length, relieving residual tensile stress.

such a weld, with a dashed line to show the final shape after shrinkage. In shrinking to the dashed line, a high tensile residual stress is introduced into the weld, since the weld material must stretch to cover the longer length of the dashed line. A much better weld shape for fatigue strength is the convex weld of Fig. 10-6. When this shape shrinks, it shrinks to the shorter length represented by the dashed line, thus greatly reducing the residual stress. The flat or concave fillet weld is entirely satisfactory for static stresses but is not successful against fatigue conditions.

10-15 WEAR

If a machine part does not fail by fatigue or corrosion, perhaps its life will be determined by wear. The loss of only a few grams of metal due to wear in an engine weighing several hundred pounds is cause for the whole engine to be scrapped, even though 99.9 % of the engine may be in good condition. Similarly, the loss of only a few thousandths of an inch on the diameter of a 2-in. shaft by wear may render the shaft useless.

The detachment of particles from a metal surface may be the result of contact with another metal surface or with a foreign material such as grit. The first case is called *adhesive wear*, also called galling or seizing, and the second case is called *abrasive wear* or abrasion. Both types of wear can be reduced or prevented by proper lubrication. But wear can rarely be avoided completely, and the designer therefore designs against it. Since worms rub against worm wheels, the wheel is usually of bronze and the worm is steel to reduce adhesion. Two unlike metals are less likely to adhere or seize than two similar alloys. Similarly, a bearing is not the same alloy as the shaft. The softer material is normally employed for the part that is the more economical to replace.

Resistance to wear is usually improved by hardness and, in the case of steels, by high carbon content and the addition of carbide-forming alloying elements such as chromium, tungsten, and molybdenum. A smooth surface also improves wear resistance, since any projections which may seize will then be minute in area. Various surface treatments are also used to protect against wear.

Electroplating of steels with hard chromium provides a surface of great hardness and low friction, as well as corrosion resistance. Decorative chromium-plating deposits only about 0.0001 in. of chromium; hard chrome is a different electroplating process that deposits several thousandths of an inch of chromium. The rods of hydraulic cylinders are often hard chrome-plated to protect against scoring by dirt. The related method for aluminum is hard anodizing, which produces a few thousandths of an inch of hard aluminum oxide on aluminum alloys.

Steels may also be carburized or case-hardened. Low carbon steels such as 1020 or 8620 are used for this purpose. Carbon is diffused into the surface of the steel in a furnace. Longer residence times in the furnace result in deeper cases. A case depth as high as 0.100 in. is possible, though the final grinding operation on the part will reduce the case depth by 0.010 in. or more. A case of 0.040 in. (after grinding) is satisfactory for most purposes. Case-hardening results in the highly useful combination of a very hard case with the very tough core of the original low-carbon steel.

An alternate method is induction hardening. This is a very fast electrical heating method which employs the part to be hardened as the secondary of a transformer. Steels of 0.35–0.45 % carbon are used and induction-hardened to the desired depth. The method has a number of advantages, such as speed (the induction-hardening operation is performed in seconds, compared to hours for case-hardening), low distortion, minimal scale, and often about 5 Rockwell hardness points above the level obtainable by furnace hardening.

Still another method of hard-surfacing is metal spraying (flame spraying), in which a metal wire is melted and sprayed onto the surface of the part. The sprayed deposit is rough and porous and must be finished by a grinding operation. The wear resistance of sprayed metals is excellent, but resistance to impact is poor.

Wear-resistant deposits may also be produced by arc and gas welding methods, as is done for crusher jaws and the rolls of crawler tractors. The welding operation is termed hard-facing. Oxyacetylene welding usually gives better results than arc welding because the heating and cooling rates are slower, resulting in less tendency to crack, but arc processes are often used, especially for automatic operations in hard-surfacing.

Besides fatigue and wear, there are a variety of reasons for failure of machine parts, including the occasional unpredictable failure. Errors in design and specification are occasionally found, usually arising from the designer's improper conception of the problems to be met by the machine part in service. Sometimes the material used is at fault, or more often, the wrong material may be selected. Improper manufacturing techniques, such as faulty electroplating, grinding, or heat-treating, also result in failures. There are types of failure due to high- or low-temperature effects or to corrosion, topics that are too complex for discussion here.

This chapter serves as an introduction to some of the considerations that preoccupy the machine designer in coordinating service conditions, stress conditions (including residual stress), and manufacturing considerations and selecting a suitable material to meet these many requirements at reasonable cost. To produce a successful design, the designer must be knowledgeable in both materials and manufacturing processes as well as the design of machine parts.

QUESTIONS AND PROBLEMS

P10-1 What methods are available to minimize seizing?

P10-2 The balls of trailer hitches are made of machinery alloys such as 4140, which rust. To prevent rusting, such balls are sometimes chrome-plated, but the plating flakes off. Would a stainless steel ball be more suitable, and if so, which stainless alloy would you recommend?

P10-3 Using handbook values, compare the ultimate tensile strength of standard mild steel (low-carbon steel) and the structural (weldable) aluminum alloys.

P10-4 Using the Goodman diagram in Fig. 10-4, state the maximum safe stress for the following cases:
  (a) butt-welded plate, 2×10^6 cycles, half-tension to full tension
  (b) as-rolled plate, 2×10^6 cycles, zero stress to full tension
  (c) Assuming that a fillet weld is half as strong as a butt weld in fatigue, what maximum stress level would you accept for a fillet weld for 100% reversing load at 2×10^6 cycles?

P10-5 Why is bronze used for the worm gear if the worm is of steel?

P10-6 Why are (a) induction hardening and (b) case hardening, preferred methods for hardening gears?

P10-7 Why must an impact tool be neither too lightweight nor too heavy?

P10-8 Explain why a convex fillet weld is superior in fatigue strength to a flat fillet weld.

P10-9 Why are the plastics not used for precision gears?

P10-10 What surface qualities does chrome-plating supply?

P10-11 Why are machine parts usually made of medium-carbon steels rather than steels of higher- or lower-carbon content?

P10-12 State the carbon content of the following AISI–SAE alloys: 5160, 8620, 4130, and 10120.

P10-13 From handbooks or the data provided in this chapter on carbon steels, state the increase in tensile strength in these steels by increasing the carbon content by 0.1%. Such data apply to the as-received or annealed condition; heat-treating will increase these strength levels.

P10-14 What are the limitations of plain carbon steels?

P10-15 You are using plain carbon steels for hardened shafts but find that they warp excessively from the heat-treating operation. What change in alloy selection would you make in general?

P10-16 Why are higher-carbon steels difficult to weld?

P10-17 Define a residual stress.

P10-18 A 2 × 2 in. square steel tube with a 16-gage wall has been used in a frame. It is desired, for corrosion resistance, to substitute a rectangular tube of fiberglass-reinforced thermosetting polyester for the steel tube. The polyester tube must have approximately the same stiffness (EI) as the steel tube, but it must retain the same width of 2.0 in. The E-value of the polyester tube is 3×10^6 and its specific gravity is 1.75.

(a)   Design a suitable hollow section in reinforced polyester.

(b)   Which is the heavier tube?

SUGGESTED DESIGN PROJECT

DP10-1 The following design requires the selection of suitable fasteners, bearings, speed reducers, belts, and chain drive. Catalogs such as that of the Boston Gear Company should be consulted for power transmission components. However, no special alloys or materials will be required.

Design a mechanical walker for exercising horses.

The walker consists essentially of a spider of four arms each about 15 feet long attached to a vertical shaft and rotated at a speed of about 4 miles per hour at the maximum radius of the arms. At the end of each arm there is a ring for attachment of a 1/2″ lead rope which is tied to the halter of the horse. The walker causes the horse or horses being exercised to walk a circle.

The four arms should slope upward to a height of 6 feet at the ends and must be braced with a light tube as shown in Fig. DP10-1. A lightweight chain, tightened with a turnbuckle, connects the end of one arm to the end of the adjacent arm.

The rotating arms should be driven by a 1/3 hp, 120-volt capacitor-start motor. Despite the large speed reduction required, the motor does not provide sufficient torque to pull the horse; the horse actually follows the lead rope attached to the arm. Since any horse has a tractive effort equal to that of a truck, the horse can stop the walker from rotating if allowed to do so.

The final drive to the vertical spindle should be a chain drive. For reasonable flexural stiffness, the vertical spindle should not be lighter than a 2″ tube of 1/4″ wall. The motor is mounted with the shaft horizontal. A right-angle gear reducer or worm reducer may be used. All bearings should be of roller or ball type protected with shields. To reduce the amount of design work, it is recommended that the design of protective covers over the machinery be omitted.

To prevent an excitable horse from overturning the walker, the base should measure about 5 feet by 5 feet. Because of the size of the walker, 30 feet in diameter, it must be made of components that can be conveniently packaged for shipment and easily assembled on site without field welds. Since it is not practical to design the walker to resist the strength of horses, no stress analysis is performed on the components.

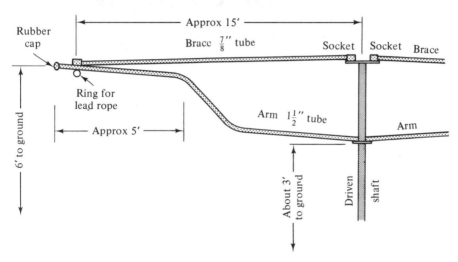

Figure DP10-1

THE SI SYSTEM

There have been several metric systems of units developed. One which enjoyed considerable popularity a few decades ago was known as the CGS system, and was based on the centimeter, the gram, and the second. More recently, the MKS metric system used the meter, the kilogram, and the second as fundamental units. The metric system currently being introduced is known as the SI system. Its name is, in French, Systeme International, hence the abbreviation SI. The British system, on the other hand, is based on the foot, the pound, and the second.

The SI system has six basic units, and all other units—for power, torque, energy, etc.—can be derived from these six:

the unit of length — the meter (m)

the unit of mass — the kilogram (kg)

the unit of time — the second (s)

plus the ampere for electrical engineering, the candela for illumination, and the kelvin for thermodynamics. Only the first three units are of concern in machine design practice.

To convert from British to SI:

$$1 \text{ inch} = 25.4 \text{ mm}$$

$$1 \text{ yard} = 0.9144 \text{ m}$$

$$1 \text{ pound} = 0.453592 \text{ kg}$$

Multiples or decimal fractions of the basic SI units are used where convenient—for example, the millimeter or one-thousandth of a meter, and the gram or one-thousandth of a kilogram. The standard multipliers for multiples and decimal fractions are given in Table A1-1.

TABLE A1-1
STANDARD MULTIPLES OF THE SI SYSTEM

MULTIPLIER	NUMBER	PREFIX	ABBREVIATION
1 million million	10^{12}	tera	T
1 billion	10^{9}	giga	G
1 million	10^{6}	mega	M
1 thousand	10^{3}	kilo	k
1 hundredth	10^{-2}	centi	c
1 thousandth	10^{-3}	milli	m
1 millionth	10^{-6}	micro	μ
1 billionth	10^{-9}	nano	n

Thus, a thousandth of a millimeter is a micrometer or μ meter.

Areas are given in mm² or in m², volume in mm³ or m³. Speed is preferably given in meters per second, m/s. Density is expressed as g/mm³ or kg/m³. Note that 6 square kilometers is abbreviated 6 km². The abbreviation means 6 (km)², not 6 k × m². Similarly, 6 square millimeters is abbreviated 6 mm², which does not mean 6 m³.

Unlike previous metric systems, the unit of temperature used in the SI system is the degree Celsius, not Centigrade, though the temperature scale is still the same (Centigrade) scale; only the name is changed.

Combined units, such as kilogram–meters, must be abbreviated with care. Consider a torque of 5 kg × 3 m. This product is expressed as 15 kg m, with a space between kg and m. It must not be written 15 kgm. If a length must be multiplied by time, consider that 10 m × 10 s = 100 meter–seconds or 100 m s. If written as 100 ms, this would mean 100 milliseconds.

EXAMPLE A1-1

As an example of conversion to SI units, suppose we convert 60 miles per hour to kilometers per hour.

To convert 1 mile/hour to km/hour, we must multiply by km/mile; that is,

$$1 \text{ mile/hour} \times \text{km/mile} = \text{km/hour}$$
$$1 \text{ mile} = 1760 \text{ yards}$$
$$1 \text{ yard} = 0.9144 \text{ meters}$$
$$1 \text{ mile} = 1760 \times 0.9144 \text{ meters}$$
$$= 1609.344 \text{ meters}$$
$$= 1.609 \text{ km}$$
$$60 \text{ miles/hour} \times 1.609 \text{ km/mile} = 96.6 \text{ km/hour}$$

THE NEWTON

The SI unit of force is the newton (N). The newton is defined as the force that will accelerate a mass of 1 kg at a rate of 1 meter per second per second; that is,

$$1 \text{ N} = 1 \text{ kg m/s}^2$$
$$1 \text{ pound force} = 4.448 \text{ N}$$

Stress is expressed in newtons per square meter (N/m^2).

Gravitational acceleration at sea level in the SI system is 9.81 m/s^2.

EXAMPLE A1-2

A structural tube with an O.D. of 80 mm and an I.D. of 60 mm supports a compressive load of 60,000 N. Determine the compressive stress in the tube in N/m^2.

$$\text{Area } (A) \text{ of the tube} = (\pi/4)(80^2 - 60^2) \text{ mm}^2$$
$$= 2200 \text{ mm}^2$$
$$= 0.0022 \text{ m}^2$$
$$\sigma = P/A = 60{,}000 \text{ N}/(0.0022 \text{ m}^2)$$
$$= 27.27 \times 10^6 \text{ N/m}^2$$

EXAMPLE A1-3

Calculate the number of MN/m² corresponding to 1 ton/sq. in.

$$1 \text{ inch} = 25.4 \text{ mm}$$

$$1 \text{ in.}^2 = (25.4)^2 \text{ mm}^2$$

$$1 \text{ pound force} = 0.4536 \text{ kg force}$$

$$1 \text{ kg} = 9.807 \text{ N}$$

$$1 \text{ pound} = 0.4536 \times 9.807 \text{ N}$$

$$\sigma = \frac{P}{A} = \frac{2000 \times 0.4536 \times 9.807}{(25.4)^2 \times 10^{-6} \text{ m}^2}$$

$$= 15 \times 10^6 \text{ N/m}^2$$

EXAMPLE A1-4

Determine the extension of a square steel bar with a cross section of 40 mm × 40 mm and a length of 8 meters when loaded in tension with a force of 64,000 N. The modulus of elasticity of steel is closely 200×10^9 N/m².

$$\text{Cross-sectional area} = 1600 \text{ mm}^2 = 1600 \times 10^{-6} \text{ m}$$

$$\sigma = \frac{P}{A} = \frac{64,000}{1600 \times 10^{-6}} \text{N/m}^2$$

$$\text{Strain} = \frac{\sigma}{E} = \frac{64,000}{1600 \times 10^{-6} \times 200 \times 10^9} = 0.0002 \text{ mm/mm}$$

$$\text{Total extension} = 0.0002 \times 8000 = 1.6 \text{ mm}$$

ENERGY AND POWER

Work is the product of force times the distance through which the force acts. In the SI system,

$$1 \text{ unit of work} = 1 \text{ newton} \times 1 \text{ meter} = 1 \text{ newton–meter} = 1 \text{ N m}$$

This unit might be stated as 1 meter–newton, similar to 1 foot–pound in English units. This latter designation is a hazardous practice, however, because 1 m N could be interpreted as a millinewton.

The SI system, however, gives the meter–newton a specific name, the joule (J):

$$1 \text{ joule} = 1 \text{ N m}$$

The SI unit of power is the newton–meter per second, which is 1 watt (W):

$$1 \text{ watt} = 1 \text{ newton–meter/second}$$

$$1 \text{ W} = 1 \text{ N m/s} = 1 \text{ J/s}$$

$$1 \text{ horsepower} = 746 \text{ watts or closely } 3/4 \text{ kW}$$

EXAMPLE A1-5

A force of 2500 N is exerted over a stroke of 225 mm in a machine tool in 9 seconds. What is the power requirement in the cutting tool?

$$\text{Power} = \text{work/time}$$

$$= \frac{2500 \text{ N} \times 0.225 \text{ m}}{9 \text{ s}}$$

$$= 62.5 \text{ watts}$$

EXAMPLE A1-6

A constant force of 30 N is applied to a body to move it 800 mm in the direction of the force. What is the work done?

$$\text{Work} = 30 \text{ N} \times 0.8 \text{ m} = 24 \text{ N m.}$$

Until properties of materials in metric units are widely available in this country, and until standard metric sizes for pipe, tube, bar, and other products are known, only a limited range of calculations in the SI system will be possible. Of course, standard sizes in inches can always be converted to metric units. The problems in SI units given at the ends of the chapters serve as an introduction to the SI system of units.

APPENDIX 2

PROPERTIES OF PURE METALS AND ALLOYS

PROPERTIES OF PURE METALS AND ALLOYS

METAL	E-VALUE (psi)	G-VALUE	COEFFICIENTS OF THERMAL EXPANSION (in./in./°F)	SPECIFIC WEIGHT (lb/in³)	ULTIMATE TENSILE STRENGTH (psi)
Al	10×10^6	3×10^6	0.0000131	0.0975	12,000
Cu	16×10^6	6×10^6	0.0000092	0.324	34,000
Fe	29×10^6	12×10^6	0.0000065	0.285	42,000
Mg	6.35×10^6	2.4×10^6	0.0000150	0.063	22,000
Ni	30×10^6	—	0.0000074	0.322	55–80,000
Ti	16.8×10^6	—	0.0000047	0.1628	59,000
Zn	—	—	0.0000220	0.2577	41,000

ULTIMATE TENSILE STRENGTH OF ALLOYS

	ALLOY	CONDITION	TENSILE STRENGTH (*psi*)
Stainless	301	Annealed	110,000
	302	Annealed	90,000
	304	Annealed	85,000
	403, 410	Annealed	75,000
	403, 410	Hardened	190,000
Aluminum	2024	Annealed	27,000
	2024	Hardened (T6)	70,000
	6061	Annealed	18,000
	6061	Hardened	45,000
Cartridge Brass		Annealed	47,000
		Hard Drawn	78,000
Muntz Metal		Annealed	48,000
		Hard Drawn	110,000
Silicon Bronze		Annealed	58,000
		Hard Drawn	108,000
Magnesium	AZ92A	Annealed	25,000
		Hardened	40,000
Monel			70–85,000

ANSWERS TO SELECTED QUESTIONS AND PROBLEMS

CHAPTER 1

P1-3 (a) $\sigma_{min} = 8000$ psi, $\sigma_{max} = 16,000$ psi
(b) $\sigma_{min} = 6060$ psi, $\sigma_{max} = 13,950$ psi

P1-5 $\sigma_{30°} = 5500$ psi
$\tau_{30°} = 4330$ psi

P1-9 (a) $\sigma_p = 1000$ psi, $\tau_{max} = 0$
(b) Principal stresses are $+5780$ and -2780 psi, $\tau_{max} = \pm\,4280$ psi
(c) $\sigma_p = \pm\,5000$ psi, $\tau_{max} = 5000$ psi
(d) Principal stresses are $+8450$ and -5450 psi, $\tau_{max} = \pm 6950$ psi

P1-13 $\theta_8 = 0.000341\,T$ radians, $\theta_{10} = 0.0000268\,T$ radians

P1-17 Section A: $R = 0.4$; Section C: $R = 0.541$

P1-19 If longitudinal and circumferential stresses are taken as 5000 and 2500 psi, and torsion stress is 710 psi, then the principal stresses are 2310 and 5190 psi.

P1-20 Strain $= 0.121$

P1-21 Loads must be proportional to E-values in the rods. $y = 0.638$

P1-23 Stress in square section, 10,000 psi; in middle section, 452 psi; in 5/8 dia. section, 19,500 psi

P1-24 Note that yielding of the right support will reduce the stress in BC and increase the stress in AB (with no reaction at C, there would be no stress in BC). $F_C = 2550$ lb and $F_A = 17,450$ lb approximately.

P1-25 Using $\alpha = 0.00001$ in./in./°F, $\Delta T = 19.75$°F.

P1-26 $L = 3.14$ in.

P1-27 Force in each bar $= 951$ lb, shear stress $= 1900$ psi

P1-28 Stress is relieved in the brass bar for $\Delta T = 29.4$°F, in the steel bar for $\Delta T = 25.6$°F.

P1-29 Stress in bolts $= 225,300$ psi

P1-30 Poisson's ratio reduces the clearance by 0.00003 in./in.

P1-31 Using $\alpha = 0.000009$ in./in./°F, a temperature change of 1.65°F will close the contacts.

P1-32 $F = 2500$ lb

P1-34 Tee up: compressive stress $= 28,900$ psi; tee down, compressive stress $= 53,600$ psi

P1-36 Maximum w for permissible shear stress $= 6250$ lb per ft

P1-37 At b–b, outer fiber $\sigma = -13,300$ psi, inner fiber $\sigma = +13,800$ psi

P1-38 $F = 87.7$ lb

P1-39 Maximum stress (at support) $= 33,600$ psi

P1-40 $F = 11,000$ lb, $e = 0.273$ in. for $\sigma_t = 1000$ psi and $\sigma_c = 10,000$ psi

P1-41 $F = 5240$ lb

P1-42 $\sigma_{max} = 10,400$ psi, $\tau_{max} = 289$ psi

P1-43 $F = 9600$ lb

P1-44 (a) $\sigma = 4100$ psi
       (b) Notch: $\sigma = 1020$ psi; hole: $\sigma = 1160$ psi

P1-45 $K_e = 59,600$

P1-46 $K_e = 1.385 \times 10^6$

P1-47 32,000 psi

P1-48 $M = 51,800$ lb-in.

CHAPTER 2

P2-5 B.M. of triangular loading about point $c = \dfrac{120 \times 18}{2} \times \dfrac{18}{3}$

P2-8 The trapezoidal loading can be taken as a combination of a uniform load plus a triangular load. $\theta = 0.051$ rad, $\delta = 0.95$ in.

P2 9 $\delta EI = \dfrac{Fl^3}{3} - \dfrac{5l2^3}{24}$ where $R =$ central reaction

P2-10 $\theta = 0.00053$ rad, $\delta = 0.0084$ in.

P2-13 $F = 2.4$ lb

P2-14 $F = 0.367$ lb

P2-15 $\rho M = EI$, $\sigma = 20,900$ psi

P2-16 $F = 590$ lb; for torsion $D = 1.5$ in., for deflection $D = 4.56$ in., for bending $D = 1.76$ in.

P2-17 0.06 in.

P2-18 $F = 19.6$ lb

P2-19 $y = 4.62$ in., $x = 1.735$ in.

P2-20 1.146 lb-in. total

P2-22 Euler. $F = 8680$ lb

P2-23 Johnson. $F = 19,100$ lb

P2-25 Euler. $F = 1490$ lb

P2-28 $\Lambda T = 61.5°F$

CHAPTER 3

P3-2 Cam layout uses prime circle 2.500 in. diameter instead of base circle. Follower lifts 1.000 in. in 180°. The angular values of Table 3-1, from 0° to 120°, are used, over a range of 180° (120° in table corresponds to 180° in the cam).

P3-3 Draw radial lines at 15° increments between 0° and 90° (six angular divisions). Use values of Table 3-1 for 20°, 40°, 60°, 80°, 100°, and 120° for radial distances. At each radial distance draw a line at right angles to the radial line to represent the face of the follower. Sketch in a suitable cam profile to touch all these follower lines.

CHAPTER 4

P4-2 max $\tau = 156{,}500$ psi, $\delta = 40.1$, $K = 9.96$

P4-3 Force in each spring $= 74.8$, max $\tau = 32{,}000$ psi

P4-6 1.36 in.

P4-7 (a) $K_e = \dfrac{1}{\dfrac{1}{K_1 + K_2} + \dfrac{1}{K_3}}$

   (c) $K_e = K_1 + K_2$

P4-9 Average $R = 1$ in., $K = 24.4$

P4-12 20 turns $= 943$ in. For a maximum stress of 56,000 psi, torque $= 2.71$ in.-lb and $\theta = 6.74$ revolutions. After unwinding 4 revolutions, remaining torque $= 1.3$ in.-lb and stress $= 27{,}000$ psi.

P4-13 Stiffer and less resilient spring; less energy absorption

P4-16 0.114 lb/in.

P4-18 Parallel system

P4-22 Max $\delta = 0.316$ in.

P4-23 0.877 cycles/sec

P4-24 0.412

P4-26 1.286×10^4 lb/in.

P4-27 $k = 69.6$ lb/in.

P4-28 (a) 14.5%
   (b) 29%

P4-29 $P = 33.3$ N, $\delta = 32.6$ mm

CHAPTER 5

P5-4 The high value of stress here would not be realized; the threads would deform and relax the stress.

P5-6 $K_e = 52.4 \times 10^6$

P5-7 2680 lb

P5-9 (a) Only the two upper bolts carry the load.
   (b) 7450 lb

P5-10 Shear stress $= 12{,}900$ psi

P5-11 There is twice the load in the 2-in. leg as in the 1-in. leg; therefore L_2 should be twice L_1. Total minimum weld length, $L_1 + L_2 = 12.5$ in.

P5-16 Total elongation $= 0.0044$.

CHAPTER 6

P6-1 $\mu = 0.292$

P6-2 $T_2 = 289$ lb

P6-4 $T_1 = 915$ lb, $T_2 = 197$ lb

P6-5   Torque = 1608 in.-lb, B.M. = 3856 lb-in.

P6-7   (a)   17 times greater
          (c)   about 80% reduction
          (d)   over double the life

P6-8   Approximate length = 198.9 in.

P6-10 Belt length = 140.8 in.

P6-11 $T_2 = 195$ lb, $T_3 = 237$ lb

P6-13 (a)   42 hp
          (b)   110 hp

P6-15 Tensile stress = 332 psi

P6-16 (a)   5.6 hp per belt
          (b)   2.2 hp
          (c)   25.5 hp
          (d)   14 hp

P6-17 (a)   *C*
          (b)   *B*
          (c)   *B*
          (d)   two *B* belts

P6-18 174.6 in.

P6-20 Three *D* belts

P6-21 Two *C* belts

P6-22 $C = 15$ in. At idle $T_1 = T_2 = 107$ lb.

P6-24 $C = 19.45$ in. At zero load $T_1 = T_2 = 121.4$ lb.

P6-25 (a)   Pitch diameter = 1.33 in., MHD = 0.875 in.

P6-26 (a)   38.9 hp

P6-27 Chain length = 120 pitches

P6-28 Center distance = 80 in.

P6-29 Pitch diameter = 9.58 in., chain length = 84 pitches, transmitted hp = 24.8, type II lubrication

CHAPTER 7

P7-2 $D = 0.917$ in., addendum $= 0.0833$ in., $O.D. = 1.084$ in.

P7-8 11.8°

P7-9 5.45 in.

P7-10 (a) 19 hp
(b) 10.55 hp
(c) 5.7 hp

P7-11 Pinion allowable stress $= 8000$ psi, gear allowable stress $= 3200$ psi. For $p_d = 4$, for the gear, $y = 0.129$, $b = 2.7$ in.

P7-12 Pinion controls the design. $p_d = 4$, 16 teeth in pinion

P7-14 $p_d = 14$, $b = 3/4$ in., 168 tooth gear

P7-15 Force at bearing 3 $= 386$ lb

P7-16 $W_r = 62.5$ lb, $W_a = 46$ lb

P7-17 $W_r = 151$ lb

P7-18 (a) 2.55 in.
(b) 2380 lb-in.
(d) 204 lb

P7-21 (b) 884 lb

P7-22 (a) 2.34 in.
(b) max hp $= 3.44$
(e) 5570 lb
(f) 2070 lb

P7-24 1033 rpm

P7-25 $\dfrac{18}{81} \times \dfrac{18}{80}$

P7-26 $\dfrac{24}{54} \times \dfrac{18}{54} \times \dfrac{24}{39}$

P7-27 Pinion has 20 teeth, output speed $= 693$ rpm

P7-28 (b) Greater torque and gear forces

P7-29 $p_d = 14$ using 30 and 68 teeth

P7-30 Best choice is 48 or 72 p_d. Driven speed is 1820 rpm.

P7-31 (a) $\dfrac{26}{78} \times \dfrac{24}{80}$ with center distance 3.25 in.

(b) $\dfrac{31}{93} \times \dfrac{24}{100}$ with center distance 5.167 in.

P7-32 $\omega_R = 166.7$ rpm counterclockwise

P7-33 $N_S = 40$

P7-34 $\omega_S = 92$ rpm

CHAPTER 8

P8-3   Symmetrical load distribution, use of full circumference of the shaft, less severe stress concentration

P8-4   (a)   Chain coupling, universal joint
      (b)   Chain coupling, universal joint
      (c)   Chain coupling, magnetic fluid coupling
      (d)   Hydraulic coupling, magnetic fluid coupling, rotating field clutch

P8-5   (c) $D = 1.0$ in.
      (d) $0.57°$

P8-6   (c)   15/16 in.
      (d) $1.54°$

P8-7 $D = 1.14$ in. Use $1\frac{1}{8}$ in.

P8-8 $D = 0.83$ in. Use $\frac{7}{8}$ in.

P8-9   1840 rpm

P8-10   For single shear $D = 0.25$ in.; for double shear $D = 0.176$ in.

P8-11   Width $= \frac{3}{4}$ in., depth $= \frac{1}{2}$ in., minimum length $= 4\frac{1}{2}$ in.

P8-13   70%

P8-14   19 hp

P8-17 $D = 0.283$ in.

P8-18   Bearing stress $= 896$ psi

P8-19   32 passes

P8-20 $D = 0.317$ in.

P8-21   375 lb-in.

P8-22   231 lb-in.

P8-23   324 lb-in.

P8-24   17 pairs

P8-26   3000 lb-in.

P8-27   234 lb

P8-28   462 Btu

P8-29   (a)   80,200 lb-in.
      (c)   Sum of start-up and steady torques $= 84,380$ lb-in. Torque at high speed end of drive $= 242$ lb-in.

P8-30   38.5 Btum

P8-31   K.E. $= 21,000$ ft-lb, braking torque $= 167.4$ lb-ft

P8-32   (a) $N = 666.6$ lb

   (c)   274 lb
   (d)   260 lb

P8-33   268 lb-in. torque and 90 Btum

P8-34   1920 and 640 lb

P8-35   311 and 142 lb.

CHAPTER 9

P9-1

	Bearing 1	Bearing 2
(a)	ball	ball
or		
	ball	roller
or		
	roller	ball
(b)	angular contact	angular contact
(c)	roller	ball
(d)	needle	needle
(e)	precision	precision

P9-2   Rolling contact produces arcing.

P9-4   Cast bronze

P9-5   More energy is needed to shear the oil film

P9-6   Like materials are prone to seize

P9-7   (a)   Bore No. 11

P9-8   2220 lb

P9-10   875 and 2480 lb

P9-12 193×10^6 revolutions or 129,000 hours

P9-13   956 lb

P9-14 422×10^6

P9-15   1820 lb

DP9-1   (a)   207,000 on two bearings
          (d)   Approximately 23.2×10^6 revolutions

CHAPTER 10

P10-2   410,431

P10-3   Mild steels, 55–65,000; weldable aluminums, 40,000

P10-6   Hard surface with a softer but tough core

P10-8   Lower residual stress after cooling and shrinking

P10-10   Hardness and low friction

P10-11 Lower carbon is not hardenable; higher carbon is not tough enough.

P10-14 Warpage in heat treatment; will harden to a depth of only $\frac{1}{8}$ in.

P10-15 Use low-alloy steels with small amounts of nickel, manganese, or chromium.

P10-16 They harden when welded.

INDEX